ZigBee Wireless Networking

ZigBee Wireless Networking

Drew Gislason

ELSEVIER

AMSTERDAM • BOSTON • HEIDELBERG • LONDON
NEW YORK • OXFORD • PARIS • SAN DIEGO
SAN FRANCISCO • SINGAPORE • SYDNEY • TOKYO

Newnes is an imprint of Elsevier

Newnes

Newnes is an imprint of Elsevier
30 Corporate Drive, Suite 400, Burlington, MA 01803, USA
Linacre House, Jordan Hill, Oxford OX2 8DP, UK

∞ Recognizing the importance of preserving what has been written, Elsevier prints
its books on acid-free paper whenever possible.

Library of Congress Cataloging-in-Publication Data
Application submitted.

British Library Cataloguing-in-Publication Data
A catalogue record for this book is available from the British Library.

ISBN: 978-0-7506-8597-9

For information on all Newnes publications,
visit our Web site at: www.books.elsevier.com

Working together to grow
libraries in developing countries

www.elsevier.com | www.bookaid.org | www.sabre.org

ELSEVIER   BOOK AID
International   Sabre Foundation

Contents

Foreword

Back in 1999, we saw the need for an organization with a mission to define a complete, open, global standard for reliable, cost-effective, low-power, wirelessly networked products addressing monitoring and control. While there were other standards that addressed higher data rates or battery-powered networks for a very small number of devices, none of these truly met the needs of this market. Instead, what we needed was something focused on:

- Large networks (a large number of devices and a large coverage area) that could form autonomously, and operate very reliably and securely for years without operator intervention

- Very long battery life (years of use from a pair of AA cells), very low infrastructure cost (low device and setup costs), very low complexity, and small size

- A relatively low data rate

- A standardized protocol, allowing multiple-vendor, interoperable products for the global market

Thus, in 2002, the ZigBee Alliance was born.

Now, with over 225 member companies, we are able to draw on a wealth of experience in every aspect of the business. Allowing so many companies the opportunity to have input is not as fast-paced as adopting a single proprietary system and declaring it a standard. Even so, the combined wisdom and the vetting process enable a much better solution to be built, one which meets all of the needs mentioned above. Getting it right is extremely important if a standard is to have a long track record of success.

The ZigBee Alliance slogan, "Wireless Control That Simply Works," is clearly what is needed for end users and implementers, but achieving that result places a heavy burden on the developers and OEMs. Drew Gislason has the ability to take complex topics and present them in a manner that is cogent and easily digestible to OEMs and developers. This book goes a long way in helping to explain ZigBee concepts, and in explaining how to implement ZigBee in a wide variety of products.

Many people do not realize that a quiet revolution is already under way in business, and in our homes. Great emphasis is being placed on reducing natural resource waste and in becoming exponentially more efficient. To achieve global efficiency goals, we must have an extensive, intelligent, easy-to-deploy, low-cost infrastructure to monitor and manage our surroundings. ZigBee is a solid technological solution for this monumental task, and brings with it other benefits and lifestyle improvements, such as increased security and convenience.

The momentum is gathering, and we are about to take a leap that brings remarkable savings, not only of resources, but of money and time. History repeatedly demonstrates that companies at the forefront set the mark for their industries. Indeed, many current industry leaders who take a "wait and see" approach are likely to be supplanted by new leaders willing to make an early investment in learning and implementing solutions such as ZigBee into their product lines. This book will help companies move up the learning curve rapidly.

Bob Heile, Ph.D.
Chairman, ZigBee Alliance

Preface

Intended Audience

ZigBee Wireless Networking is for developers who are interested in learning more about ZigBee. The developer who actually has a project at hand may benefit the most from the text and examples, but managers considering using ZigBee on a project will also benefit, especially from Chapter 1, "Hello ZigBee," and Chapter 2, "Deciding on ZigBee."

No prior experience with embedded programming, 802.15.4, or networking is required, but a working knowledge of the C programming language is helpful.

Most of the examples use the Freescale ZigBee platform, but the ideas apply to all ZigBee platform vendors. Occasionally, an example or solution is specific to the Freescale platform. Where this occurs, the text makes it clear that the solution is Freescale-specific, and not a ZigBee requirement, in general.

Formatting Conventions

Various elements in this book are specially formatted for easier identification while reading. Code samples are printed in a different style on a light gray background. Variables in the text are printed in **bold**. Each section ends in a brief summary, indicating the key points. For example, Code samples look like this:

```
void BeeAppDataIndication(void)
{
  apsdeToAfMessage_t  *pMsg;
  zbApsdeDataIndication_t  *pIndication;
  zbStatus_t  status = gZclMfgSpecific_c;
  while(MSG_Pending(&gAppDataIndicationQueue)
```

The Book's Structure

This book is designed to be read from cover-to-cover, tutorial style. Each chapter introduces concepts that are used in later chapters. However, the reader is encouraged to skip ahead if the concepts of any section are already familiar. To keep the reader oriented in each section, some overlap is necessary.

While it is not required, it is helpful to have the ZigBee and IEEE 802.15.4 specifications available when reading this book. I'll sometimes refer to a section or a concept in those specifications for further reading.

To obtain the ZigBee specification, go to http://www.zigbee.org and click on "Download the Specification." It is free, and comes in PDF format.

Go to http://standards.ieee.org/getieee802/802.15.html and select "IEEE 802.15.2-2003" to obtain the 802.15.4 standard in PDF format. There is a new draft standard, IEEE 802.15.2-2006, but ZigBee does not currently use that specification.

The chapters are organized as follows.

Chapter 1, "Hello ZigBee," lays out the basics of ZigBee and its intended use. It describes the ZigBee Alliance, which is the standards body which defines and promotes the ZigBee standard worldwide. This chapter also provides the developer with several ZigBee networking examples, complete with source code.

Chapter 2, "Deciding on ZigBee," helps the developer make technical and marketing choices about ZigBee, and even helps determine whether ZigBee is the right solution for any given problem. It provides a concise checklist, with all of the ZigBee factors that must be considered throughout the entire product life cycle, from inception through development, deployment, and maintenance.

Chapter 3, "The ZigBee Development Environment," covers the basics of the ZigBee development environment, and walks the reader through an example, step-by-step, to help gain a full understanding of what's involved in the development and debug phase. This chapter contains information necessary if the reader plans to follow along with the examples using actual hardware.

Chapter 4, "ZigBee Applications," goes in-depth into application development, including the fundamentals of ZigBee networks, nodes, addressing, Application Profiles, and the features provided to the application by the Application Framework (AF) and Application Support Sub-layer (APS).

Chapter 5, "ZigBee, ZDO, and ZDP," describes the ZigBee Device Object (ZDO) and how it interacts with and is used by applications, including how to achieve maximum battery life from ZigBee nodes.

Chapter 6, "The ZigBee Cluster Library," covers the library of common clusters used for profile and device development. It describes Home Automation in some detail.

Chapter 7, "The ZigBee Networking Layer," goes in-depth into how ZigBee actually delivers packets from one node to another, including mesh and tree networking. It also discusses some of the table management that must occur for ZigBee nodes to last for years (and decades) in the field with no required maintenance. This chapter also describes security in detail.

Chapter 8, "Commissioning ZigBee Networks," describes the commissioning process with ZigBee. This topic is critical to the successful deployment and maintenance of ZigBee networks.

Chapter 9, "ZigBee Gateways," introduces gateways and describes techniques for retrieving information from a ZigBee network, as well as controlling and configuring sensors and actuators from outside the ZigBee network.

Appendix A, "ZigBee 2007 and ZigBee Pro," is a quick reference to the ZigBee application API.

Appendix B, "ZigBee Quick Reference," is a quick reference to ZigBee architecture and commands.

Appendix C, "ZigBee Cluster Library Quick Reference," covers some of the new features in ZigBee to be found in the upcoming ZigBee 2007, a specification that was not quite ready for publication at the time of this writing.

Example Source Code

Each chapter contains examples designed to enhance understanding of the ZigBee concepts introduced. Only partial source code is included in the text of this book.

For full source code, including Freescale CodeWarrior project files, go to the web site:

http://www.zigbookexamples.com/code

The home page of this site is just an advertisement for the book, so remember to type in **/code** after the site name. The sample code may be freely used in applications. A standard author's disclaimer applies:

> *No warrantee is implied or expressed and it's probably not suitable for anything other than instructional purposes, and maybe not even that. If you use the source code, keep the copyright text intact.*

The code on the Web site is organized by chapters. Each example, except for the two in Chapter 1, contain only the application source code, and assume that you use Freescale's BeeKit to generate the project. The whole process of taking small source code samples and incorporating them into the Freescale solution is described, step-by-step, in Chapter 3, "The ZigBee Development Environment."

Acknowledgments

There are so many people that contributed to this book. First of all, I'd like to thank my wife, Alicia. Without her countless hours of encouragement and solid support of this arduous process called writing a book, this would not have been possible. And to my daughters, Lita, Genevieve, and Elanor, I love you.

To Rachel, my editor at Elsevier, a big thank you, especially for your patience. There is a White Chocolate Frappuccino® waiting for you at Starbucks®, my treat.

I'd like to thank all the staff at Freescale, in what's now called Wireless Commercial Operations (WCO), especially Brett Black for believing in and championing this technology, Darrel Simms for introducing me to Freescale, Mark Williams for always supporting his partners, Walter Young whose command of the English language is greater than Strunk and White, Mads Westergreen—a Danish man who knows more about ZigBee than just about anybody on the planet, Matt Maupin, Ryan Kelly, Shannon Reid, and so many others as well.

Also, ZigBee Alliance members deserve many thanks for this book: Bob Heile, Zachary Smith, Ian Marsden, Don Sturek, Phil Jamieson, and Skip Ashton, all of whom have made the ZigBee standard possible. I really appreciate all the time you've spent explaining the concepts to me, and discussing the edge cases. You have taught me so much. And to others within the Alliance who have patiently answered my questions: Spiro Sacre, Monique Brown, Wally Burnham, Claudio Borean, Rudi Belliardi, Greg Henry, Silviu Chiricescu, and Bill Wood, thank you.

To my good friends and colleagues, Masaru and Blanca Natsu, Christian Garcia, and Dalila Pinedo, thank you for all your help, suggestions, corrections, and insight. And to

Tim Gillman who was the idea man behind many of the stories which open each chapter. Tim, you are a funny guy.

Special thanks to Mads Westergreen and Allan McDaniel for the iPod example. I think the readers will love this one.

Any mistakes, inaccuracies, or omissions in this text are purely my own.

CHAPTER 1

Hello ZigBee

Wireless networking standards are everywhere. You've heard of WiFi™ and Bluetooth™ and cellular technology. Perhaps you've heard of Active RFID, Wibree, WiMAX™, or Wireless USB. So, why ZigBee™?

The ZigBee wireless networking standard fits into a market that is simply not filled by other wireless technologies (see Figure 1.1). While most wireless standards are striving to go faster, ZigBee aims for low data rates. While other wireless protocols add more and more features, ZigBee aims for a tiny stack that fits on 8-bit microcontrollers. While other wireless technologies look to provide the last mile to the Internet or deliver streaming high-definition media, ZigBee looks to control a light or send temperature

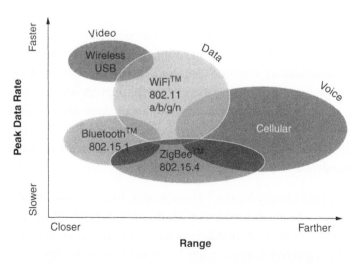

Figure 1.1: Wireless Technologies Compared

data to a thermostat. While other wireless technologies are designed to run for hours or perhaps days on batteries, ZigBee is designed to run for years. And while other wireless technologies provide 12 to 24 months of shelf life for a product, ZigBee products can typically provide decades or more of use.

The market category ZigBee serves is called "wireless sensor networking and control," or simply, "wireless control." In fact, the slogan for ZigBee is, "Wireless Control That Simply Works."

I'll get to that in a moment (the first code example in this book, "Hello ZigBee," simply works). The wireless control market has a number of unique needs for which ZigBee is ideally suited, because ZigBee is:

- Highly reliable

- Cost-effective

- Able to achieve very low power

- Highly secure

- An open global standard

To achieve the low power and low cost criteria, ZigBee added a constraint to the technology:

Low data rate

Now, why would a networking standard want to encourage low data rates? Because ZigBee is all about wireless monitoring and control. Think of a light switch. How often does it need to communicate? Ten times per day? Or perhaps, not at all, on a given day? How large is that data set? By keeping focused on the appropriate technology for wireless control, ZigBee achieves an elegant solution that makes sense and fits a set of markets very well. Just don't expect to stream voice or video over ZigBee. Yes, I have seen companies stream voice over ZigBee, but that's not the intended use, and it consumes way too much bandwidth for most ZigBee networks.

"Wireless Control That Simply Works." Those words, the slogan of the ZigBee Alliance, are oh-so-easy to say and oh-so-hard to do. ZigBee comes pretty close, as you'll see over the course of this book. Unfortunately, making wireless control that simply works from an end-user perspective requires a fair bit of work from the developer and system designer, with some thought taken on how the application is to be solved. ZigBee has the necessary networking primitives to make it possible, and in many cases also has the necessary Application Profiles to make compatibility with other products possible. In this book, I'll show you how to apply ZigBee appropriately to achieve wireless control that simply works.

> ZigBee is a standard networking protocol aimed at the wireless control market.
>
> The ZigBee protocol fits on 8-bit microcontrollers, with 16- and 32-bit solutions available.

1.1 What's a ZigBee?

Before I delve into ZigBee in detail, I'd like to address one question that seems to come up every time I speak about ZigBee in public. Where did the name ZigBee come from?

Well, the way I first heard it was at the ZigBee Open House in Seattle, back in 2004.

This story was told during a time when ZigBee was often confused with Bluetooth™. Just so you don't confuse the two: Bluetooth™ is great at point-to-point (as seen in many headsets and cell phones). ZigBee is great at wireless control, where anywhere from two to thousands of nodes are all connected together, in multi-hop mesh network.

Bob Heile, Chairman of the ZigBee Alliance explained the origin of ZigBee something like this:

A Norwegian legend speaks of a little troll by the name of ZigBee, who lived in the village of Vik far inland on the fjord of Sogn. Now, Norwegian trolls aren't the big, nasty and smelly, hard-as-rock variety often told of in other tales, at least not always. ZigBee was a kindly, quiet little troll, who didn't speak much, but when he did speak it was always reliable. A person could count on ZigBee.

One time, ZigBee sensed that a decomposing pile of hay stacked up against a barn had become too hot and had begun to smolder. ZigBee, in no time at all, sounded the alarm to every house in the village, and the villagers were able to put out the fire before the barn was lost.

Another day, a grandfather left the port of Vik i Sogn in his small fishing boat to catch Salmon with his granddaughter Brita. This day, unlike other days, Brita was not being careful. Bestefar (that is the name Norwegians give to their grandfathers) didn't notice when Brita fell overboard as he was busy hauling in a net full of fish off the stern of the little boat. ZigBee, sensing immediately that Brita had fallen, alerted Bestefar who was able to save her from drowning.

Yet another time ZigBee saved the whole village of Vik. And the way I heard it was this. A local villager by the name of Haarold Bluetooth, was far up in the snow-capped mountains, tending his flock of sheep in the early spring. It had been a warm spring that year, following a particularly hard winter.

The shepherd, Bluetooth, brought his flock to a stream he knew well, but this year he couldn't approach it. The steam had turned into a flooding river from the rapidly melting snow. Alarmed, Bluetooth now wished he could let the villagers know about the flood before it reached the village, but the village was too far away for him to be heard. Bluetooth simply didn't have the range to help the village.

ZigBee, sensing there was trouble, saw the flood as well. And ZigBee, like Bluetooth, realized he was too far away for a single shout to be heard. So he immediately began to hop down the mountain, ledge by ledge until he reached the village. He automatically opened the dam and the flood passed through without harm to the village.

It was a very lucky thing for Vik i Sogn that they had ZigBee, and that ZigBee knew how to multi-hop.

1.1.1   ZigBee Is Highly Reliable

Wireless communication is inherently unreliable. Prove this to yourself by walking around with your cell phone, then step into an elevator. Anyone who has used a cell phone has experienced dropped calls or poor reception. It's all because radio waves are just that: waves. They run into interference patterns, can be blocked by metal, water or a lot of concrete, and vary depending on many complex factors including antenna design, power amplification, and even weather conditions.

Wireless control, however, doesn't usually have the same remedies normally associated with a cell phone call, of moving to find better reception, or waiting to try back later. The ZigBee Alliance understands this, and so the ZigBee specification reflects this need. ZigBee achieves high reliability in a number of ways:

- IEEE 802.15.4 with O-QPSK and DSSS
- CSMA-CA
- 16-bit CRCs
- Acknowledgments at each hop
- Mesh networking to find reliable route
- End-to-end acknowledgments to verify data made it to the destination

The first is relying on a very reliable, low-range wireless technology, the IEEE 802.15.4 Specification. This specification is a very modern, robust radio technology built on over 40 years of experience by IEEE. It uses what's called Offset-Quadrature Phase-Shift Keying (O-QPSK) and Direct Sequence Spread Spectrum (DSSS), a combination of technologies that provides excellent performance in low signal-to-noise ratio environments.

ZigBee uses what's called Carrier Sense Multiple Access Collision Avoidance (CSMA-CA) to increase reliability. Before transmitting, ZigBee listens to the channel. When the channel is clear, ZigBee begins to transmit. This prevents radios from talking over one another, causing corrupted data. CSMA-CA is similar to what people do in conversations. (Or at least should!) They wait for the other speaker to finish, and then talk.

ZigBee uses a 16-bit CRC on each packet, called a Frame Checksum (FCS). This ensures that the data bits are correct.

Each packet is retried up to three times (for a total of four transmissions). If the packet cannot get through after the fourth transmission, ZigBee informs the sending node so something can be done about it.

Another way that ZigBee achieves reliability is through mesh networking. Mesh networking essentially provides three enhanced capabilities to a wireless network: extended range through multi-hop, ad-hoc formation of the network, and most importantly automatic route discovery and self healing.

With mesh networking, data from the first node can reach any other node in the ZigBee network, regardless of the distance as long as there are enough radios in between to pass the message along (see Figure 1.2). Node 1 wants to communicate to node 3, but is out

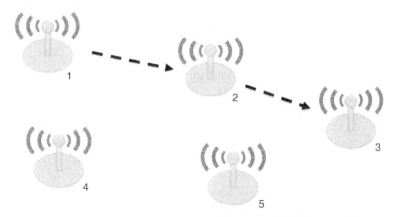

Figure 1.2: ZigBee Mesh Networking

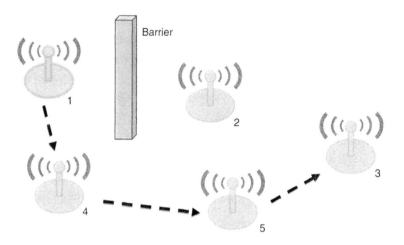

Figure 1.3: ZigBee Mesh Rerouting

of radio range. ZigBee automatically figures out the best path and so node 1 sends the information to node 2, which forwards it on to node 3.

Now suppose that over time, something happens to this route. Perhaps node 2 has been removed or dies, or some barrier intervenes, such as a concrete wall or a large tank of water. This doesn't bother ZigBee at all. ZigBee will automatically detect the route failure and route around the obstruction (see Figure 1.3).

In addition to mesh networking, ZigBee provides reliable broadcasting, a technique for distributing a message to many nodes in the network. ZigBee also provides multicasting, which can send a message to any given group of nodes. And, as a back-up routing technique, ZigBee provides tree routing to augment ZigBee mesh networking in RAM limited systems.

ZigBee also provides automatic end-to-end acknowledgments. Your application can know whether a particular packet was received by the other node. With all of these retries, ZigBee filters out any duplicate received packets, so your application doesn't have to.

All of these concepts are explained in detail in Chapter 7, "The ZigBee Networking Layer."

> ZigBee enhances reliability through mesh networking, acknowledgments and use of the robust IEEE 802.15.4 standard.

1.1.2   ZigBee Is Cost-Effective

Obviously, a book is not the place to look for current prices, but at the time of this writing, ZigBee is pretty cheap. And it looks like it is getting even less expensive over time. A quick search for ZigBee chips (I used http://www.DigiKey.com) yielded inexpensive prices (see Table 1.1).

Most of the entries in the table are the price for a combination MCU and radio. That's right, a single chip solution yields prices in the $3 to $5 range, on average. One of the entries, the Freescale MC13203, is a stand-alone radio that can be combined with any MCU or CPU.

The final two entries, the Digi International XBee and the Panasonic PAN802154, are complete, ready-to-ship, precertified modules. Connect a couple of AA or AAA batteries and a sensor or actuator to these boards, put it in some plastic enclosure, and ship your product.

These prices are for low-volume sales. OEM and higher volume (10,000+ units) knock off about a dollar, as a rule of thumb. And, as the competition for the 802.15.4 market stiffens, the prices will likely be brought down even further. Some have predicted that the 802.15.4 radio market will see prices around $1 in quantity over the next three to five years.

But the low cost of ZigBee isn't just about low silicon costs. In addition to low MCU and radio costs, developing applications for ZigBee is cheap in other ways:

- It uses the 2.4 GHz spectrum for worldwide distribution.

- There are certification houses with expertise in 802.15.4 and ZigBee.

Table 1.1: ZigBee Chip and Module Prices

Chip	Price	Type
Freescale MC13213	$4.43	MCU + radio
Freescale MC13203	$3.64	Radio only
Texas Instruments CC2420	$4.09	MCU + radio
ST Microelectronics SN260QT	$7.25	MCU + radio
Ember EM2420	$4.70	MCU + radio
Ember EM260	$5.77	MCU + radio
Microchip MF24J40	$3.72	MCU + radio
Digi International XBee	$19.00	Module
Panasonic PAN802154HAR00	$27.40	Module

- There are module vendors who provide ready-to-go ZigBee boards.

- Its core technology is free of patent infringements.

- It requires only low cost development environments.

- ZigBee experts are available for consulting or custom engineering.

- There are Application Profiles for ready-made vendor interoperability.

The 802.15.4 radios come in both 2.4 GHz and 915 MHz flavors. Radios in the 2.4 GHz band may be shipped worldwide, license-free (that's governmental RF spectrum license-free, not license-free from the standpoint of vendor licenses). The 915 MHz band is available license-free in the Americas. In either case, new RF boards must be certified by the government to ensure they don't interfere with other products in the intended markets.

Some companies offer easy and relatively low-cost governmental RF certification, including FCC certification in the U.S., IC in Canada, and CE Marking in Europe. Others offer ZigBee certification, such as NTS and TUV.

Another cost saver, available because ZigBee is an international standard, is the availability of modules. A module is a single board computer complete with an MCU and ZigBee RF solution. Module vendors, such as Digi International, Panasonic, and LS Research, provide precertified ZigBee modules, ready to communicate out-of-the-box. Using this approach, the only development cost is software, and even that can be reduced by using the sample programs which come with the ZigBee kits, or the built-in software in the modules.

One other aspect that keeps the cost of ZigBee low is the ZigBee Alliance's careful choice of using patent-free technologies. ZigBee uses AES-128 bit encryption for security, a standard that has no patents associated with it and is freely available worldwide. ZigBee also uses AODV, a mesh networking algorithm in the public domain. New members to the ZigBee Alliance must sign a disclosure statement regarding patents in the area of ZigBee to help prevent members from slipping a patented technology into the specification.

The development environment is generally inexpensive, and kits are available for around $1,500, including the compiler, the ZigBee stack, some ZigBee boards, and all the software and documentation necessary to develop ZigBee applications.

Also, because of the ZigBee Alliance, a lot of expertise is available in the form of consulting companies, training companies, or custom engineering services, saving time for you or your development staff. Developer time can be the single largest expense in a project, and reducing that time can save significant costs. San Juan Software, the company I work for, offers ZigBee services. So does LS Research, Develco, Software Technology Group, and others.

The ZigBee Alliance, which is the alliance of companies who developed and support ZigBee as a standard, has also defined what are called Application Profiles. These profiles describe how ZigBee devices interact, specifically between products of certain types and within certain markets. For example, the Home Automation profile describes how switches can control lights, how a temperature sensor sends its data to a thermostat and how that thermostat controls a heating or cooling unit (heater or air conditioner). A well-defined standard saves effort and money when developing products that need to interoperate with other vendors.

Development costs are explored in more detail in Chapter 2, "Deciding on ZigBee."

> Multiple silicon and stack vendors, ZigBee modules and many available resources all contribute to low development costs for ZigBee devices.

1.1.3 ZigBee Is Low-Power

What does the ZigBee Alliance mean when it says ZigBee is low-power? The simple answer is that devices in a ZigBee network can operate for years on a pair of AA batteries. Depending on the application, these devices can last the entire shelf life of the batteries, which means if you placed the batteries on a shelf at room temperature with nothing connected to them, they would run out of juice just as quickly as they would if they were participating in a ZigBee network, about five years later. No, that doesn't mean ZigBee radios do not consume power, it just means that if managed correctly, devices can last a long time.

Compare this to the time you get on your laptop battery: hours at best. Now, granted, your laptop is powering a bright screen in addition to WiFi™. Compare this to the life of your cell phone battery between recharging, especially one with Bluetooth™ enabled: days at best.

The real secret to low power consumption in ZigBee, in addition to radios and microcontrollers that can sleep, is low duty cycle. A node on a ZigBee network does not need to keep in constant contact with the network to remain on the network. In fact,

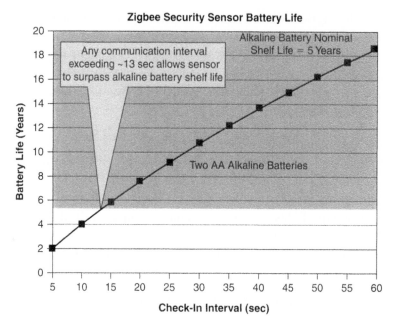

Figure 1.4: ZigBee Battery Life

ZigBee networks are often quite silent. A temperature sensor may report only once per hour, unless the temperature suddenly changes. A light switch may be toggled 6 or 10 times per day, possibly less.

Performing a simple power calculation (see Figure 1.4), demonstrates that a ZigBee node, communicating at once every 13 seconds or less, achieves (or exceeds) the shelf life of the battery. I've provided a spreadsheet which will allow you to do the calculation yourself on the ZigBee Wireless Networking Web site at http://www.zigbookexamples.com.

I do a demonstration in my training classes where during one part of the lecture, we set up a room-wide ZigBee network with 30 or so nodes. Then, 20 minutes later after I'm finished explaining some ZigBee concepts including the low-duty cycle, I show the class the sniffer that has been running that entire time, capturing all of the over-the-air packets which have been transmitted by the ZigBee nodes. To their surprise, no ZigBee packets have been exchanged over-the-air. Then the class presses buttons to wake the application (switches and lights), and the packets fly. For those experienced with other networking protocols, they are really impressed.

How to implement low power in ZigBee devices, and some of the caveats and tricks, are described in Chapter 5, "ZigBee, ZDO, and ZDP."

> ZigBee devices can achieve 5 years on a pair of AA batteries.

1.1.4   ZigBee Is Highly Secure

For securing the network, ZigBee uses the National Institute of Standards and Technology (NIST) Advanced Encryption Standard (AES). This standard, AES-128, is a block cipher that encrypts and decrypts packets in a manner that is very difficult to crack. It's one of the best-known and well-respected standards. The reason it was adopted by ZigBee was for the following key reasons:

- It's an internationally recognized and trusted standard.

- It's free of patent infringements.

- It's implementable on an 8-bit processor.

From the NIST Web site, I found this text describing the use of AES (also called Rijendael), with regards to patents:

> *I, Joan Daemen, do hereby declare that to the best of my knowledge the practice of the algorithm, reference implementation, and mathematically optimized implementations, I have submitted, known as Rijendael may be covered by the following U.S. and/or foreign patents: none.*

> *I do hereby declare that I am aware of no patent applications which may cover the practice of my submitted algorithm, reference implementation or mathematically optimized implementations.*

ZigBee provides both encryption, which means that packets cannot be understood by listening nodes who are not aware of the key, and authentication, which means a malicious node cannot inject false packets into the network and expect the ZigBee nodes to do anything with them other than throw them away. ZigBee has been very careful to ensure the security solution.

To learn more about ZigBee security, see Chapter 7, "The ZigBee Networking Layer."

> ZigBee uses AES 128-bit security for encryption and authentication.

1.1.5   ZigBee Is an Open Global Standard

Go ahead, go to http://www.zigbee.org, and download the ZigBee specification right now.

Or go to http://www.freescale.com/zigbee and download BeeKit, the Freescale ZigBee solution. Prefer TI? Use the Texas Instruments ZigBee solution instead at http://www.ti.com/zigbee. Prefer a European company? Try http://www.st.com/zigbee. Prefer an Asian company? Try http://www.japan.renesas.com/zigbee.

However you look at it, ZigBee is global, ZigBee is open, and ZigBee is standardized.

ZigBee uses as its foundation the IEEE 802.15.4 specification for the lower MAC and PHY layers. IEEE defines a reliable radio standard in the 2.4 GHz band that may be used worldwide. This standard may be freely downloaded from www.ieee.org.

The ZigBee specification was developed by the ZigBee Alliance, a standards body with over 250 member companies from every continent in the world (except for Antarctica, and the ZigBee Alliance is still looking for a company to participate from there as well).

The ZigBee Alliance also specifies a test suite, with test cases that enforce compatibility among vendors. For vendors to receive the coveted ZigBee Compliant Platform (ZCP) certification, they must pass this rigorous test suite. The ZigBee Alliance regularly hosts ZigFests that encourage ZigBee vendors to get together and verify that their stacks and applications work together.

In addition to stack-level compatibility, ZigBee also defines application level compatibility through Application Profiles. Application Profiles describe how various application objects connect and work together, such as lights and switches, remotes and televisions. The Application Profiles also specify a test suite for verifying compatibility among applications from various vendors.

In fact, a ZigBee network is expected to have products from many vendors that all interoperate. A home network may look like the one shown in Figure 1.5.

The remote control, with one touch, dims the Philips lights (and adjusts them from a cool to a warm tone), lowers the Somfy Glystro automated shades, turns on the Panasonic wide-screen television and Sony DVD player, and starts the movie. Each of these devices contains ZigBee radios from different vendors: Freescale, Texas Instruments, Renesas, Ember, and ST Microelectronics. Your wife comes in, says, "Hello," and touches the light switch. The lights brighten, and she finds her purse and leaves. As she leaves,

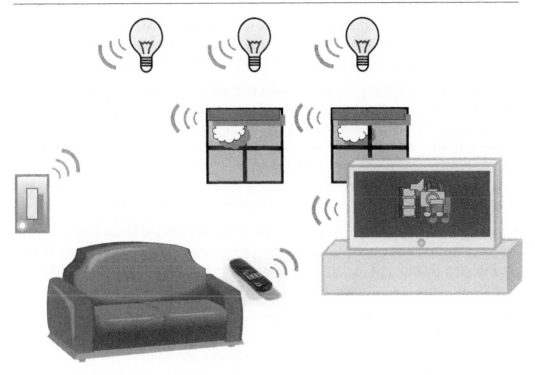

Figure 1.5: Multiple ZigBee Vendors in Home Automation

she touches the light switch again which dims the lights back to your favorite settings for movies.

Later, both of you are headed out to dinner. A single press of the "We're going out" button on the lower left of the light switch turns everything off 30 seconds after sensors detect the last motion in the house, all automatically. The convenience and power savings are all possible due to ZigBee wireless devices.

> Multiple vendors offer ZigBee stacks, silicon and application solutions.
>
> The ZigBee specification can be downloaded for free from http://www.zigbee.org.

1.1.6   ZigBee Is Low Data Rate

In order to achieve low cost and low power, and considering the application space and markets ZigBee is aiming for, the ZigBee Alliance decided to keep the protocol designed for a low data rate environment.

ZigBee resides on IEEE 802.15.4 transceivers, which in the 2.4 GHz space communicate at 250 kilobits per second (kbps), but by the time retries, encryption/decryption, and the fully acknowledged mesh protocol is applied, the actual through-put is closer to 25 kbps.

The transceivers are half duplex, which means they are either transmitting or receiving, but not both at the same time, which also is a factor in reducing the effective through-put from 250 to 25 kbps. The radios are also expected to share channels, perhaps with other ZigBee networks or other wireless technologies. ZigBee networks enjoy upto 16 channels in the 2.4 GHz space, separated by 5 MHz each. Each channel is physically separated from other channels, but ZigBee is expected to operate in environments where the number of networks might be fairly dense (think of apartments and condominiums).

ZigBee transceivers share the 2.4 GHz spectrum with other wireless technologies, including WiFi™, Bluetooth™, some cordless phones and even microwave ovens. To be good spectral citizens, and co-exist with these other technologies, ZigBee has chosen to keep the data rates in applications low. ZigBee is a very quiet protocol.

Currently, there are no ZigBee stacks that run on the 802.15.4 915 MHz radios, but that will probably be up and coming when 802.15.4b is adopted, which has higher data rates than the 40 kbps at which the 915 MHz radios currently communicate.

> ZigBee through-put is typically 25 kilobits per second.

1.2   The ZigBee Alliance

The ZigBee Alliance itself was formed in 1997 by eight Promoter companies to "enable reliable, cost-effective, low-power, wirelessly networked monitoring and control products based on an open global standard." Since that time, the ZigBee Alliance has swelled to over 200 member companies, many of them the market leaders in their respective markets (see Figure 1.6). It continues to grow.

Some of you may remember the early days of IEEE 802.11, a wireless networking standard touted as the next great thing for connecting laptop computers to the office network. Only it didn't work, at least not at first. An 802.11 card purchased from one vendor was likely not to recognize an 802.11 router purchased from another vendor. Security didn't work well. All that changed when the WiFi Alliance formed and brought the vendors together to standardize their product. Now you can't throw an iced latté without hitting a WiFi™ hot spot, all due to standards.

Figure 1.6: The ZigBee Alliance Members—Over 200 Companies and Growing!

Having learned the past lessons of other standards, the ZigBee Alliance is making sure to roll out the technology in a way that enhances the probability of both technical and market success. First of all, the standard is focused on an appropriate technology for wireless control networks, the IEEE 802.15.4 radio standard with ZigBee networking built on top. The ZigBee Alliance is also controlling the inevitable early hype until the technology is mature enough to meet the market needs and is proven in the field. And the standard is released publicly to allow comments and critical review from those outside the standards body.

To obtain a copy of the ZigBee Specification, simply download it from ZigBee Alliance Web site at http://www.zigbee.org (see Figure 1.7). Member companies of the ZigBee Alliance enjoy early access to the specification, and have the opportunity to provide input into the specification to make sure it is right for their markets and products.

The ZigBee Alliance maintains a web site with the latest ZigBee news and events, and the latest publicly available technical specification. By early 2008, the ZigBee specification had been downloaded over 73,000 times.

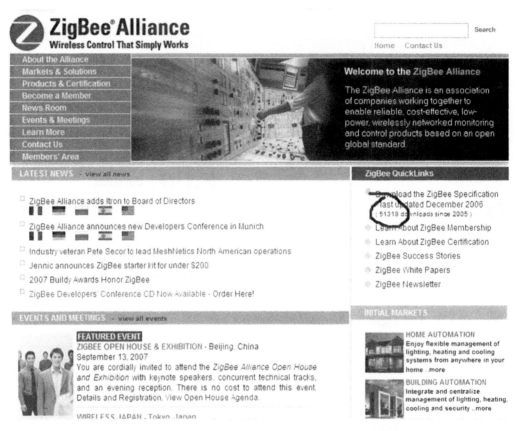

Figure 1.7: The ZigBee Alliance Web Site (http://www.zigbee.org)

Each quarter, the ZigBee Alliance hosts an open house that allows potential ZigBee developers a firsthand look at the products and services offered by various ZigBee vendors, as well as a sneak peak at upcoming end-user products from Original Equipment Manufacturers (OEMs). As an international organization, ZigBee hosts these events around the world. Each year, two are in North America, one is in Asia, and one is in Europe. For example, in 2008, one open house was scheduled in Tokyo, Japan, and another was scheduled in Atlanta, GA.

The ZigBee Alliance also hosts booths at the many wireless and embedded shows worldwide. If you're independently wealthy, love ZigBee, and have too much time on your hands, you could spend every week of the year at some ZigBee event!

Twice a year, the ZigBee Alliance hosts the ZigBee Developers Conference. This conference is for the serious developer who really wants to understand ZigBee in-depth.

Developers get hands-on experience developing a variety ZigBee platforms, and even get a chance to deploy networks around the conference area using standard commissioning techniques.

1.2.1 The ZigBee Alliance Ecosystem

Probably the single most important reason to use ZigBee in a product is that the technology is standardized. There are already lots of proprietary wireless products that cover various markets, from home automation to industrial automation. There are proprietary mesh networks, even ones that run on the 802.15.4 radios. But standards bring two things that proprietary systems cannot:

- Many diverse companies using and defining the standard

- Economies of scale

The ZigBee Alliance actively seeks an entire ecosystem of companies These range from silicon vendors (such as Freescale, ChipCon, Atmel, Ember, and others) to original equipment manufacturers (OEMs) which are researching and developing products designs using ZigBee companies (such as Philips Lighting, a world leader in lighting products, Invensys, a world leader in smoke alarms, Mitsubishi, Johnson Controls, Trane, and Eaton, all leaders in heating, ventilation, and air-conditioning (HVAC)). The ZigBee Alliance also comprises of tool vendors, test houses, NRE and consulting companies, training companies, and ready-made ZigBee module vendors which provide precertified modules ready to be incorporated into products.

So what does the ZigBee Alliance offer to you? Choices. You have a choice of radio vendors, a choice of stack vendors, and a choice of tool vendors. It also means a broader and deeper technology. Someone out there has probably already written a large portion of the software needed for your product, or a hardware vendor has created a platform which can reasonably meet your field trial needs, allowing you to get to market much more quickly and less expensively than could be done with a proprietary solution.

In addition, standards often lead to surprising technologies and markets. Who would have thought of the World Wide Web when TCP/IP was first invented? Now you have email, browsers, Google, Amazon.com, eBay, Voice over IP (VoIP), and blogs. Imagine what might happen when wireless sensor and control networks are cheap and connected to the world via the Internet? What applications might be possible?

As one industry pundit put it:

> *Just as the personal computer was a symbol of the '80s, and the symbol of the '90s was the World Wide Web, the next nonlinear shift, is going to be the advent of cheap sensors.*
>
> *—Paul Saffo, Institute for the Future.*

I just ordered my Apple iPhone™ online. When it arrived, I activated it in minutes over the Web, synced it with iTunes® from my laptop and now have a phone that does everything but wash my dishes.

What changes can we expect with wireless sensor and control networks? With ZigBee?

1.2.2   ZigBee Alliance Certification

The ZigBee Alliance is also the gatekeeper who enforces compliance and compatibility. Through its test houses, the ZigBee Alliance provides certification, both for ZigBee platforms (radio and stack) and for ZigBee end products. The two ZigBee Alliance approved test houses for ZigBee certification are NTS and TUV.

ZigBee Certified Platforms (ZCP) mean that the combination of radio and stack have been certified to interoperate with other stack vendors. ZigBee uses four "Golden Units," each from a different platform vendor to ensure compatibility. The four Golden Units are currently from Freescale, Texas Instruments, Ember, and Integration. There are also ZigBee stacks out there that are not certified, but if you use one of these, don't count on interoperability with other stacks.

Applications interoperate through what is called an Application Profile. With each Application Profile, test suites are defined and used by test houses to ensure that products interoperate with each other if they are within the same application domain, or that they at least do no harm to other ZigBee networks if they are proprietary applications.

When an application completes certification, it is then approved to use the ZigBee logo on products. I'll describe more about the (current) fees and process of certification in Chapter 2, "Deciding on ZigBee."

1.2.3   ZigBee Alliance Membership

I can't say it better, so I'll just repeat what you see when you look at the "Become a Member" page on the ZigBee Alliance Web site:

> *Help create the future of wireless data networking. As a ZigBee Alliance member, you can actively shape and take advantage of the evolution in wireless monitoring and control*

markets worldwide. By participating in member meetings, committees, and working groups, your company can help shape the ZigBee standard for reliable, secure, and low-power wireless communications. Companies from virtually every industry and every country are joining the ZigBee Alliance.

Becoming a member of the ZigBee Alliance is easy: simply fill out the online form, sign the license agreement, and pay some money. The price depends on the class of membership. For OEMs that usually means the least expensive Adopter class. There are three classes of ZigBee Alliance membership:

- Promoter

- Participant

- Adopter

The Promoter membership class must be approved by vote of the other Promoters within the ZigBee Alliance. Promoters get first-pick of booth location at trade shows, are offered the first chance at sponsorship for various events, and receive the right to vote on all proposals. They can even veto a decision made by the Participants within the Alliance, although I've never seen this happen. At the time of this writing there are 12 Promoter members.

Participants, along with Promoters, are ZigBee Alliance members who develop the ZigBee specification. They vote on all the specifications produced by the ZigBee Alliance, including the main ZigBee specification, the test suite specifications, the ZigBee Cluster Library and Application Profiles.

Adopters make up the bulk of the membership. These members are typically OEMs. They gain early access to the specifications and may participate in the application-working group meetings, but they do not receive voting rights, nor do they actively participate in the ZigBee specification itself, only in the Application Profiles.

ZigBee Alliance membership is required in order for OEMs to ship products with ZigBee technology inside. ZigBee can be used free of charge by universities and other non-profit organizations.

The ZigBee Alliance is very much a volunteer organization. Not only do members pay to be part of the Alliance, but they pay their own way to the quarterly face-to-face meetings and spend their own time developing and reviewing specifications. The ZigBee Alliance hires Global Inventures to maintain the Web site and market and promote ZigBee as a standard worldwide. The chairman of the ZigBee Alliance is a paid position as well. But the rest of us work for free. And it's worth it.

The real benefits of membership include early access to the specifications (get that jump on your competition) and ready access to experts who have many years of experience in wireless control, which can help you avoid costly pitfalls.

If your company wants to drive a particular market, providing input into the ZigBee Application Profiles can make sure the profile covers your customer's needs. The ZigBee Alliances encourages multiple vendors in a given market to provide their input into the specifications. The Application Profiles then become stronger and are better suited to their intended markets. In fact, ZigBee won't approve a specification until at least three vendors provide interoperable products or platforms. Just don't expect it to happen quickly; it's usually a year before an Application Profile matures from conception to specification.

If your company is planning to provide products or services for the wireless control space, becoming a ZigBee Alliance member can be a good investment.

> ZigBee Alliance membership is required in order to ship ZigBee technology in products. Alliance membership provides early access to specifications.

1.3   ZigBee in the Marketplace

ZigBee is beginning to be found in many, many applications.

The uses of ZigBee span so many markets it is difficult to categorize them in a single paragraph. Think of the devices around you. How many of them currently have wires to control them? Think of lights and light switches. (Hint: Philips Lighting is heavily involved in ZigBee.) Think of heating, ventilation, and air-conditioning. (Another hint: Schneider Electric, Mitsubishi, Eaton, Trane, and Johnson Controls are also involved in ZigBee.) Think advanced electrical, gas, and water metering (Itron, CellNet), irrigation systems that water on demand rather than on a schedule, industrial plants that monitor and control production, medical devices that free a patient to go home early from a hospital. A week doesn't go by when I don't hear of another product concept which is using or could be using ZigBee.

As of this writing, ZigBee has not yet started to appear in the Home Depot or Fry's Electronics, but it has been successfully deployed in both industrial and commercial products, operating in hotels, office buildings, and factories (see Figure 1.8). Over the next few years you can expect to see consumer devices with the ZigBee logo on them in

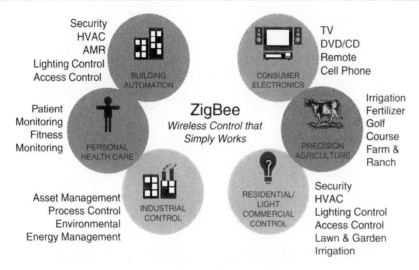

Figure 1.8: ZigBee Markets

cell phones and home automation products. The ZigBee Alliance targets a set of markets that make good sense for this technology.

While ZigBee cannot (or should not) send video over-the-air, it can control the AV system and in-home theater. ZigBee can also control pool pumps, lights, and HVAC equipment. Wires are notoriously difficult in agriculture and on golf courses. ZigBee fits well with automated control and monitoring in these environments. ZigBee is highly secure, and so works well for markets that target secure access.

The ZigBee specification that you download describes how ZigBee works as a networking protocol. What is not included (at least at the time of this writing) is how applications interact through what are called Application Profiles. To access the Application Profile specifications, you must become a member of the ZigBee Alliance.

Application Profiles describe the over-the-air behavior of devices in each of the wireless control domains. They include, at the time of this writing:

- Home Automation
- Commercial Building Automation
- Industrial Plant Monitoring
- Telecommunications Applications

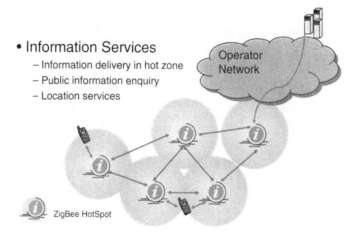

Figure 1.9: Telecom Application Information Services

- Automatic Metering Initiative

- Personal Home and Health Care

In Telecommunications Applications (TA), for example, companies are creating products that can introduce location-based services (see Figure 1.9). These services could advertise specials, services, or sales (at your option) based on the location of your cell phone. The cell phone company already knows where your phone is from a cell tower perspective, but that can be a 15-mile radius or more, far too broad for location-based services.

ZigBee, as both a low-power and shorter-range protocol, can provide local shops and restaurants a way to reach you as you walk a city or village street. Perhaps you are hungry and your cell phone informs you that a few blocks away, a cozy bistro is open, tucked into an interesting street corner. Normally, you would not have seen this restaurant unless you knew someone locally. Now, you have the most memorable meal of the year due to ZigBee.

In the Personal Home and Health Care (PHHC) arena, ZigBee can monitor a patient's condition, including heart rate, movement (to detect a fall), blood pressure and other medical information and can communicate this data securely to the hospital. The doctors and nurses can be kept apprised of the status of the patient, sending a staff member to the house immediately in the rare case of an emergency. This frees up the hospital bed early on, making medical care both less expensive and more comfortable for the patient.

Home Automation, one of the early ZigBee Application Profiles, covers all of the devices ZigBee expects to see in the home (see Table 1.2).

Table 1.2: ZigBee Home Automation Devices

Main Structure (194)

- 14 Window sensors (battery)
- 7 External doors (battery)
- 4 Door latches/deadbolts (battery/solar)
- 11 Internal doors (battery)
- 40 Internal and external light fixtures (mains)
- 26 Circuit breaker monitors (mains)
- 15 Occupancy/motion sensors (battery)
- 6 Thermostats (battery)
- 10 Control dampers (battery or mains)
- 1 Heat pump (mains)
- 1 Air blower (mains)
- 4 Pipe/Water monitors (battery)
- 1 Oven/Burner sensor (mains)
- 3 Ceiling ventilator/heater control/monitors (mains)
- 5 Smoke detectors (battery or mains/battery)
- 3 CO detectors (battery)
- 6 Controllable outlets (mains)
- 2 Standing water detectors (battery)
- 1 Earthquake shutoff sensor (battery)
- 1 Electric meter monitor (battery)
- 1 Gas meter monitor (battery)
- 1 Gas main shutoff (battery)
- 1 Water meter monitor (battery)
- 1 Water main shutoff (battery)
- 10-in wall mold/moisture monitors (battery)
- 2 Fireplace flue/damper monitors (battery)
- 2 Attic ventilators (mains)
- 4 Attic temperature/moisture sensors (battery)
- 5 Under-house crawl space moisture monitors (battery)
- 2 Under-house crawl space ventilators (mains)
- 3 Ceiling ventilator/heater control/monitors (mains)
- 5 Smoke detectors (battery or mains/battery)
- 3 CO detectors (battery)
- 6 Controllable outlets (mains)
- 2 Standing water detectors (battery)
- 1 Earthquake shutoff sensor (battery)
- 1 Electric meter monitor (mains)
- 2 Doorbell buttons (battery), 1 doorbell (mains)
- 2 Video Snapshot security cams (battery)

Garage (16)

- 2 Door sensors (battery)
- 2 Door latches/deadbolts (battery)
- 1 Garage door opener (mains)
- 2 Garage door/home remotes (battery)
- 2 Occupancy/parking space/position monitor (battery)
- 3 Light fixtures (mains)
- 3 Remote light switches (battery)
- 1 Driveway monitor (battery/solar)

Landscaping (53)

- 4 Low-voltage light strings (mains)
- 20 Sprinkler heads (battery)
- 3 Pool pump, filter, heater (mains)
- 10 Spot watering devices (battery)
- 4 Sets semi-permanent holiday lights (mains)
- 1 Pool monitor (battery)
- 1 Gate monitor (battery)
- 10 Landscape moisture monitors (battery)

Other (30)

- 6 Exterior hose bib monitors (battery)
- 1 Mailbox monitor (battery/solar)
- 1 Water softener monitor (battery/mains)
- 1 Pet door control (battery)
- 2 Pet ID tags (battery)
- 1 "Change-the-litter" monitor (battery)
- 1 Water bowl monitor (battery)
- 3 Televisions (mains)
- 5 Major appliances (mains/battery)
- 3 Sets remote speakers (mains/battery)
- 1 Cable settop box (mains)
- 1 Sound system (mains)
- Many remote controls (battery)
- 1 Perimeter fence monitor (battery/solar)

At first blush, this list looks ridiculous. Is it realistic to have nearly 300 wireless devices in the home? But read down the list. Many of these devices you already have in your home (and probably wired). What if you were making a new home or doing a remodel and could save thousands of dollars? What if you just want the added convenience of a one-touch "I'm going to work" or "I'm going on vacation" button that would set all the devices in your home to a preset condition, such as turning out all the lights but a few that do a random pattern, or turning down the air-conditioning or heating in all rooms but the one with your prize orchid collection? What if this also saved you money on your energy bill and paid for itself in a few years?

What if you wanted the convenience of a TV remote control that with one touch would turn down the lights to your preferred setting, lower the blinds and turn on your surround-sound system, television, and DVD player, and even turn the telephone ringer off? Better yet, what if the act of just inserting the DVD into the player did all that?

What if the entire state of your home was accessible to you from the Web? What if you could let your friend or neighbor into the house from your cell phone?

Many of these systems are realities today in higher-end systems (just visit http://www.crestron.com for some examples). But, like the cell phone which once was only for the elite, these features will become commonplace in the not-too-distant future. And many of them will be on ZigBee.

The next couple of examples are based on the ZigBee Home Automation Profile. They start with the basics: a simple light and a switch. While not all that sophisticated, lights and switches use nearly all the commands that any other device might use, and they provide instant feedback to the user (just press the button, and watch the remote light shine).

> ZigBee covers many markets, including home, commercial and industrial automation, medical and location-based services.

1.4   Hello ZigBee (A First ZigBee Network)

It wouldn't be a programming book without a "Hello World" program. And in ZigBee terms, "Hello World" is an on and off light and switch. In this example, you'll get your first experience running a ZigBee network and get your first quick first look at ZigBee source code.

The concept of this example is very simple: two boards collectively form a ZigBee network, one of which is a switch, and the other, a light (see Figure 1.10). The switch is able to remotely, wirelessly, toggle the light on and off. The networked is formed and a connection is made between the two devices automatically: Simply turn them both on.

To find the source code to this example (Example 1-1), go to http://www. zigbookexamples/code.

The boards used for this example are Freescale Sensor Remote Boards (SRBs), which are available in a variety of Freescale ZigBee development kits. The kit I used was the Freescale 1321xNSK (Network Starter Kit), which can be found from the Freescale ZigBee home page http://www.freescale.com/zigbee.

Look on that page for "Hardware Development Kits," and it will lead you right to it. I'm making the assumption, of course, that Freescale hasn't changed the site significantly since I've written this. But even if they have, the kit won't be hard to find anyway.

The original version of this demo ran in October 2006 during the development of the Freescale ZigBee Home Automation applications. For fun, rather than blinking an LED

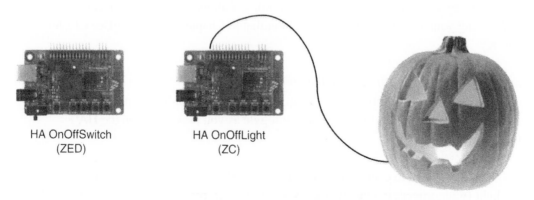

HA OnOffSwitch
(ZED)

HA OnOffLight
(ZC)

Figure 1.10: Hello ZigBee

on and off, the developers controlled a lighted pumpkin (it was close to Halloween) and his nickname was, of course, "Jack." The developer of the application, Dalila, dutifully made C macro constants for these commands. There was **gCmdJackToggle_c** to toggle the light, **gCmdJackOn_c** to turn the light on, and of course **gCmdJackOff_c** to turn the light off. Correct naming convention or not, our development staff still hasn't let Dalila live it down.

You'll notice the figure includes the terms (ZED) and (ZC) in parentheses underneath the two nodes. These are ZigBee acronyms which stand for ZigBee End-Device and ZigBee Coordinator. ZigBee End-Devices (ZEDs) are designed to be low-power boards, able to run for years on a pair of AA batteries. A stick-on-the-wall light switch is a classic example of this type of device. ZigBee Coordinators (ZCs) are the only ZigBee device-type that can form (as opposed to join) a network. In this small two-node network, the light, which is probably mains powered, is a natural choice for the ZigBee Coordinator.

You'll also notice the figure includes the terms HA OnOffSwitch and HA OnOffLight. The ZigBee Home Automation (HA) Application Profile describes a large set of standard devices that are expected to be wireless in the home, including switches and lights. Think of an HA OnOffSwitch as the type of switch you put into your wall to control lights, or anything else that has two states.

ZigBee is generally like that, defining general-purpose over-the-air interfaces. This same switch could be bound (connected) to a ceiling fan to cool the room, or to your water main to shut off water with the flick of a switch, or to even set your entire home into "vacation" mode: turning down the heat, turning off all the lights but a select few, and arming your security system. More about that in Chapter 6, "The ZigBee Cluster Library."

So go ahead, turn on both devices, then press switch one (SW1) on the HA OnOffSwitch and watch the remote-controlled pumpkin light up! If you don't have a pumpkin to light, LED2 on the HA OnOffLight will toggle as you press SW1 on the HA OnOffSwitch.

If you actually want to control a pumpkin (or any other external light), simply connect the light (with proper circuitry) to the 3V GPIO Port B, pin 6, which physically translates

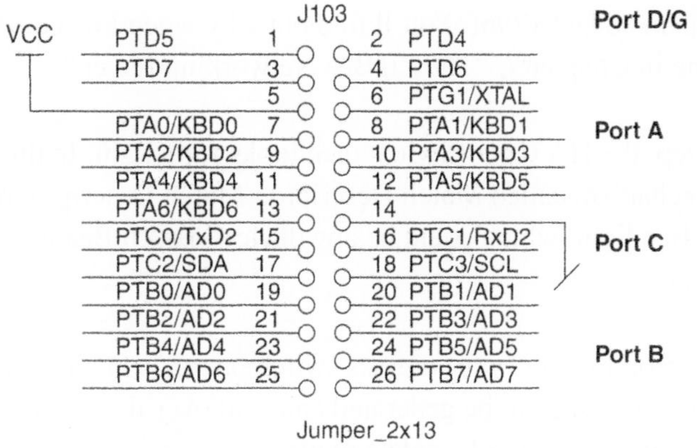

```
                              J103              Port D/G
     VCC    PTD5       1   ○  ○   2   PTD4
       ┐    PTD7       3   ○  ○   4   PTD6
       │              5   ○  ○   6   PTG1/XTAL
            PTA0/KBD0  7   ○  ○   8   PTA1/KBD1   Port A
            PTA2/KBD2  9   ○  ○  10   PTA3/KBD3
            PTA4/KBD4  11  ○  ○  12   PTA5/KBD5
            PTA6/KBD6  13  ○  ○  14
            PTC0/TxD2  15  ○  ○  16   PTC1/RxD2   Port C
            PTC2/SDA   17  ○  ○  18   PTC3/SCL
            PTB0/AD0   19  ○  ○  20   PTB1/AD1
            PTB2/AD2   21  ○  ○  22   PTB3/AD3
            PTB4/AD4   23  ○  ○  24   PTB5/AD5    Port B
            PTB6/AD6   25  ○  ○  26   PTB7/AD7
```

Jumper_2x13

Figure 1.11: SRB External Pins

to pin 25 on the external 26-pin connector located at the back of the SRB board. The illustration of the pumpkin shows where to connect it (see Figure 1.11). The schematic to the SRB and all other Freescale reference boards are available for download from the Freescale Web site. Also, see the Freescale MC9S08GT60 Data Sheet for more information about GPIO pins on this 8-bit microcontroller.

So what happens from a ZigBee perspective?

- The HA OnOffLight (ZC) forms the network.

- The HA OnOffSwitch joins the network.

- The HA OnOffSwitch finds the light and binds to it.

- The HA OnOffSwitch can toggle the light with SW1.

During the first step, the HA OnOffLight, which is the ZigBee Coordinator forms the network. In this case, the light forms the network on channel 25, one of the 16 channels available to 2.4 GHz 802.15.4 ZigBee radios. Channels are numbered 11–26, in ZigBee terms. The network has a particular PAN ID, in this case 0x3bab.

During the second step, the HA OnOffSwitch then joins the network, and it receives a short address (also called a network address) from the HA OnOffLight when it joins.

In this case it happens to be 0x796f. You'll find out why when I discuss CSkip and ZigBee tree routing in Chapter 7, "The ZigBee Networking Layer."

During the third step, the HA OnOffSwitch then looks for a light. In this case it's using a ZigBee mechanism called Match Descriptor. Match Descriptor returns the address of the ZigBee light, which could be any device in a ZigBee network with hundreds of devices.

During the fourth step, the user (you) presses switch one (SW1) on the HA OnOffSwitch. This causes a ZigBee message to be generated and sent over-the-air. This process is called an APSDE-DATA.request, or simply, a data request.

The packets generated over-the-air are shown in the screen capture. If you look at packet sequence number (Seq. No.) 5, you'll see the HA OnOffSwitch joining the network with an Association Request. There is no line to show the HA OnOffLight forming the network as ZigBee is silent when it forms a network, other than issuing a Beacon Request to make sure there are no conflicts.

Look at Seq. No. 11 and you'll see the HA OnOffSwitch sending out the Match Descriptor request. It receives the response on Seq. No. 17, and now knows where the light can be found in the network. Granted, this is overkill for a network of two devices. ZigBee could have just as easily sent directly to node 0x0000, the light, but this code works in the more general case as well with a network of any size.

Then, in Seq. No. 22, you can see the HA OnOffToggle command go over-the-air to toggle the remote HA OnOffLight. Really, not very many packets at all considering that ZigBee has formed a two-node network, commissioned the devices to speak to each other, and controlled a remote device:

Seq No.	MAC Src	MAC Dest	Nwk Src	Nwk Dest	Protocol	Packet Type
1		0xffff			IEEE 802.15.4	Command: Beacon Request
2		0xffff			IEEE 802.15.4	Command: Beacon Request
3		0xffff			IEEE 802.15.4	Command: Beacon Request
4	0x0000				IEEE 802.15.4	Beacon:
5	0x0050c211dc051801	0x0000			IEEE 802.15.4	Command:Association Request
6					IEEE 802.15.4	Acknowledgment
7	0x0050c211dc05180	0x0000			IEEE 802.15.4	Command: Data Request
8					IEEE 802.15.4	Acknowledgment
9	0x0050c210080c9c13	0x0050c211dc051801			IEEE 802.15.4	Command: Association Response
10					IEEE 802.15.4	Acknowledgment
11	0x796f	0x0000	0x796f	0xfffd	Zigbee APS Data	ZDP:MatchDescReq
12					IEEE 802.15.4	Acknowledgment
13	0x0000	0xffff	0x796f	0xfffd	Zigbee APS Data	ZDP:MatchDescReq
14	0x0000	0xffff	0x796f	0xfffd	Zigbee APS Data	ZDP:MatchDescReq
15	0x796f	0x0000	0x796f		IEEE 802.15.4	Command: Data Request
16					IEEE 802.15.4	Acknowledgment
17	0x0000	0x796f	0x0000	0x796f	Zigbee APS Data	ZDP:MatchDescRsp
18					IEEE 802.15.4	Acknowledgment
19	0x0000	0xffff	0x796f	0xfffd	Zigbee APS Data	ZDP:MatchDescReq
20	0x796f	0x0000			IEEE 802.15.4	Command: Data Request
21					IEEE 802.15.4	Acknowledgment
22	0x796f	0x0000	0x796f	0x0000	Zigbee APS Data	HA:On/off
23					IEEE 802.15.4	Acknowledgment
24	0x796f	0x0000			IEEE 802.15.4	Command: Data Request
25					IEEE 802.15.4	Acknowledgment
26	0x796f	0x0000	0x796f	0x0000	Zigbee APS Data	HA:On/off
27					IEEE 802.15.4	Acknowledgment
28	0x796f	0x0000			IEEE 802.15.4	Command: Data Request
29					IEEE 802.15.4	Acknowledgment
30	0x796f	0x0000			IEEE 802.15.4	Command: Data Request
31					IEEE 802.15.4	Acknowledgment

There are a lot of other packets shown in the capture, and all of them will become clear to you by the end of this book, but one thing at a time. First, examine the toggle command in a bit more detail. The packet is 30 octets in length.

```
Frame 22 (Length = 30 bytes)
   Time Stamp: 14:00:55.739
   Frame Length: 30 bytes
   Capture Length: 30 bytes
   Link Quality Indication: 122
   Receive Power: -58 dBm
IEEE 802.15.4
   Frame Control: 0x8861
        .... .... .... .001 = Frame Type: Data (0x0001)
        .... .... .... 0... = Security Enabled: Disabled
        .... .... ...0 .... = Frame Pending: No more data
        .... .... ..1. .... = Acknowledgment Request: Acknowledgement
required
        .... .... .1.. .... = Intra PAN: Within the PAN
        .... ..00 0... .... = Reserved
        .... 10.. .... .... = Destination Addressing Mode: Address field
contains a 16-bit short address (0x0002)
        ..00 .... .... .... = Reserved
        10.. .... .... .... = Source Addressing Mode: Address field
contains a 16-bit short address (0x0002)
   Sequence Number: 57
   Destination PAN Identifier: 0x3bab
   Destination Address: 0x0000
   Source Address: 0x796f
   Frame Check Sequence: Correct
ZigBee NWK
   Frame Control: 0x0048
        .... .... .... ..00 = Frame Type: NWK Data (0x00)
        .... .... ..00 10.. = Protocol Version (0x02)
        .... .... 01.. .... = Discover Route: Enable route discovery (0x01)
        .... ...0 .... .... = Multicast
        .... ..0. .... .... = Security: Disabled
        .... .0.. .... .... = Source Route
        .... 0... .... .... = Destination IEEE Address: Not Included
        ...0 .... .... .... = Source IEEE Address: Not Included
        0. .... .... = Reserved
   Destination Address: 0x0000
   Source Address: 0x796f
   Radius = 10
   Sequence Number = 190
```

```
ZigBee APS
    Frame Control: 0x00
        .... ..00 = Frame Type: APS Data (0x00)
        .... 00.. = Delivery Mode: Normal Unicast Delivery (0x00)
        ...0 .... = Indirect Address Mode: Ignored
        ..0. .... = Security: False
        .0.. .... = Ack Request: Acknowledgement not required
        0... .... = Reserved
    Destination Endpoint: 0x08
    Cluster Identifier: On/off (0x0006)
    Profile Identifier: HA (0x0104)
    Source Endpoint: 0x08
    Counter: 0xd8
ZigBee ZCL
    Frame Control: 0x01
        .... ..01 = Frame Type: Command is specific to a cluster (0x01)
        .... .0.. = Manufacturer Specific: The manufacturer code field
shall not be included in the ZCL frame. (0x00)
        .... 0... = Direction: From the client side to the server side.
(0x00)
        0000 .... = Reserved: Reserved (0x00)
    Transaction Sequence Number: 0x42
    Command Identifier: Toggle (0x02)
```

As is typical of many ZigBee application-level commands, there are four frames in this packet:

- The IEEE 802.15.4 (or MAC) frame

- The ZigBee network (NWK) frame

- The ZigBee application (APS) frame

- The ZigBee Cluster Library (ZCL) frame

The MAC frame is responsible for the PAN ID and the per-hop information. The NWK frame is responsible for multi-hop communication. The APS frame is responsible for the applications and the ZCL frame contains specific application commands.

Notice that the Profile Identifier (0x0104) in the ZigBee APS frame is called HA, which stands for Home Automation. This is the same profile used by all devices in the Home Automation specification. The Cluster Identifier indicates which set of commands this particular command belongs to, in this case, the OnOff Cluster. Then, looking in the ZigBee Cluster Library (ZCL) frame, you'll see the Command Identifier: in this case toggle.

The following code is the (abbreviated) source code for the ZigBee "Hello World," on the switch side. Notice that it follows the same set of steps as in the preceding outline by using a small-state machine. It starts out by indicating that it wants to join a network. Then, once the network is started, it moves on to finding the light. Then it goes to the application-ready state so SW1 can be used to toggle the remote light.

Something that's interesting, that is not obvious at first with ZigBee, is that it's irrelevant which device you turn on first. If the HA OnOffSwitch is started first, it will keep trying to join until the HA OnOffLight forms the network, then the switch will join it. This behavior is adjustable, of course, under application control, but it's a sensible default, and is typical of ZigBee systems. They can be put together in a very ad hoc (random) fashion, and they just work.

```
void BeeAppInit( void )
{
 /* start with all LEDs off */
 ASL_InitUserInterface("HaOnOffSwitch");
 /* tell the network to join automatically */
 TS_SendEvent(gAppTaskID, gAppEvtStartNetwork_c);
}

void BeeAppTask( event_t events )
{
 /* join the network automatically */
 if(events & gAppEvtStartNetwork_c)
   ASL_HandleKeys(gKBD_EventSW1_c);
 /* go find the light */
 if(events & gAppEvtNetworkStarted_c)
   ASL_HandleKeys(gKBD_EventSW3_c);
 /* found the light, go to application mode */
 if(events & gAppEvtFoundLight_c)
   ASL_HandleKeys(gKBD_EventLongSW1_c);
}
```

The concept of BeeAppInit() and BeeAppTask() are Freescale BeeStack concepts which I'll describe in some detail in Chapter 3, "The ZigBee Development Environment." Freescale BeeStack is the ZigBee stack portion of their ZigBee Compliant Platform. Basically, all ZigBee systems operate under some multitasking kernel, because networking and multitasking fit well together.

The ASL_HandleKeys() function takes keys as input, and is normally called from the keyboard callback function, BeeAppHandleKeys(). Freescale uses a standard keyboard, LED and LCD interface for all of their sample applications and they call this the

Application Support Library (ASL). This application takes advantage of ASL and feeds it keys to perform the various functions needed (joining the network, finding the light, etc.).

If you were following along with hardware (or even if you weren't), pressing SW1 on the HA OnOffSwitch generated a ZigBee toggle command over-the-air to the HA OnOffLight. The code for this begins where all keyboard input begins, in BeeAppHandleKeys().

```
void BeeAppHandleKeys ( key_event_t events )
{
 /* Application-mode keys */
 if( gmUserInterfaceMode == gApplicationMode_c )
{
 switch ( events )
 {
  /* Sends a Toggle command to the light */
  case gKBD_EventSW1_c:
   OnOffSwitch_SetLightState(
   gSendingNwkData.gAddressMode,
   aDestAddress,EndPoint,gZclCmdOnOff_Toggle_c,
   0);
  break;
```

Now, BeeAppHandleKeys() is not a ZigBee construct, but is specific to the Freescale platform. ZigBee does not deal at all with concepts such as user interface. It is enough for ZigBee that the system has some way of initiating the HA toggle command if the device supports turning things on or off. ZigBee is concerned exclusively with over-the-air behavior.

Notice the OnOffSwitch_SetLightState() function in BeeStack takes as its last parameter gZclCmdOnOff_Toggle_c. This is the same value as the toggle command (0x02) observed from the over-the-air capture decoded earlier. The third parameter could also be gZclCmdOnOff_On_c or gZclCmdOnOff_Off_c, both legitimate ZigBee commands for an OnOffSwitch.

Following the code down through the common code of the stack, this eventually ends up at an AF_DataRequest() function, which takes as its parameters a payload (which in this case is the ZCL frame seen earlier) and addressing information, which includes among its parameters the cluster identifier, endpoints and, indirectly, the profile.

I'll be examining that function in detail in Chapter 4, "ZigBee Applications."

> ZigBee networks can be put together in a very ad hoc (random) fashion, and they simply work.

1.5   ZigBee Home Automation

This second example is similar to the first, but rather than simply binding (connecting the wireless wire) to any light arbitrarily, it gives the installer (you) the option of which light(s) to bind to (see Figure 1.12). It also uses three nodes, rather than two, one of each ZigBee node type: A ZigBee Coordinator (ZC), which is used to form the network, a ZigBee Router (ZR) which can route packets, and a ZigBee End-Device (ZED), which can sleep and go into low-power modes.

What's also interesting about this example is that it uses the same ZigBee setup as standard off-the-shelf Home Automation devices. The devices could be plugged into a real ZigBee network, within an actual home or small office and the lights could be controlled by switches in the network and the switch could control other lights in the network, regardless of vendor. Now, there is certainly more to making a commercial product than just this example, but it goes a long way toward it.

The Freescale boards don't actually look like the lights and switches pictured in the figure. After all, they are general purpose development boards, intended to be used with

Figure 1.12: HA Lights and Switch

almost any application. The three boards that actually come with the NSK are actually displayed in Figure 1.13.

One of the Freescale ZigBee boards, the Network Control Board, will be both an HA OnOffLight and the ZigBee Coordinator for the network. One thing that often confuses those new to ZigBee is that the ZigBee node type (ZC, ZR, or ZED) is not directly related to the application that is running in that node. A switch could be mains powered (and be a ZR) or it could be battery-operated (and be a ZED). The same goes for a light. Any of those nodes could be a ZigBee Coordinator (ZC).

One of the things I like about the Freescale boards is that they come in a protective plastic case. This makes them easy to handle at trade shows or in the lab, and the plastic case also contains a compartment for two AA batteries, making battery-operated demonstrations easy. The NCB (see Figure 1.13, left) also contains a 2×16 LCD screen which can be used for application display (we'll use this for a thermostat display later on) or for debugging. The SRB boards also contain a 3-axis accelerometer, and a temperature sensor, as well as exposed GPIO pins for external hardware.

In this example, however, the NCB and one of the SRBs will be lights, the other SRB will be a switch.

Like the previous example, these nodes will form and join the network automatically. However, they will not **bind** automatically. For the binding process SW3 will be used.

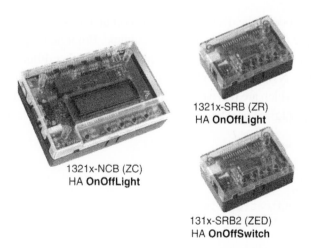

1321x-SRB (ZR)
HA **OnOffLight**

1321x-NCB (ZC)
HA **OnOffLight**

131x-SRB2 (ZED)
HA **OnOffSwitch**

Figure 1.13: Freescale NSK Boards

The way it works is this. Start by downloading the three images to the three boards. Chapter 3, "The ZigBee Development Environment," explains this process step-by-step. For now, just assume that it has happened. You can come back to this section after reading Chapter 3 if you are following along with actual hardware, but the concepts are what are important here.

Once all three boards have become part of the network (the flashing lights chasing each other will have stopped), press SW3 on each of the lights. The lights will start flashing LED3 to indicate they are in "Identify" Mode. The NCB will even display this fact on the LCD screen. Identify mode is a standard part of the ZigBee Cluster Library and is used to identify (usually visually) which node will be commissioned. Identify mode can be used in a lot of creative ways, but for now we'll use it to tell the switch which light(s) to control.

While the HA OnOffLights are still in Identify Mode (it lasts about 10 seconds in this case), press and hold SW3 on the HA OnOffSwitch. It will add those lights which are in Identify Mode to what is called a group. The switch and lights are now commissioned and ready to talk to each other.

Now press SW1 on the switch and it will toggle all lights in the group. As simple as that. A group can control all devices in the network or any subset of devices. For example, if the network contained 100 lights, 50 could be in one group, 50 could be in another, and all of them could be in an "all of the lights" group. The Home Automation specification allows any given node to belong to up to 16 groups.

Take a look at the decode of the over-the-air toggle command. Notice how it looks nearly identical to the toggle decode from the previous example. But it no longer contains a destination endpoint. A group identifier is used instead. One other difference is the destination address in the ZigBee NWK frame. In the former case, it was unicast to node 0x0000. But groups are broadcast which are received by all nodes but only sent to applications which belong to that group. ZigBee uses the special address of 0xffff to indicate a broadcast:

```
Frame 51 (Length = 31 bytes)
   Time Stamp: 21:13:44.106
   Frame Length: 31 bytes
   Capture Length: 31 bytes
   Link Quality Indication: 141
   Receive Power: -52 dBm
IEEE 802.15.4
   Frame Control: 0x8861
        .... .... .... .001 = Frame Type: Data (0x0001)
```

```
            .... .... .... 0... = Security Enabled: Disabled
            .... .... ...0 .... = Frame Pending: No more data
            .... .... ..1. .... = Acknowledgment Request: Acknowledgement
required
            .... .... .1.. .... = Intra PAN: Within the PAN
            .... ..00 0... .... = Reserved
            .... 10.. .... .... = Destination Addressing Mode: Address field
contains a 16-bit short address (0x0002)
            ..00 .... .... .... = Reserved
            10.. .... .... .... = Source Addressing Mode: Address field
contains a 16-bit short address (0x0002)
        Sequence Number: 93
        Destination PAN Identifier: 0x0bef
        Destination Address: 0x0000
        Source Address: 0x796f
        Frame Check Sequence: Correct
ZigBee NWK
    Frame Control: 0x0048
            .... .... .... ..00 = Frame Type: NWK Data (0x00)
            .... .... ..00 10.. = Protocol Version (0x02)
            .... .... 01.. .... = Discover Route: Enable route discovery (0x01)
            .... ...0 .... .... = Multicast
            .... ..0. .... .... = Security: Disabled
            .... .0.. .... .... = Source Route
            .... 0... .... .... = Destination IEEE Address: Not Included
            ...0 .... .... .... = Source IEEE Address: Not Included
            000. .... .... .... = Reserved
        Destination Address: 0xffff
        Source Address: 0x796f
        Radius = 10
        Sequence Number = 235
ZigBee APS
    Frame Control: 0x0c
            .... ..00 = Frame Type: APS Data (0x00)
            .... 11.. = Delivery Mode: Group Addressing (0x03)
            ...0 .... = Indirect Address Mode: Ignored
            ..0. .... = Security: False
            .0.. .... = Ack Request: Acknowledgement not required
            0... .... = Reserved
        Group Address: 0x0001
        Cluster Identifier: On/off (0x0006)
        Profile Identifier: HA (0x0104)
        Source Endpoint: 0x08
        Counter: 0x91
ZigBee ZCL
    Frame Control: 0x01
            .... ..01 = Frame Type: Command is specific to a cluster (0x01)
```

```
     .... .0.. = Manufacturer Specific: The manufacturer code field
shall not be included in the ZCL frame. (0x00)
     .... 0... = Direction: From the client side to the server side.
(0x00)
     0000 .... = Reserved: Reserved (0x00)
  Transaction Sequence Number: 0x44
  Command Identifier: Toggle (0x02)
```

The ZigBee Home Automation specification recommends that if more than four nodes will be the target of a communication that a group should be used rather than unicasting to each individual node. Group casting (my term, not ZigBee's) is a very bandwidth-efficient way to control a larger set of nodes.

Now reset the boards, or power them off. Power them on again. Notice that they retain the group information and can still be controlled by the HA OnOffSwitch. ZigBee as a specification does not indicate how persistent data should be stored, but the Home Automation Application Profile does specify it must be present. In this case, flash memory is used on the Freescale boards for this purpose.

When they are powered on again, there is no flurry of information required over-the-air. In fact, if you look at a decode it is silent until you press SW1 again to toggle the group of lights. The nodes simply come up and continue to operate as they did before they were powered off. Usually nodes also include a way to reset themselves back to factory defaults in case something goes seriously wrong and they no longer function with the current settings.

> ZigBee can communicate to individual nodes or groups.
>
> ZigBee devices remember their settings across resets and power outages.

1.6 ZigBee Speak

ZigBee, like all technologies, has its own language. Those of you familiar with IEEE networking specifications may be familiar with some of them already (e.g., data indications, service access points). You've been exposed to a variety of them in the previous sections as well. I've included an abridged glossary of terms (see Table 1.3), enough for you to go on in this book. A complete list of terms can be found in the ZigBee 2007 specification, Section 1.4 "Glossary."

Table 1.3: ZigBee Glossary Terms

Term	Description
802.15.4, or simply 15.4	The IEEE 802.15.4-2003 MAC and PHY specification used as the foundation for all ZigBee communications.
AES 128-bit Security	An application of the Advanced Encryption Standard to ZigBee authentication and encryption.
Attribute	A data item within a cluster, for example the state of an on/off light (is it on or off) or the current level of a dimmable light.
Authentication	Verifying that a packet came from the node it claims it did.
Bandwidth	Use of the RF spectrum. When the full bandwidth is used, no more messages may be transmitted at that time.
Broadcast	A mechanism to send the same packet to some or all of the nodes in a network.
Cluster	A collected set of commands and attributes. The ZigBee Cluster Library specifies common clusters.
Command	A single command within a cluster causing some specific action.
Commissioning	The process of connecting everything together in the network so that the normal application can run. For example, connecting a switch to a light, wirelessly.
Data Confirm	A confirmation that a data request has been delivered.
Data Indication	An application (or other portion of the ZigBee stack) is receiving data from another node. Basically, the other side of a data request.
Data Request	An application (or other portion of the ZigBee stack) is requesting to send data to another node in the network.
Descriptor	A description of some object in the node. For example, a simple descriptor describes an endpoint, a node descriptor describes a node.

(continued)

Table 1.3: *continued*

Term	Description
Encryption	Encoding a packet in such a way it cannot be understood by nodes without the security key.
Endpoint	A virtual (wireless) wire connects an endpoint on one node to an endpoint on another node. A node may contain multiple endpoints.
Frame	An over-the-air sequence of bytes (octets) defined by the ZigBee or IEEE standards.
Groupcast	Not a ZigBee term. My term used when sending a command via groups which is broadcast across the network.
Hop	Or single hop. Passing a message from one node to a neighboring node.
MAC	Media Access Control. This layer knows about networks' a node addresses, but not much else. Defined by IEEE.
MAC address	A 64-bit number uniquely identifying this node from all other nodes in the world. Also called a long address, IEEE address or EUI.
Mesh	The term used for ad-hoc multi-hop networking.
Mesh Network	A network which discovers routes to other nodes in the network.
Multi-hop	The ability for the network to pass messages from one node to another, ultimately to the final destination node.
Neighbors	Other nodes within radio range.
NHLE	Next Higher Layer Entity, basically the layer above the current layer being discussed. For example, the network layer in ZigBee passes information to the NHLE, also known as the APS layer.
Node	A device that contains a single 802.15.4 radio. A node may control or monitor multiple things. For example, a bank of four light switches would need only one 802.15.4 radio.
OUI	Organizational Unique Identifier, the top 24 bits of the MAC address.

Table 1.3: *continued*

PAN	Literally a personal area network. However, ZigBee often encompasses a whole building or campus. In some profiles, ZigBee discusses NANs, or neighborhood area networks, or HAN, a home area network. PANs, HANs, NANs, it's all the same thing to ZigBee: a single network.
PAN ID	A unique 16-bit number identifying the PAN.
PHY	The physical layer. This is basically the radio which converts a binary sequence into RF and back again. Defined by IEEE.
Profile	Generally referring to an Application Profile, or an agreed upon set of messages so applications from different vendors can interact.
Route	The path during a unicast from one node to another.
SAP	Service Access Point, the API between layers within the ZigBee stack. ZigBee is layered similarly to the common OSI networking layered model.
Sequence Number	A rolling number to uniquely identify a message. Also called a counter or transaction ID.
Unicast	A bandwidth-efficient mechanism to send a packet from one node to another node, even if it multiple-hops away.
ZC	ZigBee Coordinator. A special ZigBee node type that can form networks.
ZCL	ZigBee Cluster Library. A library of common clusters.
ZED	ZigBee End-Device. A special ZigBee node type that can sleep, extending battery life to years.
ZigBee	A networking protocol standard aimed at wireless control.
ZR	ZigBee Router. A special ZigBee node type that can route packets and understands mesh networking.

1.7   ZigBee Architecture

Before I end this chapter, I'd like to describe to you the basics of the ZigBee stack architecture. The classic ZigBee architecture diagram, seen in nearly every technical presentation on ZigBee looks something like Figure 1.14.

Each layer in the stack knows nothing about the layer above it. The layer above it could be considered a "master" that commands its "slave" below it to do the work. Each layer builds more and more sophistication upon the foundation of the lower layers.

ZigBee does not exactly fit the OSI 7-layer networking model, but it does have some of the same elements, including the PHY (physical), MAC (link layer), and NWK (network) layers. Layers 4–7 (transport, session, presentation, and application) are wrapped up in the APS and ZDO layers in the ZigBee model.

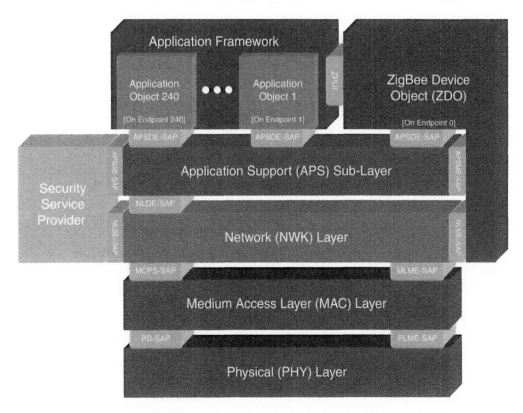

Figure 1.14: The ZigBee Architecture

Between the layers are Service Access Points (SAPs). SAPs provide an API that isolates the inner workings of that layer from the layers above and below. Like the IEEE 802.15.4 specification, ZigBee uses a two-SAP approach per layer, one for data and one for management. For example, all data communications to and from the network layer go through the Network Layer Data Entity Service Access Point (NLDE-SAP). Why the data SAP for the MAC layer is called MAC Common Part Service SAP (MCPS-SAP) instead of the MLDE-SAP is confusing to me, but it is. You'll see commands in the ZigBee specification that look like APSDE-DATA.request. Now you know that is a request to send data out the radio, but initiated just above APS layer.

The two lowest layers, the MAC and PHY, are defined by the IEEE 802.15.4 specification. The PHY layer simply translates packets into over-the-air bits and back again. The MAC layer provides the concept of a network, including a PAN ID, and networking discovery through beacon requests and responses. It also provides per-hop acknowledgments and some of the commands for joining and forming a network. The MAC does not multi-hop or mesh.

The NWK layer is responsible for mesh networking, which includes broadcasting packets across the network, determining routes for unicasting packets, and generally making sure packets are sent reliably from one node to another. The network layer also has a set of commands for security purposes, including secure joining and rejoining. ZigBee networks are all secured at the NWK layer, and the entire payload of the NWK frame is encrypted.

The APS layer is responsible for application meaning. It acts as a filter for the applications running above it on endpoints to simplify the logic in those applications. It understands what clusters and endpoints mean, and checks to see if the endpoint is a member of the Application Profile and (if present) group before sending the message on up. The APS layer also filters out duplicate messages that may have been sent up by the NWK layer. The APS layer keeps a local binding table, a table which indicates the nodes or groups in the network that this node wishes to speak to.

The ZDO layer (which includes the ZigBee Device Profile, ZDP), is responsible for local and over-the-air management of the network. It provides services to discover other nodes and services in the network, and is directly responsible for the current state of this node on the network.

The Application Framework contains the ZigBee Cluster Library and provides a framework within which applications run. Endpoints are the mechanism used to distinguish one application from another.

The security services are used by a variety of layers, and can be used by ZDO, APS, or the NWK layer, hence it's off to the side.

All layers have what is called an information base. At the MAC layer, this is called a PAN Information Base (or PIB). At the network layer it's called a Network Information Base (or NIB), and of course there is an AIB for the APS layer. All "information base" means is the current settings of that layer. How many retries are required? What is the current PAN ID or network address of a particular node? The fields in the information base are generally either set by higher layers or through the use of management commands through the management SAPs.

Notice there is nothing in here about interaction with any hardware within a given ZigBee device other than the radio. There is nothing that talks to LEDs, the LCD, speaker, GPIO ports, non-volatile or flash memory. ZigBee is exclusively concerned with the networking protocol and over-the-air behavior. The ZigBee test suites reflect this fact. Since all the over-the-air messages can be interpreted correctly by any other ZigBee node, this approach allows vendors to innovate while still providing complete compatibility between vendors.

ZigBee layers build on each other.

ZigBee is exclusively concerned with over-the-air behavior.

CHAPTER 2

Deciding on ZigBee

One of the first questions I get from an OEM is, "How does ZigBee compare to wireless technology *X*?" So before I get into the technical details of how ZigBee works, perhaps it would be good to spend a bit of time on what ZigBee is really good for, what it is not good for, and how it compares to some other popular, competing networking technologies. I have seen plenty of companies trying to shoe-horn ZigBee into a problem space that just doesn't make sense for the technology, and I'd like to help you avoid that.

Another common question put to me is, "What does ZigBee cost?" That is a complicated question. Obviously a book is not the right medium to keep current on prices, but it can help you ask the right questions, and describe "hidden" costs, some of which you may not have considered yet. I love ZigBee. ZigBee is great! But I am also an engineer, and I like to consider a variety of factors before deciding on the solution that is right for any particular problem.

And yet another question I am often asked is, "So what does ZigBee stand for?" Well, I'll tell you.

There are many people who expect that ZigBee is an acronym and they are actually correct, even though the acronym has been forgotten over time. The acronym represents the attributes associated with the original ZigBee objectives, which sort-of evolved into a mission statement for the technology. "ZIGBEE" originally stood for a "Zonal Intercommunication Global-standard, where Battery life was long, which was Economical to deploy, and which exhibited Efficient use of resources."

Two other objectives were added a short time later: "Reliable" and "Secure." But by then, of course, it was too late to change the name. In addition there was concern among the promoter member companies that ZigBee might not get taken seriously, and that their respective companies might not pay for travel to the exotic locations hosting the ZigBee Alliance quarterly meetings if they added the other two letters to the original acronym, forming the new one: "ZigBeers."

2.1   Deciding on the Right Technology

This is a book about ZigBee. But ZigBee isn't always the right solution for a given problem. This section discusses some other possibilities for networking sensors and control devices, and some of the advantages and disadvantages of each approach.

2.1.1   Wired Versus Wireless

Wireless is inherently unreliable. Does this sound strange coming from a proponent of ZigBee? It's true. There are so many factors that affect wireless; RF noise coming from machinery, changes in the physical environment, even the vagaries of the atmosphere. In the 2.4 GHz spectrum, which is the ISM band used by ZigBee, there is even more interference. Water, people, concrete, metal, foliage can all change the wireless characteristics, causing packet delivery to fail. Now granted, ZigBee, with its mesh networking and per-hop and end-to-end retries and acknowledgments, turns what is an essentially unreliable medium into a very reliable network, but ZigBee is not the only solution to network devices.

A wired network doesn't have any of the above problems. But wires do have problems of their own. Wires have connectors, and connectors get broken over time, especially if they are plugged in frequently. The wires themselves get caught on things and get cut. One major white-goods manufacturer I spoke with was changing over the entire production line to wireless to avoid just this problem. A wireless system can test and inspect the washing machines, dryers and refrigerators coming down the line without any of the drawbacks of wires.

Salt and water corrode wires. A wire on a ship corrodes or chafes over time, and so becomes less reliable than a wireless solution in that same environment.

Wired solutions can be more expensive per foot. At a recent building construction show, the number used was $10 US per foot (2007 dollars) for wired control. That $10 is not in just the cost of the wire, but the cost of conduit, installation, and maintenance. In a similar situation, ZigBee wireless is generally considered to cost less than $0.10 per foot, including the cost of the wireless boards and installation.

But there are many good, prominent wired protocols and hardware out there. Don't just jump into wireless because it's the next greatest thing. Determine first if wireless is a great thing for your application.

> Wires don't have the interference problems of wireless, but wires need connectors.

2.1.2   Other Wireless Technologies

802.15.4 is a great MAC and PHY specification, and ZigBee, which relies on the 802.15.4 MAC and PHY, is a great networking protocol. ZigBee is well-targeted at the wireless sensor and control network space. But 802.15.4 is not the only radio which targets this space, and ZigBee is not the only networking protocol which runs on the 802.15 MAC and PHY:

- *Non-ZigBee 802.15.4 radios* are not supported. One thing that is not obvious to the casual observer is the fact that ZigBee runs only on 2.4 GHz (ISM band) 802.15.4 radios. The 802.15.4 specification allows for 900 MHz, and 868 MHz radios as well, but ZigBee doesn't run on those radios, due mainly to data rate. ZigBee requires a higher data rate than the 40 kbps offered by the 900 MHz standard and 20 kbps offered by the 868 MHz standard. The 2.4 GHz 802.15.4 specification operates at 250 kbps, a rate sufficient for the ZigBee protocol. The newer 2006 802.15.4 specification offers higher data rates in the sub-1 GHz RF spectrum, but those have not yet been adopted by ZigBee. The other reason ZigBee chose the 2.4 GHz spectrum is that products can interoperate worldwide in this RF space.

- *Wireless USB* is emerging as a great technology for small, battery-operated devices. For example, the Cypress CYWUSB6934 is used in a number of point-to-point devices such as a wireless mouse. This technology is very inexpensive, great on batteries, but is not meant to function on a scale as large as ZigBee can, and doesn't have the security. Wireless USB is based on Ultra Wideband (UWB) which supports 480 Mbps data rate over a distance of two meters (about 6.6 feet). If the speed is lowered to 110 Mbps, UWB will go a longer distance (up to 10 meters or about 30 feet). This technology generally focuses on the PC peripherals, but could be adapted to the sensor and control space.

- *WiFi*™ is also starting to see more use in sensor and control networks. WiFi is a ubiquitous, well understood technology that previously targeted the PC and laptop space. It's a bit more expensive than ZigBee, typically requiring a larger CPU to run the full protocol. WiFi can mesh like ZigBee and if you Google "WiFi mesh" you'll see a number of vendors that serve this space, but there are no interoperable standards for meshing WiFi. WiFi is not as good with batteries, using up batteries in hours, rather than the months to years of ZigBee, but not all control networks require batteries.

- *Bluetooth*™ is very secure, and is used in headsets and cell phones everywhere. Bluetooth is relatively inexpensive, but due to its frequency-hopping technology, does not have the battery life of a ZigBee device. Think of Bluetooth battery life as days, not the months to years like ZigBee. Also, Bluetooth doesn't scale well and is typically limited to networks of seven devices or less, whereas ZigBee networks can contain thousands of devices.

- *Wibree* falls under the Bluetooth Special Interest Group, and is aimed at watches and body sensors. It's designed to be very, very low power, but again is limited in the number of nodes in a network. At the time of this writing, Wibree was not in production, but still in the specification stage.

- *Z-Wave,* by Zensys (see http://www.zen-sys.com), is useful if you are planning home automation products. Z-Wave is aimed specifically at the home automation space, and while not a standard (Zensys is the only manufacturer of Z-Wave), it is used by some of the large home automation manufacturers, including Danfoss, Leviton, and Universal Electronics. The new ZW0301 contains a 32 K flash 8051 CPU, and a radio that operates in the 868 MHz space for Europe and the 908 MHz space for the U.S., at 9 kbps (compared to ZigBee's 250 kbps).

  Zensys is naturally worried about ZigBee, because these two technologies compete directly, and so Zensys has formed a group called the Z-Wave Alliance, and regularly releases white papers promoting Z-Wave over ZigBee.

- *Proprietary radios* are still the norm today. I remember this used to be the case in operating systems, too. Everyone had their own system, but while there are still quite a number of proprietary operating systems, most OEMs realize it is cheaper in most cases (all things considered) to go with a commercial operating system. I believe this will happen to radios over time in this space.

  There are so many good radios in various RF frequencies. Texas Instruments makes one called the Dolphin. Nordic Semiconductor makes a variety of low power radios, such as the nRF2401A. The biggest issue with the proprietary radios is the lack of a protocol. So much of RF communication is in the software. If you have to write your own, it can cost significant amounts of development effort, which equals time and money.

  Which brings me to software protocols.

> ZigBee operates in the worldwide 2.4 GHz ISM band.
>
> Proprietary radios are the primary competitor of ZigBee today.

2.1.3 Other Software Protocols on 802.15.4 Radios

The 802.15.4 radio (MAC & PHY) specification provides the foundation for the ZigBee networking protocol, but ZigBee is not the only protocol that runs on these radios:

- *802.15.4 MAC* may work for you. Yes, it's possible to build a network with only the 802.15.4 MAC and PHY. Most 802.15.4 silicon vendors offer an application development solution on their 802.15.4 MAC. Granted, it's only point-to-point or a star network. Granted, it doesn't have device or service discovery or interoperable applications like ZigBee. But, a MAC-only solution does have a smaller footprint than ZigBee and might be enough networking protocol for your application. The MAC uses the same per-hop retry mechanism that ZigBee uses, the same network associate commands, and the same sleepy end-device polling commands. What the MAC-only solution is missing is the application-level compatibility and the higher reliability and multi-hop nature of mesh networking.

- *SimpliciTI*™ may work, too. This Texas Instruments protocol is a simple, point-to-point, or repeating protocol for use with their 802.15.4 radios and solutions. This protocol is extremely small (4 K flash), and they tout it as being extremely simple. There are only 6 API calls.

- *Synkro,* originally by Freescale Semiconductor, is now an emerging standards-based protocol targeted specifically at the consumer electronics space. This protocol uses 802.15.4 radios and solutions to control televisions, DVD players, stereos, and the like. Sony is one of the early adopters of this technology and it looks as if the standard might take off. It's still too early to know at the time of this writing, but the consumer electronics manufacturers are looking to avoid the IR fiasco (too many proprietary control methods), and instead focus on standardized RF for the remote. Synkro is point-to-point, not meshing, but does offer a form of frequency agility to avoid potential interference problems.

- *MiWi,* by MicroChip, is a peer-to-peer and mesh protocol specifically targeted at the Microchip MRF24J40 radio. This protocol has no cost associated with it if used with the MicroChip solutions.

- *PopNet*™, by San Juan Software, is a small-footprint full-mesh solution that could be considered "ZigBee Lite." Like ZigBee, PopNet supports broadcasting, unicasting, and mesh route discovery. Unlike ZigBee, PopNet fits into 16 K of flash and does not require ZigBee certification.

- *TinyOS* is a mesh networking technology written in a language called NesC. This C-like language allows sensors to be treated as objects. TinyOS was even adopted by one of the ZigBee stack vendors, MeshNetics, as the foundation for their ZigBee solution.

- *6LoWPAN,* standardized by IETF.org, is an 802.15.4 networking protocol that looks similar to IPv6. Companies such as ArchRock (http://www.archrock.com) are promoting this emerging standard for interfacing sensors to the World Wide Web.

All of these protocols have their own strengths and weaknesses. And there are plenty more protocols! With the exception of PopNet, and a vanilla 802.15.4 MAC, I have not personally written software or produced products using the other protocols, so I won't try to give advice about which one is best for your particular application. I did, however, want you to be aware that ZigBee isn't the only networking protocol available on 802.15.4.

But ZigBee is the primary protocol based on 802.15.4, so if standards or interoperability are important to your application, ZigBee is the route to go.

> ZigBee is the primary networking protocol based on 802.15.4 radios, but many other protocols are also commercially available.

2.2 Deciding on a ZigBee Solution

Okay, so you've decided ZigBee is the right solution for your product. Now, which ZigBee solution is best? Sorry. I can't answer that. But I can give you a set of questions to think about.

At the time of this writing, some of the 802.15.4/ZigBee radio vendors (also called ZigBee silicon providers or platform vendors) include:

- Atmel

- Ember

- Freescale

- Integration Associates

- Jennic

- Microchip

- NEC

- Oki

- Radio Pulse

- Renesas

- ST

- Texas Instruments

In general, the easiest way to learn the latest details of the 802.15.4 offerings from these vendors is to go to the World Wide Web. All the vendors include PDF data sheets and various white papers to show why their solution is best. The Web sites for larger corporations (which have vast and often confusing Web sites) tend to include a simple way to find information on their ZigBee products, such as http://www.freescale.com/zigbee, or http://www.ti.com/zigbee, or http://www.st.com/zigbee. For smaller companies, the ZigBee information tends to be found somewhere on their home page.

Below I've compiled a list of questions you should ask yourself when evaluating each platform vendor:

- Is the platform certified?

- Will the product use an integrated MCU/radio or a stand-alone radio?

- Does the platform have enough memory for the application?

- Does the price-point make sense for the product?

- Is the part low-power enough for the power supply?

- What voltage range does the part support?

- Does the part support all the required physical characteristics?

- Are you familiar with the silicon vendor?

- How good is the stack?

- What tools are available to support compiling and debugging?

- What reference designs are available?

- What is the availability of the radio?

- Is an alternate supplier a requirement?

Is the ZigBee platform certified?

A combination of the radio, MCU, and ZigBee stack software is called a *platform* by the ZigBee Alliance. Some ZigBee platforms are certified by the ZigBee Alliance, and some are not. Is this important to you?

If your product will interact with products from other companies, such as Home Automation, Smart Energy, or Commercial Building Automation, I would say the answer is a resounding, "Yes!" The platform you choose *must* be certified. If your product will be a stand-alone network, then the range of vendors opens up.

Be sure to check when evaluating a vendor. A non-certified stack may not work with other ZigBee stacks.

Will the product use an integrated MCU/radio or a stand-alone radio?

OEMs using ZigBee often choose an integrated MCU and radio. This combination can be very cost-effective, allowing both the ZigBee networking protocol and the application to reside in a single chip, 8-bit solution.

The single chip solution offers the advantage of containing everything required for the application except a power supply and sensors (see Figure 2.1). The MCU will contain memory, UARTs, even LCD, LED, or keyboard controllers, and (hopefully) enough RAM and flash (ROM) to host your application. Most single-chip solutions are built on an 8-bit processor, often an 8051 (although some of the newer parts, such as Freescale's MC13223 contain a 32-bit ARM7). Check with your vendor for their latest offerings.

The single chip solution works fine for many applications. But sometimes, an 8-bit MCU just doesn't have the memory or raw processing power that is required. It may surprise you (it did me), but variable airflow valves (VAVs), those vents you see in the ceiling of every office building, use a 32-bit CPU with a very complex application to control the opening and closing of the vent.

If you are choosing a stand-alone 802.15.4 radio with some host CPU, the main question is whether ZigBee will need to run on that host processor. ZigBee normally runs as software (including a small OS or kernel), much like Windows or MAC OS X runs on

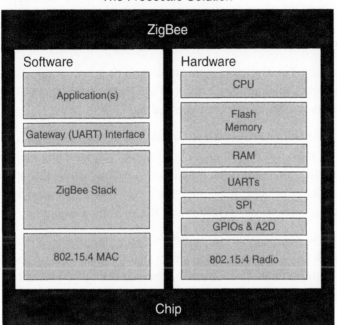

Figure 2.1: ZigBee Chip Solutions Contain Everything Needed for the Application

your PC or laptop. If you plan to use, for example, an ARM 9 with a stand-alone 802.15.4 radio, does a ZigBee stack run on that combination? The answer is often, surprisingly, "No." Most ZigBee implementations run really well on the integrated MCU/radio solutions, but may not be portable enough to run on another MCU, or the silicon vendor who supplies the ZigBee stack may not be willing to release full source code. Most don't.

Another way to use whatever CPU you need for your application, and still use ZigBee for the networking communications, is to treat the ZigBee integrated chip as a communication "module" (see Figure 2.2). The ZigBee stack and a small serial

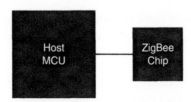

Figure 2.2: ZigBee as a Communications Module

"gateway" application run in the single chip solution, while the main application runs on the host processor, talking to ZigBee through a UART.

Does the platform have enough memory for the application?

It might surprise you, but ZigBee is quite large. As a protocol designed for 8-bit devices, ZigBee often stretches the limit, using most of the CPU's RAM and ROM, leaving only a few kilobytes of flash left over for your application.

When evaluating vendors, make sure you first do a code-size estimate for your application. Then, using the development kit from the vendor, build the Home Automation On/Off Light application (nearly every vendor offers this application) for a ZigBee Router device. Check in the .map file to see how much room is left for your application, both in terms of flash (ROM) and RAM. Is there enough?

Because there are so many options in ZigBee, it can sometimes be a pretty big task to configure the stack to include only those pieces that your application needs. Feel free to call the vendor's support line, or to check out their online help. This will also help you evaluate the vendor's ability to support your product development.

Does the price-point make sense for the product?

The price of a solution is not just the price of the silicon. I often see customers say, "Well, this chip is $0.05 cheaper!" when the cost of the external components required for the radio to function make it significantly more expensive than the alternatives. Consider the following production costs:

- What external components are required?
- Can the board layout be two-layer, or must it be four-layer?
- What quantity will I ship? This can greatly affect price per unit.
- Do the cost of the tools make a significant difference?
- How solid is the software? Will it require significant development or debugging effort?

Only after you've evaluated all of these questions can you determine the real cost of that particular solution. First, create a bill-of-materials (BOM), and then price them out with one or more distributors.

Consider also the cost of manufacturing. Some solutions require only two layers on the board, which makes manufacturing costs less. Others require four layers, which increases manufacturing costs.

One cost that is not always considered is the cost of the software. It used to be that hardware was the major cost in any given design. Today, more and more, it is the cost of the software that really makes or breaks a product.

Is the part low-power enough for the power supply?

All 802.1.54/ZigBee radios can go into low-power modes. When evaluating low power for a particular radio or single-chip solution, consider both the wake power and the sleep power. The sleep power, for most applications, dwarfs the wake power (and makes the wake power almost irrelevant). Put together a little spreadsheet, or use the power calculator provided at http://www.zigbookexamples.com, and you'll see why this is. If a node only wakes for a few milliseconds every few minutes, then power consumption while sleeping is everything.

This may be irrelevant if your application does not require battery-operated nodes.

What voltage range does the part support?

One thing that might not be readily apparent is that battery supplies vary widely in the voltage they can produce with a given charge. Some of the longer-lasting battery technologies (some that will last 20 years or more) start out providing five volts and end up providing only two volts. Many 802.15.4 radios (or MCUs) cannot deal with such a wide range without a power regulator which completely defeats the long battery life (power regulators consume power). So, make sure to match the batteries to the radio. And don't forget all the other power consumers on the board, including sensors.

Does the part support all the required physical characteristics?

Sometimes temperature, shock, or other factors are required for a given application. Will the radio be in the tire of a car? It had better withstand Phoenix heat and Fargo cold. Most ZigBee radios are in the industrial range of −40 to +85 Celsius.

Are you familiar with the silicon vendor?

One factor that plays into both schedule and cost is familiarity. If your company (or you personally) have worked with a particular vendor before, you may prefer to stay with that vendor; you know the quirks of their parts, tools, and sales processes. This can have a significant impact on both cost and schedule. Start out with the vendors you know. Then evaluate other vendors as appropriate.

How good is the stack?

Not all ZigBee stacks are created equal. ZigBee is a complicated specification with many edge cases. I've seen many stacks not take care of even the basics, in their first incarnations. Here are a few tests for you to try before wasting your time building an application on an unstable stack:

> Build a diamond network with a light on one end and a switch on the other. Alternatively, turn off the middle nodes. Does the stack recover a route and turn the light on every time? For more on this, see Chapter 4, "ZigBee Applications."

> Turn on 20 nodes at once. Does the stack handle it gracefully, and do all the nodes join the network?

> Run a test in a secure network, with a node sending a packet via APSDE–DATA. request to another node two hops away. Put a 32-bit incrementing counter in the payload, to show the number of messages sent. Send a packet after every data confirm. Let this test run for a few days. Is the stack still sending the packet? Is the number of messages sent reasonable? Expect at least one message every 100 ms or less.

> Build a max-depth network for the stack profile (10 hops across in stack profile 0x01, 30 hops in stack profile 0x02). Can the stack send from end-to-end?

What tools are available to support compiling and debugging?

The proper tools can make a significant difference during development of a product. Are your developers already familiar with the tools? Do they allow your developers to debug efficiently or are they wasting significant time fighting tools? Do the tools provide the information needed in a clear, easily understood presentation? It can be so frustrating trying to debug an application that cannot set sufficient breakpoints, or does not detect NULL points.

What reference designs are available?

Reference designs can save time and money when building custom 802.15.4 boards. All silicon vendors offer reference designs for their radios and MCUs. A reference design will usually include a full schematic and a bill-of-materials.

A word of caution to those of you who make your own boards: you may build boards in your sleep, but building RF boards is another thing. I've seen so many projects fail because the engineering house thought it knew how to build the RF board, only to find that they couldn't build it to production quality with a sufficient radio range.

Consider outsourcing board layouts to a company that has a proven track record, or consider using modules in your design. Everyone thinks they will ship millions of copies of a given product. But why not get your initial production going using off-the-shelf, precertified boards (aka modules), and then do a cost-reduction phase later? In most cases, you'll be better off with this approach.

What is the availability of the radio?

Sometimes companies forget to consider supply carefully enough. Sometimes, a project will begin and plan to use a silicon product not yet in production. The vendor you work with says it will be in production in June, but June comes and goes, July and August come and go, and pretty soon your product is late or canceled or you must scramble to find another solution, because the vendor's silicon is just not ready. My advice is to only plan products on silicon that is already in production.

Also, make sure that the supply will be big enough for you. Sometimes there is supply now, but the silicon manufacturer won't be making another run for two more years. Perhaps the fab is used for something else in the meantime, or the company uses a third-party fab.

It's a shame to have a successful product, only to run into supply problems.

Is an alternate supplier a requirement?

Some vendors will not consider a technology without multiple suppliers of that technology. That's one of the great things about ZigBee. There are so many suppliers. But while ZigBee speaks the same language over-the-air, ZigBee does not have a common software API. If you have a product that requires multiple suppliers, design your application software so that it is portable among stack vendors.

> Carefully consider all aspects of your design when selecting hardware.
>
> See http://www.zigbee.org for a complete list of ZigBee silicon and solution providers.

2.3   ZigBee Modules

A ZigBee module is a board that is manufactured "application-ready." The module will almost certainly be government-certified, including CE Marking in Europe,

IC in Canada, and FCC in the U.S. The module will certainly have various I/O pins, including binary GPIOs, and analog-to-digital pins. The module can be battery-operated, and comes in a usable form, sometimes with a common connector to attach the module to a motherboard (see Figure 2.3).

I can't say enough good things about ZigBee modules. Modules save you money and time, and until your production reaches the millions of units, there is rarely a good reason to build your own board (other than it's fun and you've always wanted to build your own RF board, anyway).

Figure 2.3 shows a variety of modules. Simply google "ZigBee Modules" to get a sampling of what modules are available. At the time of this writing, some popular ZigBee modules are built by Panasonic, MeshNetics, LS-Research, Digi, Cirronet, AeroComm, and many others, representing all the major ZigBee vendors.

Ask your hardware vendor what ZigBee modules they can recommend.

All the ZigBee module vendors offer some sort of development kit, which includes a larger development board where power can be easily applied, sensors and other peripherals can be added, and application software can be downloaded.

Figure 2.3: A Sampling of ZigBee Modules

Other module vendors provide a "wire-replacement" or other application to use ZigBee communications without knowing anything about ZigBee. Simply connect the module to a UART on two more devices, send a few Hayes-style AT commands and you have a mesh network. Easy!

Most modules are priced for the under-100 unit OEMs. But generally, if your company will ship volume products, a better price per unit can be negotiated.

These same modules can be used with any protocol (802.15.4 MAC, ZigBee, proprietary) that runs on 802.15.4 radios. If you later replace the modules with custom boards, the software does not change, except possibly for a few low-level drivers interacting with LEDs or other pins.

Okay, one more plug for modules. I have seen many projects fail because of overconfidence from the hardware design group. They've created printed circuit boards in their sleep. They've shipped many successful products, and know how to make a board. They've looked at modules, but none of them are just perfect. So they embark on making a custom board, only to find six months later that RF is hard. Really hard!

Subtle things, such as a technical balan that matches the specification but doesn't perform as expected, can cause delays. Perhaps the range available from the radio is just not what it should be. Perhaps there is noise near the chip that interferes with the application. Perhaps the antenna design just isn't doing what the simulations say it should.

Do yourself a favor. Use a module, at the very least, through the Beta phase. For many products, modules are cheaper even in the long run. They are certainly cheaper in the short run. Save the custom RF board during the cost reduction phase of the product.

2.4   A ZigBee Checklist

This is really just a list of things to remember when building a ZigBee application. There is no right or wrong answer, and this certainly isn't a definitive list, but make sure you've thought about each item below. Hopefully they will save you some time and money, more than you've spent on this book.

Protocol Selection

How often does the application need to communicate? Think of all nodes that need to speak and develop a worst-case "bandwidth" simulation. Can the application be less "chatty?"

- Does the protocol need to speak 802.15.4? ZigBee?

- Is simplicity (a shorter learning curve) more important, or is compatibility with industry standards more important?

- How large will the expected network be? Normal case? Largest configuration? Is it better to ship more quickly with a smaller network size target?

- Will it interact with products from other OEMs?

- Will other networks be in the vicinity? Other 802.15.4 networks? Will they cause interference?

- What security is required by the application?

- Is low latency more important or low power?

- How important are gateways? To a host processor? To a PC? The Internet?

Hardware Selection

- Will a ZigBee module work for the Proof-of-concept? Beta? Production?

- What power supply will be used? Does the hardware selected support the appropriate voltage across the life of the power supply?

- Develop a power budget. What are all the power consumers on the board?

- Is size an issue? Does the package need to fit into a small space? How small?

- Is range important? What type of antenna will be used?

- Are there any MCU hardware or peripherals that are particularly important to the application? For example, is a second SPI port needed? A UART?

- Is time important to this application? Is an external crystal needed for accuracy while in low-power mode?

- Is the price range for the part acceptable?

- What other hardware is needed beyond the hardware provided by the primary 802.15.4 silicon vendor? Will they have any troubles working together?

Stack Selection

- Is the ZigBee stack certified? If not, is compatibility important?

- How easy is it to set up a 12 node network? Set up as large of a network as you can when evaluating a stack. Many problems don't appear with just a few nodes. Try 20 or more, if possible.

- Have you done the diamond network test (verify automatic route recovery)? If not, see Chapter 4, "ZigBee Applications," for a complete explanation.

- Try turning on all the nodes, to join the network at once. Does the network figure it out and eventually allow all nodes to join? This test can be tough for some stacks.

- On the selected hardware, how much room does the stack leave for your application? In ROM (Flash)? In RAM?

- Does the application use a ZigBee public Application Profile? If so, is the one from the stack vendor good enough? How much work is needed to go from a technology demo to commercial quality on this profile?

- What software is needed, beyond what the stack vendor provides, to complete this application?

Application Development

- How many developers will be on the project? Do they have sufficient tools? Are they familiar with the tools, or will there be a learning curve?

- Would ZigBee or protocol training be helpful for your staff?

- Is a protocol analyzer (such as Daintree) budgeted for? Plan at least one for the group, perhaps one per developer.

- Can the application be developed in parallel to the hardware? If so, are the silicon vendor reference boards good enough? Are ZigBee modules needed?

- Plan to build a mock-up test in the intended physical environment (at the loading docks, for example). How many nodes are needed? What test software is needed to capture the proper test data? Can a network of proper size be built ahead of time using a mock application?

- How will the software be tested? Will there be an automated nightly build?

- Think about in-field testing? How will this be conducted? How will the application be debugged in the lab? In the field?

- What debugging diagnostics are needed to ship with the product?

- Is a watchdog used? Under what conditions?

- How real-time does the application need to be?

- How will the application manage non-volatile memory (persistent storage)? Are there other ZigBee tables or memory that must be managed by the application? If so, what are they?

- Think about designing the application so that it is portable between processors and stacks. Is the application paying attention to the Endian-ness of its own application protocol? Will a mixed network be used with different Endian MCUs?

- Is any application code that is stack-specific separated from application code that is stack-neutral?

- Is it better to write the application in-house, or hire a third party already familiar with ZigBee and Networking?

Manufacturing

- Will units be manufactured in-house? If so, how will they be managed?

- If manufactured out-of-house, can they meet your volume and schedules? What tools do they have for serializing the board images?

- How will MAC addresses be assigned to each board? How will they be managed? Do they need other information assigned at manufacturing time?

- How will the nodes be packaged?

- How will they be shipped? Are the radios excluded from air shipments (which don't allow RF)?

Deployment

- Who will be the installer(s) of the network? How simple does it need to be?

- How will the installer(s) know if routers are near enough to each other?

- Does every node in the network at least have two routers within hearing range? How will the installer know?

- How will the installer verify that everything is installed correctly?

- Will the network be built in pieces, such as one hotel room at a time? Or will the network be built out from the center?

- Does the application take care of all commissioning simply and easily for the user?

- How will nodes be replaced if they break, or stop functioning over time?

CHAPTER 3

The ZigBee Development Environment

This chapter is for those who want to understand how to develop for ZigBee. It does not describe the ZigBee protocol, just the embedded development process, and in particular, how to build and run the example programs available throughout the book.

If you've already developed for wireless networks before, and have used tools to compile, download, and debug embedded programs, then some of this will be review. If you're already very familiar with the Freescale BeeKit development environment, the CodeWarrior IDE, the HCS08, and the Daintree Sensor Network Analyzer, then much of this will be review.

If, however, you are setting up a team to develop ZigBee-based products, or are new to embedded development, or if you want to follow along with the examples using actual hardware rather than simply reading about ZigBee, then this chapter contains what you need.

As usual, we begin with the *real* story about how ZigBee got its name.

> *The Aztecs were a Pre-Columbian Mesoamerican people of what is now known as Central Mexico.*
>
> *Chapultepec, "at the grasshopper hill" in the Nahuatl language, is a large hill on the outskirts of central Mexico City. In the days when Tenochtitlán was the island capital of the Aztecs, built on a series of islets in Lake Texcocoth, the city was linked to Chapultepec by a causeway. Aztec chiefs turned the hill and the surrounding forest into a royal retreat. The poet-king Nezahualcóyotl built a palace there in the 1400s, along with an aqueduct to carry spring water to the Aztec capital, and a sculpture of Moctezuma can still be seen carved into the rock of Chapultepec, not far from Huemac's cave. One of the pleasures the royalty enjoyed during this time was one we take for granted today: ice.*

Ice, as you can imagine, was very difficult to come by in the hot climate of Central America. The Aztecs would cut ice from glaciers in the nearby mountains and runners would carry the ice packed in skins, down to Chapultepec.

No single runner could make it all the way with a packet of unmelted ice, so a series of runners were stationed along the route, each to take the ice and pass it along to the next runner, eventually to end up at the final destination—Chapultepec. They called this service Zigibi.

Zigibi was similar to how ZigBee functions today: multi-hopping packets along a route from one node, through a series of other nodes to the final destination, ensuring that the packet arrives in a timely manner.

The hill of Chapultepec and the surrounding land are now Chapultepec Park, a popular spot both for locals and tourists. The park covers 1,600 acres of land, with centuries-old forest, several small lakes, landscaped areas and outdoor cafes. Chapultepec Zoo is located there, as well as an amusement park, La Feria, and it is the official residence of the President of Mexico, Los Pinos.

3.1   Development Hardware

ZigBee is a small wireless networking protocol designed for IEEE 802.15.4 radios and 8-bit microcontrollers. Together, the radio, ZigBee stack, and microcontroller make what is called a platform. If the platform has been certified by the ZigBee Alliance as compliant (and not all ZigBee platforms are), then the platform is called a ZigBee Certified Platform, or ZCP.

Developing for ZigBee is basically the same for all platforms (see Figure 3.1). A PC set of tools builds and compiles applications, which are then downloaded into target boards for debugging, usually through USB or Ethernet (although some platforms allow wireless download).

The PC tools required for ZigBee development usually include:

- An IDE for development, including compiling code into the appropriate form for the target microcontroller

- A ZigBee stack configuration or application "template" tool

- A debugger for downloading and stepping through source code lines on the target platform

- A protocol analyzer for debugging the network over-the-air

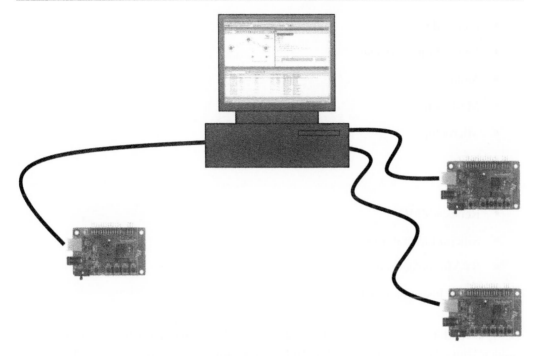

Figure 3.1: ZigBee Development Hardware

- Hardware for ZigBee development kits usually include:

- Two or more ZigBee boards

- A means of connecting target boards to the PC, usually through USB or Ethernet

- A JTAG or BDM debug device for programming the flash memory in the target boards

- A sniffer for detecting over-the-air packets and delivering them to the protocol analyzer

- Power supplies and/or batteries for powering the boards

At the time of this writing, some of the main ZigBee platform vendors include:

- Airbee

- Atmel

- Ember

- Freescale

- Integration Associates

- Jennic

- MeshNetics

- Microchip

- NEC

- Oki

- Renesas/ZMD

- Silicon Laboratories

- ST Microelectronics

- Texas Instruments/ChipCon

Of these vendors, only TI, Ember, Integration Associates, and Freescale are considered Golden Units. Golden Units form the ZigBee compatibility test suite sanctioned by the ZigBee Alliance. All other ZigBee Certified Platforms are tested against the Golden Units for compliance. Not all vendors provide ZigBee Certified Platforms, but most do. Make sure to check when evaluating a vendor, as a non-certified stack may not work with other ZigBee stacks (but could still be used as a proprietary mesh-networking stack).

Of the ZigBee Certified Platforms, I've chosen Freescale for this book, mainly because it's the one I'm most familiar with. I'll be delving into many details of that platform, but as you can expect with any book, some information will be out-of-date by the time this is in print. The ZigBee concepts will remain the same, however, as ZigBee, like IP and WiFi™, is a standard protocol designed to be available for many years to come.

Of course, the easiest way to learn the current details of each platform is to go to the World Wide Web. The Web sites for larger corporations include a simple way to find information on their ZigBee products, such as http://www.freescale.com/zigbee, http://www.ti.com/zigbee, or http://www.st.com/zigbee. For smaller companies, the ZigBee information tends to be somewhere on their home page (e.g., http://www.ember.com).

Some of the platforms are pictured in Figure 3.2. Notice the variety of form factors, antenna designs, and peripherals. And these are just development boards from the silicon vendors. There are also many, many pre-certified module designs for any application.

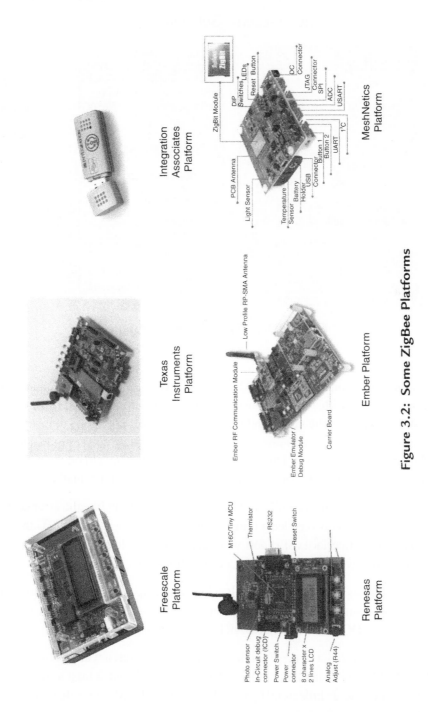

Figure 3.2: Some ZigBee Platforms

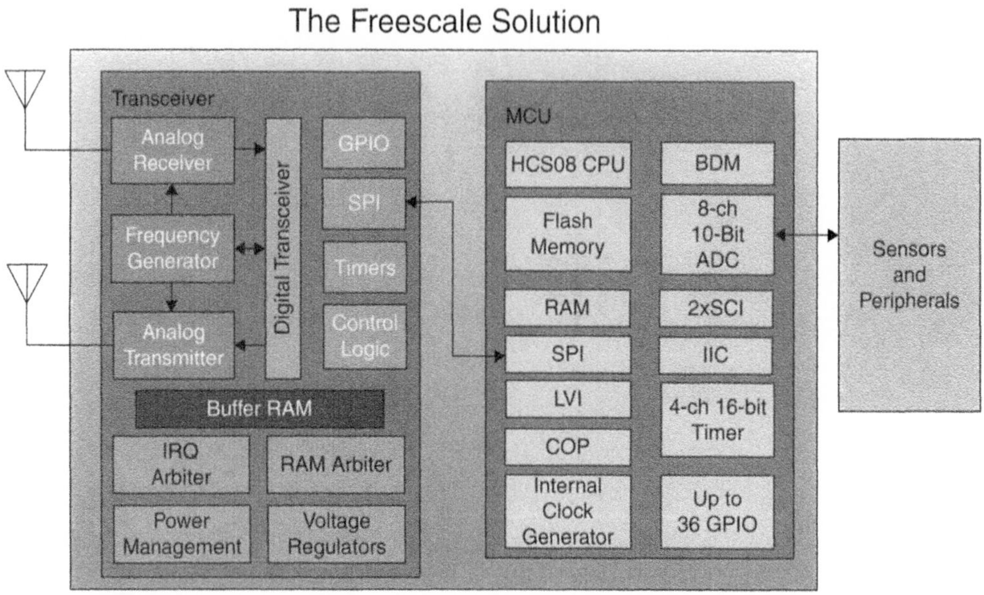

The Freescale Solution

Figure 3.3: A Typical ZigBee System

The heart of a ZigBee platform is the 802.15.4 radio. While it can be sold stand-alone, the radio is often coupled with a microcontroller inside a system-in-package (SIP) or system-on-chip (SOC). Typically the radios and microcontrollers are both good at low power, consuming something below two microAmps (μA) when in low power mode. Due to the low-cost nature of ZigBee, the microcontrollers are usually 8-bit, although 16-bit and 32-bit micros are starting to become common and with surprisingly small price tags.

A typical diagram for a ZigBee radio and MCU looks something like the arrangement in Figure 3.3. The radio (also called a transceiver) communicates to the MCU through the Serial Peripheral Interface (SPI) port with interrupts signaling events, such as when a packet has been received, or the channel is clear to transmit.

The sensors, actuators and other peripherals which make up the full application beyond the ZigBee networking portion tend to be external, and connected through an analog-to-digital converter (ADC) or SPI, IIC or UART (serial port).

The ZigBee microcontroller in these single-package systems has a wide set of peripherals and enough RAM and Flash memory to run the ZigBee stack and at least a small ZigBee application. Some have as little as 60 K of flash and 4 K of RAM; others have 128 K of

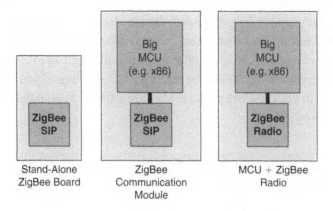

Figure 3.4: ZigBee Board Configurations

flash or more, and 8 K of RAM. Check with the vendors for their latest offerings. Generally you will see one of three silicon configurations for ZigBee boards (see Figure 3.4).

- A stand-alone board with a ZigBee single chip solution, in which the application is completely encapsulated on the board.

- A communications module in which the ZigBee single chip solution acts as a networking link, but the more complex application resides on another, usually 32-bit MCU.

- A board in which the ZigBee radio is stand-alone, and the ZigBee protocol runs on the larger host CPU.

Every silicon vendor who makes ZigBee radios offers some sort of development kit, complete with all the hardware and software necessary to develop ZigBee or 802.15.4 applications. These kits come with two or more development boards, a USB, serial or Ethernet connection to the boards, and some kind of debug connection (BDM or JTAG) to download new binary images into the boards and to debug applications. The kits also come with sample applications to get you started.

These kits vary quite a bit in pricing, so check the Web for the latest information. They can be purchased directly from the silicon vendor's site, or through distributors such as DigiKey (http://www.digikey.com).

I won't recommend a particular ZigBee kit, radio or stack here because that landscape changes quickly, and different components are better suited for different projects. A simple, broad recommendation is that you can trust the products from the big name silicon vendors. They tend to have large distribution channels, worldwide support, and strong internal testing, and tend to stay in the market for long periods of time. The smaller companies, however, can often offer more personalized and flexible support. Because nearly every 8-bit silicon vendor offers a ZigBee solution, you are probably already familiar with your favorite players. My suggestion is to read the online literature and talk to their customers.

> Go to www.company.com/zigbee to find information on a company's ZigBee offerings, for example, http://www.freescale.com/zigbee, or http://www.ti.com/zigbee.

3.1.1   Introduction to the Freescale NSK

As mentioned previously, the examples in this book are mostly made on the Freescale ZigBee Certified Platform, BeeStack. The choice was made because it's one of the prominent ZigBee solutions, and also the one this author is most familiar with.

To keep things simple, the bulk of the examples throughout this book are designed to work with a particular hardware kit provided by Freescale, the Network Starter Kit (1321xNSK-BDM). This kit comes complete with everything needed to develop ZigBee applications: three ZigBee boards (one NCB and two SRBs), a compiler (CodeWarrior), a debugger, and a ZigBee stack configuration tool (BeeKit). In addition, a trial version of a third-party ZigBee network analyzer is included (the Daintree Sensor Network Analyzer).

Look over the NSK components as shown in Figure 3.5. As you can see, the kit includes a set of ZigBee boards, power supplies and batteries for the ZigBee development boards, USB cables to connect to the boards, and the installation software on CD.

Each ZigBee board in the Freescale NSK uses the Freescale MC13213 single-chip solution (the radio and MCU in a SIP). Each board comes in a plastic case with a clear cover. Each board hosts four application-driven buttons, a reset button, four LEDs, and a serial connection via the USB port. In addition, each board exposes the GPIOs available from the microcontroller, so that sensors and actuators can be added using a simple connector.

The smaller Sensor Remote Boards (SRB) also include a low-g 3-axis accelerometer (MMA7260Q) and a temperature sensor (LM61BI), both of which will be used in examples later on in this book. The larger Network Control Board (NCB) does not include sensors, but does include a 2-line by 16-column LCD display.

Figure 3.5: The Freescale Network Starter Kit (NSK)

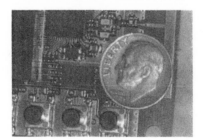

Figure 3.6: ZigBee Is as Small as a Dime

Although the development boards themselves are the size of a small PDA, it's interesting to note that the ZigBee components fit into the area of a U.S. dime (the MC13213 is in a 7 × 7 mm package). The white rectangle contains all the ZigBee components on the Freescale SRB (see Figure 3.6). This is true of every vendor. ZigBee is small.

Each Freescale board is also available as a full reference design, and can be downloaded free-of-charge from the Freescale Web site, complete with schematics and a bill of materials (BOM). Free schematics are pretty typical with the ZigBee platform vendors. They are interested in selling chips so they do everything they can think of to make your life easier.

> The Freescale Network Starter Kit (NSK) is used for the examples in this book.
>
> Full hardware reference designs for many platforms are available on the web.

3.2   The ZigBee Stack

Hardware is important, but the development environment wouldn't be ZigBee without a ZigBee stack. Every 802.15.4 silicon vendor offers at least one version of a ZigBee stack for their radio (at least in the 2.4 GHz band). Some offer more than one.

There are even ZigBee stack vendors who offer a software stack independent of the hardware. These stacks can run across multiple silicon vendors' products (both MCUs and radios). Airbee offers one such platform.

The quality of the ZigBee stack may be the single most important decision you make when selecting a ZigBee vendor. A simple test I like to perform when evaluating a new stack is this: Start with 4 nodes and spread them in a diamond pattern, as shown in Figure 3.7.

Make sure that node A is out-of-range of node D. You can test this by having node D try to join the network. It should fail (because it can't hear node A). Next, bring node B into the network and it should join A. Then node D should join node B (you may need to bring node D a little closer). Toggle the HA On/Off Light at node A using the HA On/Off Switch at node D. Use an acknowledged message so it is retried automatically by ZigBee at node D. I often use a sniffer to make sure that the packet is truly multi-hopping.

Next, turn node B off and turn node C on. Attempt to toggle the light. If the stack doesn't recover within one or two button presses, you have a problem: ZigBee mesh networking

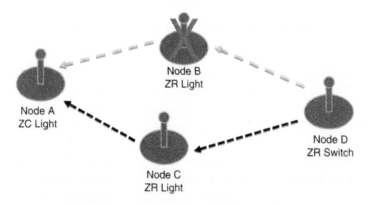

Figure 3.7: ZigBee Diamond Network

is not working for the stack. I am amazed time and again how many stacks fail this basic, simple test. I provide a working example of this test in Chapter 7, "The ZigBee Networking Layer."

Another simple, but related test works this way. Do the same thing, but use two or more different ZigBee vendors. All vendors should have some example of the Home Automation On/Off Light and Switch. The On/Off Light application is the "Hello World" of ZigBee.

The ZigBee Alliance only certifies platforms, a combination of the hardware and software. The ZigBee Alliance does not certify the individual components, such as the stack or silicon, independently. If a silicon vendor makes a new chip, they must recertify. If they make a (significantly) new stack, they must recertify.

A ZigBee stack usually runs in a small multitasking environment, complete with drivers for the on-board peripherals such as buttons and LEDs, non-volatile, storage and the like. In the 8-bit microcontrollers, these environments are usually cooperative (as opposed to pre-emptive) multitasking and are pretty lightweight.

The multitasking kernel (see Figure 3.8) found with ZigBee implementations include multiple timers to allow applications to time various events, and for ZigBee to time random back-offs, retransmissions, acknowledged packets, and other networking

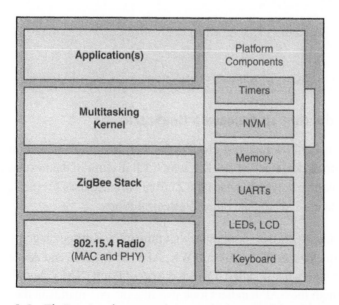

Figure 3.8: ZigBee Implementations Include a Multitasking Kernel

operations. The kernel also controls memory allocation for transmitted and received packets.

ZigBee typically comes in partial source, with the bulk of the code provided in object or library form. Some vendors offer full code for their ZigBee stacks, either for free or a price.

I don't know of any generic open-source project that is active at this time that supports multiple platforms, but perhaps someone will write one someday and certify it. ***Hint:** If you are interested in working on such a project, contact me and I'll work to get you the support you need within the ZigBee Alliance.*

I've already covered the ZigBee stack architecture at a high level in Section 1.4, "Hello ZigBee," and I'll cover it in depth in Chapters 4 through 7, but for now I'll recap the Chapter 1 information here. ZigBee is separated into layers, with the lowest layers, the MAC and PHY, adhering to the IEEE 802.15.4 specification. The next layer up is the NWK layer which keeps track of network concepts, such as mesh routing. The highest layers (below the application) are APS and ZDO, which provide application level interoperability and functionality, including the end-to-end acknowledgment of packets and the concept of Application Profiles.

So with that overview, I'll get into a few more details of a specific ZigBee Stack: Freescale BeeStack.

> Make sure the ZigBee implementation is a ZigBee Certified Platform (ZCP).
>
> ZigBee stacks also include a multitasking kernel.
>
> Verify that the stack can easily recover routes in the diamond network configuration.

3.2.1   Introduction to Freescale BeeStack

BeeStack is the name Freescale has given to their ZigBee implementation. It comes as part of a software package called BeeKit, which I'll discuss in the next section. BeeStack is Golden Unit-certified, which means the ZigBee Alliance uses Freescale BeeStack as part of the test harness to validate other ZigBee solutions.

BeeStack implements ZigBee in modules that mirror the ZigBee diagram shown in Figure 3.9, complete with PHY, MAC, NWK, APS, SSP, ZDO, and AF layers. Each layer is connected through what is called a Service Access Point (SAP). And the messages which flow through the SAP handlers look just like the commands in the ZigBee specification.

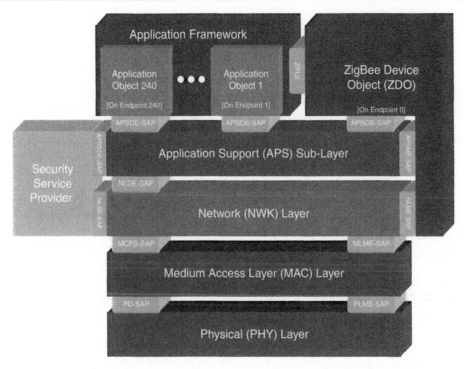

Figure 3.9: The ZigBee Architecture

For example, open up the ZigBee Specification book to page 33 and look at Table 2.2, APSDE-DATA.request. You have already downloaded the ZigBee specification, right? If not, go to http://www.zigbee.org, and download it. The specification will come in very handy for reference if you do any ZigBee development at all. The ZigBee specification is free.

Table 3.1 describes a ZigBee command, called APSDE-DATA.request, which is the command that sends data from one node to another at the application level. Since this command is in the APS layer, it starts with APS (there is also an NLDE-DATA.request, but applications don't interact with that command directly).

Now take a look at the C data type used for this command in BeeStack, as seen by the AF-APS SAP handler:

```
typedef struct zbApsdeDataReq_tag {
  zbAddrMode_t dstAddrMode;   /* indirect, group, direct-16,
                                 direct-64 */
  zbApsAddr_t dstAddr;        /* short address, long address
                                 or group (ignored on indirect
                                 mode) */
```

```
    zbEndPoint_t dstEndPoint;  /* destination endpoint (ignored
                                  if group mode) */
    zbProfileId_t aProfileId;  /* application profile (either
                                  private or public) */
    zbClusterId_t aClusterId;  /* cluster identifier */
    zbEndPoint_t srcEndPoint;  /* source endpoint */
    uint8_t    asduLength;     /* length of payload */
    uint8_t    *pAsdu;         /* pointer to payload */
    zbApsTxOption_t txOptions; /* options on transmit */
    uint8_t      radiusCounter; /* # of hops */
} zbApsdeDataReq_t;
```

Notice any similarities? That's right, they're identical. Where possible, BeeStack implements exactly what is in the ZigBee specification. See Chapter 4, "ZigBee Applications," for more discussion on this topic, and to learn about data requests in detail, including the meaning of all the parameters above.

If you see the term "higher layer," think of the data flowing upward in Figure 3.9, from the radio toward the application. If you see the term "lower layer," think of the data flowing downward from the application toward the radio.

BeeStack applications are written in the C programming language, a common language for 8-bit embedded programming. ZigBee does not mandate the C language. One vendor, MeshNetics, uses a proprietary language called NesC. But a ZigBee stack could be

Table 3.1: The APSDE-DATA.request

Name	Type	Range	Description
DstAddrMode	Integer	0x00–0xff	The addressing mode for the source address…
DstAddress	Address	As specified by DstAddrMode	The individual device address or group…
DstEndpoint	Integer	0x00–0xff	Destination endpoint…
ProfileId	Integer	0x0000–0xffff	The identifier of the profile for which this frame is intended
ClusterId	Integer	0x0000–0xffff	The identifier of the object…
SrcEndpoint	Integer	0x00–0xff	The individual endpoint from which the ASDU is being transferred
asduLength	Integer	0x00–0x50	Length of ASDU in octets
Asdu	Set of Octets		Octets comprising ASDU
txOptions	Bit-Map	0000 0xxx	The transmission options for ASDU…
Radius	Integer	0x00–0xff	Distance, in hops…

written in any language, including Java, Basic or C #. However, because most ZigBee stacks run on 8-bit or 16-bit microcontrollers, C is the natural language of choice.

BeeStack is divided into a set of tasks that run in a small multitasking kernel, one task per module. The NWK layer is contained in a task, so is ZDO, APS, and the other ZigBee modules.

Each BeeStack application has an application task. This task is initialized through a function called BeeAppInit(). The job of BeeAppInit() is to initialize any platform components that the application needs, such as LEDs, the LCD, the keyboard, sensors and so on, and to register endpoints with BeeStack so the stack can send and receive data on those endpoints, interacting with other ZigBee nodes in the network.

A common BeeAppInit() function looks as follows in BeeStack:

```
/*****************************************************************
*
*BeeAppInit
*
*Initializes the application
*****************************************************************/
void BeeAppInit( void )
{
  index_t i;
  /* initialize LED driver */
  LED_Init();

  /* register to get keyboard input */
  KBD_Init(BeeAppHandleKeys);

  /* initialize LCD (NCB only) */
  LCD_Init();

  /* initialize buzzer (NCB, SRB only) */
  BuzzerInit();
  BuzzerBeep();

  /* register to get ZDP responses */
  Zdp_AppRegisterCallBack(BeeAppZdpCallBack);

  /* flash LED1 to indicate not on network */
  LED_TurnOffAllLeds();
  LED_SetLed(LED1, gLedFlashing_c);

  /* indicate the app on the LCD */
  LCD_WriteString(2, "CustomApp");

  /* register the application endpoint(s) */
  for(i=0; i<gNum_EndPoints_c; ++i) {
  (void)AF_RegisterEndPoint(endPointList[i].pEndpointDesc);
  }
```

```
    /* remember first endpoint */
    appEndPoint=
    endPointList[0].pEndpointDesc->pSimpleDesc->endPoint;

    /* remember first cluster */
    Copy2Bytes(appDataCluster, endPointList[0].
    pEndpointDesc->pSimpleDesc->pAppInClusterList);

    /* allocate timers for use by this application */
    appTimerId = TMR_AllocateTimer();
}
```

A mechanism that is used frequently in BeeStack is the concept of callbacks. Take a look at the line of code above which calls KBD_Init(). As a parameter, KBD_Init() is passed the function to call when a key is available, in this case BeeAppHandleKeys().

Likewise, the function AF_RegisterEndPoint() is passed an endpoint description which includes, as one of its fields, a function to call when that endpoint receives data from another node in the network. The default callback function is named BeeAppDataIndication(). You may remember from the ZigBee terms discussion in Chapter 1 that a "data indication" means to receive data.

The last line in the initialization function above allocates a timer for use by the application. Timers in BeeStack are allocated before they can be used (started and stopped). Timers are used for anything that must be timed, for example to read a sensor at specific intervals or to flash an LED. The timer, when it expires, calls the call-back function provided to the start timer function.

```
void MyLedBlinkFunction(void)
{
   /* flash LED4 for 3 seconds */
   LED_SetLed(LED4, gLedFlashing_c);
   TMR_StartSingleShotTimer(appTimerId, 3000, MyTimerCallBack);
}
void MyTimerCallBack(tmrTimerID_t timerId)
{
   (void)timerId; /* timer ID not used in this case */
   LED_SetLed(LED4, gLedOff_c);
}
```

When the timer of 3,000 milliseconds (3 seconds) expires, the MyTimerCallBack() function is called, which then turns off the flashing LED. Timers can be set up to be repeating by calling TMR_StartIntervalTimer(). There is no limit to the number of timers in the system, other than RAM. Each timer requires six bytes of RAM. They all share and are controlled by a single hardware timer (TPM1, if you are familiar with the Freescale HCS08 MCU).

Another callback, the keyboard, is used extensively in the examples, but may not be used as much in commercial applications. Some commercial applications have no buttons at all on the units (not even a reset button).

A "key" may not always be a physical button press. For example, one application San Juan Software wrote used the keyboard interface to indicate if a door or window opened or closed for a security system. The keyboard interface on the HCS08 MCU can wake the processor with an interrupt. The interrupt was used to allow the unit to sleep most of the time but to wake instantly if a door or window opened. The unit then woke the radio and used ZigBee to communicate to the security system.

Below is an example of a keyboard callback from a Home Automation On/Off Switch.

```
void BeeAppHandleKeys(key_event_t events)
{
  if(gmUserInterfaceMode == gApplicationMode_c) {
    switch (events) {
      /* SW1 toggles the remote light */
      case gKBD_EventSW1_c:
        OnOffSwitch_SetLightState(
          gSendingNwkData.gAddressMode,
          aDestAddress,
          EndPoint,
          gZclCmdOnOff_Toggle_c,
          gApsTxOptionDefault_c);
        break;
      /* SW2 turns the remote light on with ZigBee ACK */
      case gKBD_EventSW2_c:
        OnOffSwitch_SetLightState(
          gSendingNwkData.gAddressMode,
          aDestAddress,EndPoint,
          gZclCmdOnOff_On_c,
          gApsTxOptionDefault_c | gApsTxOptionAckTx_c);
        break;
      /* Long SW2 turns remote light off with ACK */
      case gKBD_EventLongSW2_c:
        OnOffSwitch_SetLightState(
          gSendingNwkData.gAddressMode,
          aDestAddress,EndPoint,
          gZclCmdOnOff_On_c,
          gApsTxOptionDefault_c | gApsTxOptionAckTx_c);
      /* all other keys are handled by common ASL library */
      default:
        ASL_HandleKeys(events);
        break;
    }
```

```
  }
  /* In configuration mode. Keys are handled by ASL library */
  else
    ASL_HandleKeys(events);
}
```

The keyboard callback receives control when a key is pressed. In BeeStack, keys can be pressed for a short duration, called a short press, or for a longer duration (about one second by default), called a long press. The application can do anything it likes on a key press. In the code above, the application toggles the remote light on switch 1 (gKBD_EventSW1_c), and turns the light off with end-to-end acknowledgment using long switch 2.

In Freescale BeeStack, the Home Automation examples all use a common user interface and library of routines called the Application Support Library (ASL). One of the common UI elements is that the "screen" (which includes LEDs and the LCD), and the keyboard (which includes the four switches) both have two modes: a Configuration Mode and an Application Mode. The Configuration Mode is used for things like joining the network or choosing which channel on which to communicate. The Application Mode has application-specific keys. Think of modes as similar to the caps-lock or num-lock keys on your PC keyboard: They affect the meaning of subsequent key presses.

The two modes, and the meaning of key presses in those modes, are summarized in Table 3.2.

Probably the most interesting callback, from a ZigBee application standpoint, is BeeAppDataIndication(). This callback receives all incoming ZigBee data for the node. An implementation of this callback is shown below.

```
void BeeAppDataIndication(void)
{
  apsdeToAfMessage_t *pMsg;
  zbApsdeDataIndication_t *pIndication;
  zbStatus_t status = gZclMfgSpecific_c;

  while(MSG_Pending(&gAppDataIndicationQueue))
  {
    /* Get a message from a queue */
    pMsg = MSG_DeQueue(&gAppDataIndicationQueue);

    /* give ZCL first crack at the over-the-air frame */
    pIndication = &(pMsg->msgData.dataIndication);
    status = ZCL_InterpretFrame(pIndication);

    /* not handled by the ZigBee Cluster Library */
    if(status == gZclMfgSpecific_c)
    {
      /* insert manufacturer specific code here… */
    }
```

```
      /* Free memory allocated by data indication */
      MSG_Free(pMsg);
   }
}
```

This BeeAppDataIndication() routine is from an HA On/Off Light. In this case, the application has nothing to do other than call ZCL_InterpretFrame() because the only commands it supports are those that are already defined in the ZigBee Cluster Library (ZCL), a library of common clusters supported by the ZigBee alliance. ZCL is described in detail in Chapter 6, "The ZigBee Cluster Library."

There are a couple of important points to note. It is up to the application to freeup the memory allocated by BeeStack for the incoming over-the-air packet. This allows the application to keep the packet as long as necessary for processing. RAM is limited, so free the message as soon as possible. Data indications are described in detail in Chapter 4, "ZigBee Applications."

Table 3.2: Common ASL Keyboard Interface

Mode	Switch	Description
Cfg	SW1	Start network
Cfg	SW2	Join Enable
Cfg	SW3	Bind to another node
Cfg	SW4	Choose channel (use before starting network)
Cfg	LSW1	Switch between Cfg and App mode
Cfg	LSW2	Leave network
Cfg	LSW3	Remove all bindings
Cfg	LSW4	Start network without NVM
App	SW1	--- (application-specific)
App	SW2	--- (application-specific)
App	SW3	Toggle identify
App	SW4	Recall scene
App	LSW1	Switch between Cfg and App mode
App	LSW2	--- (application-specific)
App	LSW3	Add identifying nodes to a group
App	LSW4	Store scene

In addition to callbacks, the BeeStack multitasking system also uses events to signal tasks that there is some processing to perform. Every layer in BeeStack (MAC, NWK, APS, ZDO, etc. . . .) is defined by a task.

Events are typically used by applications to control state machines. For example, an application might signal to itself that one axis (say the X-axis) on the accelerometer has been read and it's now time to read the next axis. This code snippet comes from the "Generic Application" template in BeeStack, which uses the accelerometer available on the SRB boards.

```
void AccelerometerStateMachine(accelState_t state)
{
  /* starting a new read */
  switch(state)
  {
    /* read the X channel */
    case accelStateReadX_c:
      /* X axis value ready? */
      if((ATDSC & 0x80) != 0) {
        gaAccelDemoXYZ[accelDemoX_c] = ATDRH;
        giAccelDemoState = accelStateReadY_c;

        /* next channel (Y) */
        ATDSC = accelAdcChannelY;
      }
      /* indicate ready to read Y channel */
      TS_SendEvent(gAppTaskID, accelEventState_c);
      break;

  ...

void BeeAppTask(event_t events)
{
  /* handle accelerometer events */
  if(events & accelEventState_c) {
    AccelerometerStateMachine(giAccelDemoState);
  }

  ...

}
```

Events are sent to the BeeAppTask() callback. One or more event bits may be set (it is a bit mask), so BeeAppTask() must check each event bit it cares about. In addition to creating state machines, events are used to break up processing that might take a relatively long or unknown time to complete, so that the stack can continue with other system processing, such as routing packets. A rule of thumb is don't use more than 2 ms

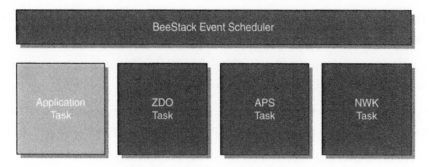

Figure 3.10: BeeStack Event Scheduler

for processing at any given time. If the computing will take longer, split it into a state machine so the other tasks are not starved for processing time.

The actual data indications from the radio (PHY data indications) are received on interrupts, so these will not be missed even if they come in during application processing. The same goes for timers and the keyboard. But for the NWK layer or other stack components to *process* the indication, that requires time from the multitasking environment.

In BeeStack, each event is a bit in the bit mask of type event_t. At the application task, bits 0 through 11 are available to the application and bits 12–15 are reserved by BeeStack.

The multitasking environment in BeeStack is cooperative, not preemptive. That means if a single task (say the application) stays in a while() loop, no other task receives control until the BeeAppTask() function exits.

A simple way to look at the event scheduler (see Figure 3.10) in BeeStack is to consider it a simple loop that looks at all of the tasks in the system and checks whether any task has an event ready for that task. If so, control is passed to the task's event handling function. In the application task, this is BeeAppTask(). The tasks are prioritized, so the NWK task, for example, gains control before any of the other layers do. The priority can be adjusted by the application at compile-time.

All the ZigBee stacks I have used for developing include a cooperative event scheduler. Networking as a problem space lends itself to a multitasking environment, and light weight, often 8-bit, MCUs do best with cooperative (rather than preemptive) multitasking.

To summarize BeeStack:

- Freescale BeeStack represents ZigBee as a set of separate tasks, mirroring the modules in the ZigBee architecture.

- The BeeStack ZigBee modules (also called layers) communicate to each other through service access points.

- Each task may receive callbacks for various physical events such as key presses, serial input, expired timers, and data indications.

- Each task may also receive logical events, represented by a bit-mask of events passed to the task's event function. Application events go to BeeAppTask().

- A common user interface is used for all demo applications.

> BeeStack is the name of the Freescale ZigBee stack.
>
> BeeStack uses a cooperative multitasking kernel.
>
> BeeStack includes a common user interface for all demo programs.

3.2.2   Introduction to Freescale BeeKit

Freescale BeeKit is a Windows PC-based application that creates and configures ZigBee applications (see Figure 3.11). BeeKit is not the compiler and IDE, but BeeKit does create project files that are used by the IDE, CodeWarrior. BeeKit's main features include:

- A set of application templates

- The ability to configure application and stack options through properties

- An easy-to-use New Project Wizard

- Full context-sensitive help for properties

- The ability to easily upgrade to a new code base

BeeKit works the same way whether using the ZigBee networking protocol or one of the other Freescale wireless protocols, such as the full IEEE 802.15.4 MAC or SMAC.

BeeKit uses the following terms: solution, project, template, code base, and properties:

- *Solution* refers to a collection of projects created through BeeKit. For example, a solution may include both a light and a switch. The light and the switch are each separate **projects**. A single file is created by BeeKit which describes the solution, and is called something like "MySolution.bksln." The solution file and any source files that have been modified for the application are all that is needed to send between developers or to back up in version control. Everything else can be recreated given the same code base.

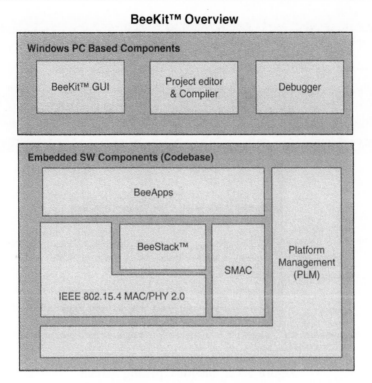

Figure 3.11: Freescale BeeKit

- *Template* is the source code to demonstration applications for developers to use as a starting point for custom applications. BeeKit copies the chosen template, which includes all the source code and libraries required to build the application to the project directory.

- *Code base* refers to a given release of BeeStack. For example, if Freescale fixes some bugs or adds new features to BeeStack, a new code base can be downloaded from the Web and applied to the application. This makes upgrading to new versions of BeeStack very convenient.

- *Properties* in BeeKit parlance are settable compile-time options that configure BeeStack or the application. For example, if you are making a ZigBee End Device (a ZigBee node that can sleep) you have a choice as to whether that node will poll its parent or not by setting the property RxOnIdle to TRUE or FALSE. Many options can be set at run-time in addition to compile-time. "Properties" refer specifically to the compile-time options settable through BeeKit.

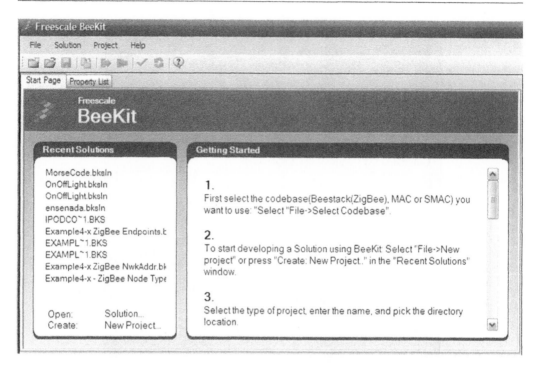

Figure 3.12: BeeKit Opening Screen

The opening screen of BeeKit is shown in Figure 3.12.

Step-by-step examples of using Freescale BeeKit, in addition to the Freescale CodeWarrior compiler and IDE are provided in Section 3.5, later in this chapter.

> BeeKit creates application templates and configures ZigBee applications.
>
> Upgrading to new versions of BeeStack is simple using a BeeKit code base.

3.3   The Embedded Compiler and Debugger

In addition to the ZigBee stack, an embedded compiler and debugger are required to build ZigBee. Some companies, like Freescale, make both the stack and the compiler: CodeWarrior is a Freescale product. Other companies use a commercially available compiler and debugger, for example, IAR in the case of Ember.

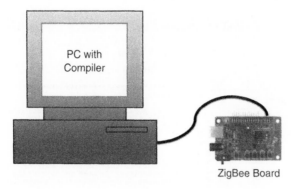

Figure 3.13: The ZigBee Download Process

A compiler takes the source code of the ZigBee application, the source code of the ZigBee stack, and combines it with ZigBee libraries to create a binary image. The binary image is then downloaded into the ZigBee development (also called target) board for debugging (see Figure 3.13).

As part of putting the binary image together, a process called "linking" is used. There is usually a file associated with linking called a locator file, which tells the linker where to place the code and data within the target MCU.

A debugger uses a JTAG or BDM connection (through USB) to program the application into flash memory in the target board. Once the download is complete, the JTAG or BDM can be disconnected from the board and the board can now run independently of the PC.

3.3.1   Introduction to Freescale CodeWarrior IDE

CodeWarrior is an integrated development environment (IDE) which contains both a C compiler and a debugger (see Figure 3.14). The compiler is your usual cross-compiler, with a module that outputs optimized code for the target MCU. In the case of the Freescale MC13213 single chip ZigBee solution, this is the HCS08.

CodeWarrior includes a fully colorized editor, but if you have a favorite editor, feel free to use it instead. CodeWarrior also creates nice hyperlinks between code and types, so it's easy to go to a given function or to the definition of a variable or type.

CodeWarrior supports multiple BDMs, so if enough USB ports are available, multiple boards can be debugged at the same time. The CodeWarrior IDE is shown in Figure 3.14.

Figure 3.14 shows CodeWarrior with the HA On/Off Light project open. The window to the left shows the project and set of files that make up the application. The window to the

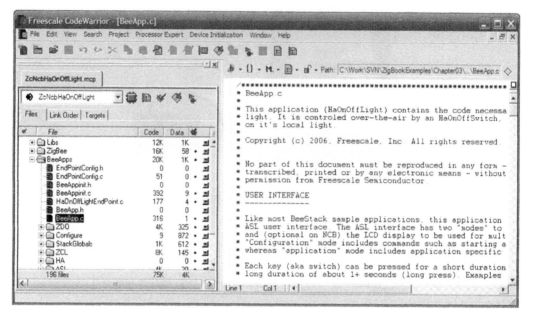

Figure 3.14: CodeWarrior IDE

right shows the source code. The little icon in the project window that looks like a hand writing on paper is the "make" button, and the little bug with the pointer (which is green in CodeWarrior) is the "debug" button.

If you are using an external editor, make sure to click the check mark icon before making the application. This will check file dates to make sure CodeWarrior knows about the external changes.

The debugger allows a developer to step through the source code, line by line, to debug the application, and includes windows to examine variables and MCU settings, including all registers.

3.3.2   CodeWarrior Tricks and Tips

All IDEs and debuggers have their quirks, and CodeWarrior is no exception. This section describes a few of the tricks and tips I've learned over the years about CodeWarrior and the debugger:

- *Hyper-links aren't available until the compiler has compiled a module.* The source code editor still has colorization, but right-clicking on a variable, type, or function will not go to the definition until *after* the source has been compiled. So first compile the original working source code, then make your changes, and it will be easier to navigate.

- *Source code is automatically pulled into the project if it has the same name and is in the project tree.* Make sure if you are keeping multiple copies of a source file around (sometimes I keep a backup handy), and place it outside the source code tree or you could be tracking down false compiler or linker errors.

- *The debugger gets stuck in an interrupt handler when single stepping (F10 or F11) through source code.* To get out of the interrupt handler, use assembly step (Ctrl-F11), then step out (Shift-F11). The debugger doesn't have this problem when running to location or stopping on breakpoints.

- *Breakpoints can only be set when the debugger has stopped code execution.* The debugger will allow setting a breakpoint through the interface when the code is running, but it won't actually stop at that location. Instead, halt execution (F6), set the breakpoint, and then continue running (F5).

- *The optimizer can make debugging confusing.* The code optimizer will fold lines of C code together or optimize out lines altogether (this is good, as it makes a smaller code image). Look at the assembly code and it will make sense. Just don't be surprised if the debugger skips around a bit in the C source code when single stepping.

3.4 Debugging the Network

The IDE-like CodeWarrior allows developers to debug individual nodes in the network, but some bugs are difficult to detect in this manner. Some bugs require looking at the network as a whole, or examining the over-the-air packets in order to solve them. To accomplish this, another type of debugger, a protocol analyzer, is used.

One bug that caught me seems like a simple one, but was very difficult to find until I saw the over-the-air behavior. In ZigBee, every application communicates data on an Application Profile. If a packet is sent on a profile which the receiving node doesn't understand, the packet is dropped by the receiving node before it reaches the application. The sending side says it sent the data. The receiving side receives nothing. Very perplexing!

During a class I was teaching once, we had modified the source code on one side of the application (the sending node), changing the profile and some code. But, on the receiving side, although the code was changed, the profile was not. We pressed the buttons, sent the data, and voila! Nothing.

Using the IDE debugger, we could see the data request being sent. All looked correct. Yet we did not see the data indication on the receiving side. A quick examination of the over-the-air data showed the problem. We could see the data going over the air, so we knew for sure the data was being sent. We could see the proper destination node profile and endpoint. All looked correct, so we knew the bug must be on the receiving side.

```
IEEE 802.15.4
   Frame Control: 0x8841
   Sequence Number: 57
   Destination PAN Identifier: 0x0f00
   Destination Address: 0xffff
   Source Address: 0x0001
ZigBee NWK
   Frame Control: 0x0048
   Destination Address: 0x0000
   Source Address: 0x0001
   Radius = 10
   Sequence Number = 121
ZigBee APS
   Frame Control: 0x08
   Destination Endpoint: 0x08
   Cluster Identifier: (0x0100)
   Profile Identifier: (0xc035)
   Source Endpoint: 0x08
   Counter: 0xd3
   APS Data: 03:48:65:6c:6c:6f:20:5a:69:67:42:65:65:00
.Hello ZigBee.
```

A quick examination of the endpoint structure in the destination node showed the wrong profile. It was still 0xc021 (Freescale's private profile ID), but it should have been 0xc035 (San Juan Software's private profile ID).

```
const zbSimpleDescriptor_t Endpoint8_simpleDescriptor =
    8,            /* Endpoint number */
    0x21, 0xC0,   /* Application profile ID */
    0xFF, 0xFF,   /* Application device ID */
    0,            /* Application version ID */
    1,            /* Number of input clusters */
    (uint8_t *) Endpoint8_InputClusterList,
    0,            /* Number of output clusters */
    (uint8_t *) Endpoint8_OutputClusterList,
};
```

Other bugs can happen because packets are received out-of-sequence. Sometimes, due to retries or other random latencies in the network, a packet sent after another packet can be received before the first. If an application must sequence packets, make sure to wait for the reply before sending another one.

Networks, especially busy ones, can be very difficult to debug without a protocol analyzer. I highly recommend purchasing one. The one I use is called the Daintree Sensor Network Analyzer.

A Network Analyzer allows debugging the entire network from an over-the-air perspective.

3.4.1 Introduction to Daintree SNA

The protocol analyzer used throughout this book is the Daintree Sensor Network Analyzer (see Figure 3.15). This analyzer does a great job of decoding ZigBee and 802.15.4 MAC packets, and displays the data in a wide variety of formats, including time line views, network-topology views, packet views, or physical-node-placement views.

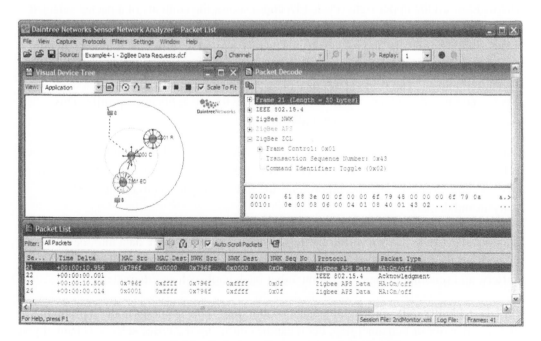

Figure 3.15: Daintree Sensor Network Analyzer

This protocol analyzer is so rich that an entire book could be written describing its features. I'll describe only a few of the highlights here.

Figure 3.15 shows some of the features of Daintree, and is my preferred layout for debugging.

The window on the upper left (Visual Device Tree) displays the nodes in the network. It may be hard to read in the figure, but on my screen I can see that endpoint 8 on the ZigBee End-Device On/Off Switch (796F) is communicating to endpoint 8 on the ZigBee Coordinator On/Off Light (0000). I sometimes use the Visual Device Layout view instead, which allows me to place nodes in their physical location on a blueprint or map.

The window on the bottom (Packet List) lists all of the packets received, and can be sorted and filtered by time, source address, destination address, network, or any combination desired. One nice feature of the Packet List window is the ability to build complex filters. All of the data is still captured, but only those packets of interest are displayed. Filters are very easy to build by simply right-clicking on fields within the packet, or by building them using C syntax.

The window on the upper right (Packet Decode) shows the contents of a packet selected in the Packet List window in a more detailed form, including the entire set of octets (bytes) sent over the air, and each ZigBee layer, separated by different colors. I can tell, for example, that packet 21 in Figure 3.15 is the Home Automation Toggle command on the On/Off cluster. The Packet Decode can be extended to understand your application's specific commands and data by creating XML code for the application profile. This makes reading decodes much easier and faster.

Daintree SNA receives its data through what is called a sniffer. Every silicon vendor, such as Freescale or Texas Instruments, offers a sniffer that plugs into a USB port. The sniffer then captures over-the-air data and presents the raw set of packets to Daintree SNA for analysis and display.

Daintree also offers its own sniffer hardware that can inject ZigBee data into the wireless network for testing, or capture data over a wider physical area by combining multiple sniffers, and sending the data over some other back haul such as Ethernet.

Sniffers generally only sniff a single channel at a time. This is not normally a problem since ZigBee doesn't channel-hop, but it can be confusing if ZigBee selects a channel you don't expect when forming the network. One solution is to compile the ZigBee nodes so they select a particular channel during work in the lab. At San Juan Software, we even assign channels to individual developers, so they don't interfere with each other.

Daintree SNA does work with multiple sniffers, so another solution is to plug in multiple sniffers, one sniffer watching each channel. I keep waiting for someone to make a sniffer than can sniff on all 16 IEEE 802.15.4 (ZigBee) channels at the same time.

The data is captured to .dcf files. These files are a simple text representation of the over-the-air octets (bytes), along with a packet sequence number and a time stamp. These files are easy to share with colleagues, and can be viewed (and even modified) with a text editor. The Daintree SNA documentation describes these features and more.

Of the packet analyzers I've used, I prefer Daintree SNA for its easy interface, its robust data views, and excellent support. Daintree SNA provides a very accurate decode of the complex ZigBee protocol and can handle millions of packets in any given capture.

> The Daintree Sensor Network Analyzer is easy to use and very robust and understands well the complex ZigBee protocol.

3.5   Example Development Sessions

This section walks you through two complete development sessions, from the out-of-the-box experience, to downloading real applications into ZigBee development boards, and then running and debugging them.

To follow along with the examples with hardware, purchase the Freescale Network Starter Kit (NSK) available at http://www.freescale.com/zigbee. This kit has everything required to follow nearly all the examples in this book. The development sessions in this section can also be followed without the hardware, to gain an understanding of how the development process works.

Alternatively, other Freescale ZigBee boards may be used, such as the SARD, EVB, SRB, or NCB boards, all available in various kits from Freescale and their distributors. To use other boards than those in the Network Starter Kit the BeeKit Solution files must be modified to support the correct hardware.

The process is very similar with any platform, so even if you plan to use another platform vendor, it is still worthwhile to read the following sections.

3.5.1   Example Session 1: The On/Off Light

This example session shows you how to build a small two-node network with a light and a switch, similar to the example in Section 1.4 of Chapter 1, "Hello ZigBee." The switch can turn the light on, off, or toggle it, both with and without end-to-end acknowledgment. The example uses the Freescale common user interface, the one used with all of the Freescale sample programs.

The development setup I used for this entire book is depicted in Figure 3.16. I used a Targus four-port USB hub to connect all the ZigBee boards to my laptop, a P&E Microcomputer USB Multilink BDM for debugging, and a Freescale Sniffer for capturing over-the-air packets. I used a Windows XP-based laptop. I haven't tried Windows Vista yet, but the initial reports aren't encouraging for compatibility with existing software.

The software installed on my laptop was Freescale CodeWarrior (the IDE for editing, compiling, and debugging the code), Freescale BeeKit (the ZigBee stack configuration and template generation tool), and the 30-day evaluation of Daintree Sensor Network Analyzer for capturing over-the-air packets. All of this hardware and software, with the exception of the Targus 4-port USB hub, comes with the Freescale NSK.

In Figure 3.16, you can see the NCB board (the larger one with an LCD screen), and three SRB boards. The NSK only comes with two SRBs (the smaller boards), and one NCB (the larger board) but some of the ZigBee concepts (such as mesh route recovery)

Figure 3.16: ZigBee Development Hardware

require more than three nodes. So I borrowed extra hardware from the San Juan Software labs as I needed it.

The steps in creating this very first example include installing the software and hardware, compiling and downloading the code, and running the example. This process is outlined in more detail below. Note that the first step, installing the software and hardware, occurs only once. The other steps are repeated for every example in the book.

First, install the software and hardware:

- Purchase the Freescale Network Starter Kit (NSK)
- Purchase the Freescale sniffer
- Install BeeKit
- Install CodeWarrior
- Install Daintree SNA
- Connect and install the BDM
- Connect and install the Freescale sniffer
- Connect and install the NCB and two SRB boards
- Download and install example code

Second, create, compile, and download each project:

- Modify and/or export the solution in BeeKit
- Import each project into CodeWarrior
- Compile the code
- Download (debug) the code

Third, run the example:

- Optionally set up the sniffer
- Boot the ZigBee boards
- Bind the application
- Press buttons and watch the fun

The detail and the lengthy explanations I include make it look more complicated than it really is. Once you've been through the process two or three times, it's really very easy. I do this all day long most every day, and I've learned a few tricks that make ZigBee development life easier. I've included these helpful hints throughout this book.

Install the Software and the Hardware

Many studies claim debugging is the longest task in any project. I say installing is longer. For whatever reason, installing always seems to be an arduous task. I recently had to set up a new laptop with all the usual tools and development programs and it took me nearly two days before it was finally set up the way I like it. This was using Windows XP. A business colleague of mine bought a new laptop with Windows Vista, and after many weeks it's still not right!

The initial thing to do is to order the hardware, as shipping can take a few days. This can be ordered directly from the Freescale Web site (http://www.freescale.com/zigbee) or from a distributor such as DigiKey (http://www.digikey.com). Order the NSK with BDM (1321XNSK-BDM). The BDM is used for downloading and debugging. Optionally, order the sniffer (FSL-ZB-SNF), which allows you to use the Daintree Sensor Network Analyzer to decode over-the-air packets. Then, wait, patiently. Perhaps read the rest of this book, and then come back to here.

To get the latest copy of Freescale BeeKit, CodeWarrior, and Daintree SNA, feel free to go to the Web site (http://www.freescale.com and http://www.daintree.net), or you can use the version included on the CDs in the NSK. I won't go through this process, as the Install Wizards are pretty straightforward. Make sure to install all three of the following software packages:

- Freescale CodeWarrior

- Freescale BeeKit

- Daintree Sensor Network Analyzer

All of these software packages are trial versions that last for 30 to 90 days. If you continue beyond reading about ZigBee and advance to actual development, they are all available for purchase, but 30 days should be enough to get you through this book.

First, unpack the hardware and plug in the NCB and SRB boards into the USB hub. *Hint: Mark which boards are plugged into which USB ports, and use the same configuration each time. USB is fussy and will require you to reinstall the hardware if you move it to a new port.* **Another hint:** *leave the NCB and SRB boards in the "off" position until after installing them. For some reason, Windows is happier about this.*

Choose to install the drivers from a specification location, as shown in Figure 3.17.

"C:\Program Files\Freescale\Drivers"

You'll have to repeat this twice (Windows needs to install two drivers per board). Windows will complain that this is not a digitally signed driver, but install it anyway.

Second, and after installing them, turn the NCB and SRB boards on. The result should be a new COM port for each board, which can be seen in Device Manager shown in Figure 3.18.

Third, install the USB Multilink BDM the same way. When installed correctly, it should have a blue light lit as seen in Figure 3.19.

Fourth, install the sniffer, shown in Figure 3.20. Now, all the hardware has been installed. If you wish to operate the ZigBee boards on batteries, install two AAs each, but it's not needed. The USB port supplies all the power required.

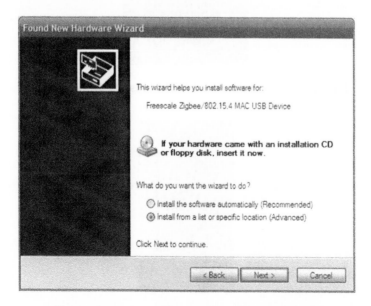

Figure 3.17: Installing SRB and NCB Boards

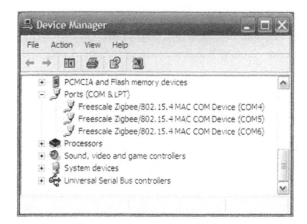

Figure 3.18: Installed NCBs and SRBs Are Seen as COM Ports

Figure 3.19: The Installed BDM Has the Blue Light Lit

Figure 3.20: Installed Sniffer Is Lit

The fifth and final step is to install the software examples from this book. I didn't include a CD with the book because you can go directly to the book's Web site http://www. zigbookexamples.com\code.

From there, download the install package "ZigBeeWirelessNetworking.exe" and save it in an empty directory on your PC. Run the installer, which will by default create a directory called C:\ZWN.

The directory is organized into chapters, and includes a .txt file that briefly describes the examples in that chapter. Complete descriptions of each example are in this book. All examples come with full source for the application.

Create, Compile, and Download Each Project

This first example demonstrates how to create a new solution in BeeKit. In this case, the solution will contain two projects: an HA On/Off Light, and an HA On/Off Switch.

From the BeeKit menus, choose "File/New Project…" which will bring up a dialog as shown in Figure 3.21. From the project types, select "ZigBee Home Automation Applications." Then, from the templates, choose "HA OnOffLight." Only then adjust the paths and files in the dialog as shown in the figure.

Figure 3.21: A New Project in BeeKit

As you can see, I always name projects based on their ZigBee node type (Zc, Zr, or Zed), the board type (Ncb, Srb), and the application name, such as "ZcNcbHaOnOffLight."

Just to keep Freescale terminology clear, a *solution* is a collection of one or more projects. A *project* is exactly one downloadable ZigBee application.

The location field is one directory above where the solution directory and file will go, in this case, "C:\ZWN\Chapter 03." The solution name is "OnOffLight." The project name is "ZcNcbHaOnOffLight."

Click "OK," and the BeeKit Project Wizard comes up, shown in Figure 3.22. This handy Wizard exposes the common ZigBee options in an easy-to-set format. In the Wizard, click "Next," to get to the "Platform Type" tab. On this page select NCB and enable the display check-box. The HA On/Off Light application will be placed into an NCB board. At this point, leave the rest at the default settings, and simply click "Finish."

Now add another project to this solution (in this case an HA OnOffSwitch) from the "Solution/Add Project…" menu. Name the project "ZedSrbHaOnOffSwitch." If you

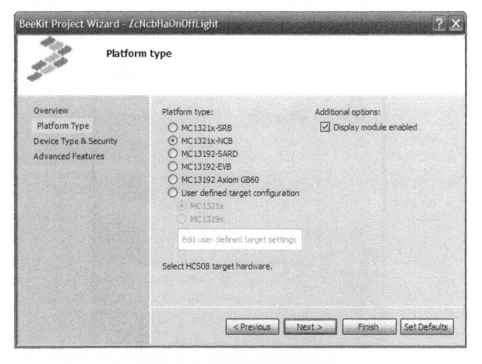

Figure 3.22: BeeKit Project Wizard

mistyped the name and have already clicked "OK," don't worry. BeeKit allows you to adjust anything through properties, including the name of the projects. In the Project Wizard, choose the SRB board for the on/off switch and click "Finish." When you're done, you should see a completed solution with two projects as shown in Figure 3.23.

Once the solution is created, save it. Then export it from the "Solution/Export Solution…" menu. Exporting takes copies of the appropriate files from the BeeStack code base, to the solution directory, in this case "C:\ZWN\Chapter03\OnOffLight." If you look at that directory, you'll see a 500 K file called "OnOffLight.bksln." This file, plus any modified application source code files, are all that is needed to archive the application, or send it to a colleague. Every other file can be generated from the code base.

Updating to a new code base, that is, to a new version of BeeStack, is as simple as selecting "File/Select Codebase…" from the menu system. Code bases make it easy to get the latest stack fixes or features and add them into your existing projects. Once the solution is exported, all the source code is available for the usual edit/compile/debug cycle.

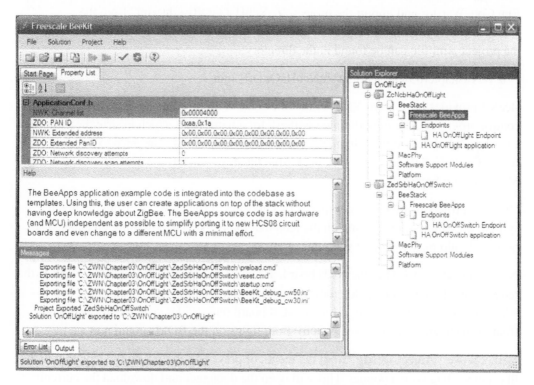

Figure 3.23: Completed Two-Project Solution in BeeKit

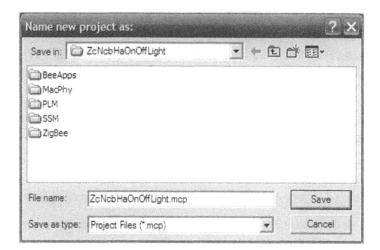

Figure 3.24: Import Project into CodeWarrior

Run CodeWarrior and import the project "ZcNcbHaOnOffLight.xml" from the CodeWarrior "File/Import Project…" menu. If you saved the file as described above, the project .xml file can be found in the directory:

C:\ZWN\Chapter03\OnOffLight\ZcNcbHaOnOffLight

BeeKit provides projects in XML so that they can be archived using text-only version control programs. The import dialog will ask for a name with which to save the project. Choose "ZcNcbHaOnOffLight.mcp," as shown in Figure 3.24. The .mcp extension stands for Motorola CodeWarrior Project, back from the days when CodeWarrior was owned by Motorola, not Freescale.

Once the project is imported, compile it by choosing "Project/Make" from the menu system, pressing F7, or clicking on the icon that looks like a handwriting on paper. The project should make, without warnings or errors.

Next, connect the BDM to the NCB board, as shown in Figure 3.25, on the left. Pin 1 (the red portion of the ribbon cable) goes to the left when looking at the back of the board where all the pins are located. Turn the NCB board on (a slider switch is located on the front of the board near the reset switch). Both lights on the BDM should now be lit.

Next, choose "Project/Debug" from the menu, or press F5, or click on the icon that looks like a green bug with a pointer. This will open the debugger and download the current application.

Figure 3.25: The BDM Connected to the NCB and SRB Boards

Next, select the "ZedSrbHaOnOffSwitch" project, and follow the same steps. Compile it, and download it.

Once both projects are downloaded into their respective boards, boot both boards by pressing the reset button on each. They will both begin blinking LED1. This indicates the Ready state (not yet on the network). As mentioned previously, all Freescale demos use a common user interface with two modes (a Configuration Mode and an Application Mode). Too see the entire list of key presses, see Table 3.2. The boards start out in Configuration Mode.

Press SW1 on each (can be either order) and the LEDs will begin chasing each other. Once the network has been formed, the NCB board will have LEDs 1 and 2 lit, and the SRB board will have LED1 lit. Next, press SW3 on each, to bind the switch to the light. This network has only one light and one switch, but if there were more than one you could select which switch is attached to which light(s). LED3 should blink for a few moments, and then become solid. The switch now knows which light to control.

Next, press long SW1 (press and hold SW1 for about one second). The application will go from Configuration Mode to Application Mode. Press SW1 on the HA On/Off Switch (the SRB board), and LED2 on the remote HA On/Off Light (the NCB board) will light.

You've now built your first application using BeeKit! Granted, this may seem like a lot of steps to create and download an application. The process is really quite easy and quick. Reading about it takes longer than actually doing it.

> The basic steps used to create an application with BeeKit are:
>
> - Install CodeWarrior, BeeKit, and Daintree (first time only).
>
> - Create project(s) in BeeKit from templates and export them.
>
> - Import projects into CodeWarrior, compile and download them.
>
> - Reset the boards to run the application (most use the Freescale Common UI).

3.5.2   Example Session 2: Morse Code

The previous example showed how to create an application from an existing BeeKit template. This example shows how to create an application which includes additional source code beyond what is in the template. In *Example 3-2 Morse Code*, the application communicates Morse Code over ZigBee.

For those not familiar with Morse Code, it was originally created by Samuel F. B. Morse for the electric telegraph in the early 1840s. This code uses a series of "dots" and "dashes" to represent the letters in the English alphabet and the numbers 0 through 9. Most of you probably know S-O-S (dot-dot-dot, dash-dash-dash, dot-dot-dot). The code can be transmitted across wires, using lights, taps, or clicks, or in this case, across a wireless ZigBee network. Morse Code is still in use by some HAM radio operators today. Go to http://www.wikipedia.org and look up "Morse Code" if you wish to see the entire set of codes and learn more about it.

All the examples in this book use either the previous method (creating the program using BeeKit only), or the method described in this section where additional source code is required for the example. The steps for an example which includes additional source code are as follows:

- Open the existing BeeKit solution file.

- Export the solution (the collection of projects).

- Copy the additional source files from the example source directory to each project directory.

- Import each project into CodeWarrior.

- Compile and download each project into the proper board.

- Run the application.

In the root folder of each chapter is a text file, which briefly describes the examples (see Figure 3.26). In addition, any capture files for the examples can be found here. Each example is contained in its own subfolder. Each subfolder includes a BeeKit solution file to save the reader the step of configuring each project in the solution.

First, open the solution file "C:\ZWN\Chapter03\MorseCode\MorseCode.bksln." This file contains two projects, both of which use the same Morse Code application. The first, "NcbZcMorseCode," is targeted for a Freescale NCB board and will be the ZigBee Coordinator for the network (the device that forms the network). The second project, "SrbZrMorseCode" is targeted for a Freescale SRB board and will be a ZigBee Router, capable of routing packets from any ZigBee node, in addition to its Morse Code functionality.

Next, export the solution from the "Solution/Export Solution…" menu.

In projects like *Example 3-2: Morse Code*, which contain special source code, at least one "example source" folder can be found inside the example's folder. In this case that folder is called:

C:\ZWN\Chapter03\MorseCode\ExampleSource

This folder contains all the necessary source code to build the example. Simply copy this folder over the ones that were created during the export process. This process will replace a single file called BeeApp.c in each project. Note that the same application is used on both the ZigBee Coordinator (NCB) node, and the ZigBee Router (SRB) node. So stated

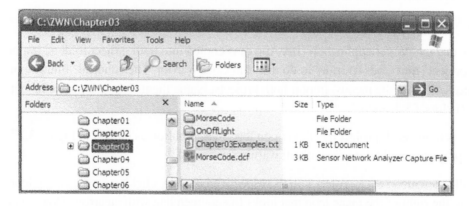

Figure 3.26: Examples Are Stored in Chapters

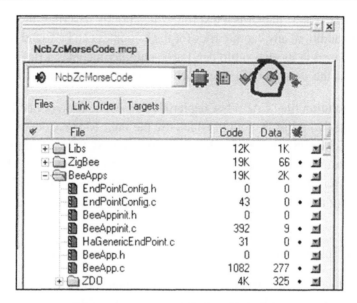

Figure 3.27: Compile Each Project

plainly, copy the ExampleSource folder (and all the files in it) into the NcbZcMorseCode and SrbZrMorseCode folders. One way to do this in Windows, is to open both the ExampleSource and NcbZcMorseCode folders. Press Ctrl-A to select all items in the ExampleSource folder. Press Ctrl-C to copy everything in that folder (including all sub-folders). Then click on the NcbZcMorseCode folder and press Ctrl-V to paste. These are the folders:

C:\ZWN\Chapter03\MorseCode\NcbZcMorseCode

C:\ZWN\Chapter03\MorseCode\SrbZrMorseCode

Now that the proper application (BeeApp.c) is copied into both project folders, import the projects into CodeWarrior. To do this, run CodeWarrior. Then choose "File/Import Project…" from the menus. Import both projects. Name them NcbZcMorseCode.mcp and SrbZrMorseCode.mcp.

Now compile each project by clicking on the "make" button as circled in Figure 3.27. The projects should compile without warning or errors.

Finally, download each project into its respective board by clicking on the debug icon, the one to the right of the make icon that looks like a green bug with an arrow. Download the NcbZcMorseCode project into the Freescale NCB board. Download the SrbZrMorseCode

Table 3.3: Morse Code UI

Switch	Description
SW1	Send "Hello ZigBee" over-the-air to the remote node
SW2	Send a dot over-the-air to the remote node
SW3	Send a dash over-the-air to the remote node
SW4	Display SOS in Morse Code locally

project into the Freescale SRB board. The previous section describes the details of installing and connecting the BDM for downloading.

To see the over-the-air packets, run Daintree SNA. In the "Source:" drop-down box, choose the Freescale sniffer. In the "Channel:" drop-down box, choose channel 25. This example, like most examples in this book, uses channel 25 and ZigBee PAN ID 0x0f00. The channel and PAN ID can be seen on the NCB board's LCD screen at startup.

If you don't have a sniffer, you can still see the over-the-air packets by looking at the capture I made. Open the file "MorseCode.dcf" from the chapter folder by selecting File/Open Capture File…" from the menus.

Example 3-2: Morse Code uses a very simple interface. The nodes automatically form a network on startup. Simply press the reset button on both. LED1 will blink for a bit, and then remain lit. At this point the nodes are ready to communicate using Morse Code.

Press SW1, SW2, or SW3 to send Morse Code over-the-air, as shown in Table 3.3. Press SW3 to display it locally. LED2 is used for the blinking light and the buzzer is used to beep out the dashes and dots in addition to the LED.

These two examples should give you enough to go on to follow the examples in the rest this book. All of them are organized and operate in the same way. The user interface changes some in each example, but that is explained along with the example.

> The steps to add example source code to a project are:
> - Export projects from the BeeKit solution.
> - Copy example source code into the projects.
> - Import projects into CodeWarrior, compile and download them.
> - Reset the boards to run the applications.

CHAPTER 4

ZigBee Applications

This chapter describes how ZigBee applications interact with other ZigBee applications. It describes what constitutes a ZigBee network, how individual nodes are addressed, and how to address application objects within a node. Terms such as PAN ID, extended PAN ID, network address, profile ID, cluster, endpoint, attribute, and command will become clear (if I have done my job). This chapter is filled with plenty of examples. If you're a developer and read only one chapter in this book, read this one.

But first, as usual, the *real* truth about how ZigBee gained its name.

ZigBee. Zig… Bee…

Bees. The origin of ZigBee came from the peculiar behavior of bees, first noted in the 1960s by Nobel Prize-winner Karl von Frisch. Bees, after zigging and zagging around in the fields, return to the hive, and perform what some call the Waggle Dance to communicate the distance, direction and type of food to others in the hive. After receiving a WAGGLE-DANCE.indication, bees fly off directly to the source of food.

"We have solved it once and for all," said Professor Joe Riley, team leader at Rothamsted Research, an agricultural research center. The team tracked a group of bees as they flew to a food source. To track bees by radar, the researchers first had to create a transponder small enough and light enough so that a bee could carry it. The transponder weighed approximately 10 to 12 milligrams, a fraction of the pollen load bees are accustomed to carrying.

Opponents to the Frisch theory have suggested that while the bees dance, it's not to convey information. They believe bees are actually guided to the food source by odor conveyed by the scout bee. Von Frisch, on the other hand, claimed that recruits understood the dance and flew directly to the food source. But "the bees take five to 10 minutes, not one minute," said Riley.

To make sure bees were not following a scent, a control group of bees was transported 250 meters after seeing a Waggle Dance. When released, the bees flew off in the direction indicated by the dance, the team at Rothamsted Research found.

The team's results show that bees do decode the dance and fly off immediately in the direction indicated. But, "they very rarely get it absolutely right," said Riley. "The mean error is about 5 to 6 meters." Once the bees get to the end of the flight, they change their flight pattern, and start zigging and zagging, looking for the food they were instructed to find. That takes time, Riley said, sometimes up to 20 minutes.

ZigBee chips do not currently fit onto the back of a Bee, but they do come in 5×5 mm packages, and consume so little power they can last longer than the lifetime of a bee on a couple of AAA batteries.

4.1   Sending and Receiving Data

The whole point of a wireless network is to send reliable data between nodes in the network, and ZigBee makes it easy. The ZigBee network automatically figures out how to route the data from one node to another with the maximum chance of success.

To send data from one node to another in the Freescale platform, for example, simply call:

```
AF_DataRequest(&addrInfo, iDataSize, pPtrToData, NULL);
```

That's it. ZigBee takes care of the rest!

ZigBee uses standard networking terms for data transmission, as defined by IEEE. This includes:

- Data Request (which means to transmit data)

- Data Confirm (which means the acknowledgment of a data request)

- Data Indication (which means to receive data)

Data requests are initiated by the application, as shown in Figure 4.1. A data confirm is the direct result of a data request: Each time an application generates a data request, it will receive exactly one data confirm to indicate the success (or failure) of that request. A data indication, however, may arrive at any time.

Data requests come in a variety of flavors. The options include:

- Unicast with end-to-end acknowledgment

- Unicast without end-to-end acknowledgment

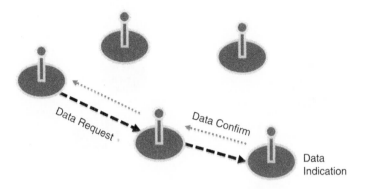

Figure 4.1: ZigBee Data Requests

- Broadcast

- Groupcast/Multicast

Unicasts are transmitted from one node to exactly one other node. Unless the nodes are neighbors (within radio range of each other), route discovery takes place the first time these nodes speak together. If end-to-end is acknowledged, the unicast will be retried up to three times, perhaps even initiating a new route discovery if the old route is broken.

Broadcasts are transmitted from one node to all of the nodes in the network, within a sender-definable radius.

Groupcasts, and the upcoming Multicast in ZigBee 2007, broadcast to a specific set of nodes. Any node (actually, any endpoint, but I'll explain it in detail later) not part of the group will discard the packet. For example, assume the dark nodes in Figure 4.2 all belong to group A. If a node groupcasts to group A, the packet will reach the applications in these specific nodes and no others.

The example in the next subsection, *Example 1-1 ZigBee Data Requests*, demonstrates how to initiate all these various data request mechanisms, and how it looks from the standpoint of data indication.

While this chapter describes how to use ZigBee unicasts, broadcasts, and groupcasts, if you would like to understand how they work from a stack standpoint, see Chapter 7, "The ZigBee Networking Layer."

ZigBee is an asynchronous protocol; any node may transmit or receive data at any time. When a user flips a light switch, the data is sent immediately, even if that light switch is a

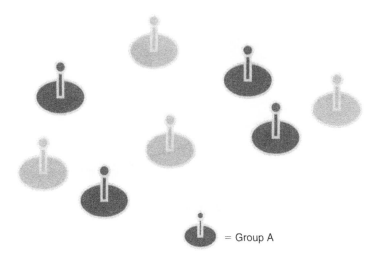

Figure 4.2: ZigBee Groups

battery-powered device designed to run for years on a couple of AAAs. If you have read the 802.15.4 specification, you may have read about beacons. ZigBee is beaconless.

Due to the nature of ZigBee, one thing that is difficult to predict is packet latency. A good rule of thumb is to assume that transmitting a packet requires about 10 milliseconds per hop. The difficulty comes in when retries or route discoveries are needed. Route discoveries require a broadcast across the network and the initiating node must wait until the results are received. Retries come in the form of both per-hop MAC retries and end-to-end APS retries.

In Figure 4.3, a node is transmitting to another node, four hops away. Expect the packet to be received 40 ms later (on average) with the acknowledgment returning 80 ms later.

When using acknowledged unicast data requests, up to three end-to-end retries occur with approximately a 1.5-second delay between each. This means the latency could be, in the worst case, nearly five seconds for a packet to successfully be delivered, or to indicate to the application that it failed to get through. Freescale BeeStack allows the duration between retries to be adjusted through the `gApsAckWaitDuration_c` property in `BeeStackConfiguration.h`.

ZigBee applications must allow for a broad range of latency for the worst-case scenarios.

The size of the packet doesn't affect latency much if the channel is fairly clean, as the random wait times dwarf the transmission times. But, if the channel is particularly noisy,

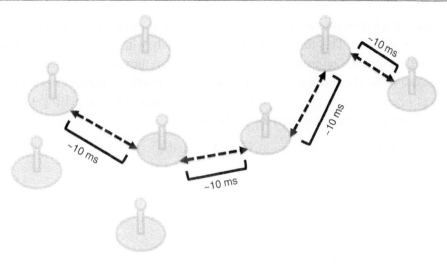

Figure 4.3: ZigBee Latency

then the length of the packet exponentially increases the chance of retries, which means an increase in latency.

A good policy is to keep the packet length as short as possible for the given transmission. Keeping packets short is good for two reasons: decreasing latency and reducing bandwidth usage, allowing more for other applications. Granted, if you are transferring bulk data, send as much as you can to reduce the number of packets, but for other kinds of data, think about how to keep it short.

> ZigBee data transmission options include: unicast, broadcast, and groupcast.
>
> Keep data transmissions as short as possible.

4.1.1   Example 1-1: ZigBee Data Requests

In Freescale BeeStack, the C functions which perform data requests, confirms, and indications mirror the ZigBee terms:

- AF_DataRequest()

- BeeAppDataConfirm()

- BeeAppDataIndication()

A data request, whether unicast, broadcast or groupcast, is initiated by a call to AF_DataRequest(). This function returns immediately with the data queued up to be sent out the radio and returns of gZbSuccess_c. A return of gZbNoMem_c means the message couldn't be queued due to lack of memory or other internal resources. A return of gZbInvalidEndpoint_c means that the source endpoint (I'll get to endpoints later) hasn't been registered with BeeStack.

The C prototype for AF_DataRequest() looks like this:

```
typedef struct afAddrInfo_tag
{
  zbAddrMode_t dstAddrMode;
  zbApsAddr_t dstAddr;
  zbEndPoint_t dstEndPoint;
  zbClusterId_t aClusterId;
  zbEndPoint_t srcEndPoint;
  zbApsTxOption_t txOptions;
  uint8_t     radiusCounter;
} afAddrInfo_t;
zbStatus_t AF_DataRequest
(
  afAddrInfo_t *pAddrInfo,
  uint8_t payloadLen,
  void *pPayload,
  zbApsCounter_t *pConfirmId
);
```

The first parameter to AF_DataRequest(), the afAddrInfo_t includes all the addressing information to indicate from where the packet is coming, and to where it is going. The afAddrInfo_t also contains transmission options, which define whether the packet is acknowledged or not, and the radius to indicate how far the packet should propagate in the network.

The field dstAddrMode in afAddrInfo_t has the same meaning as this field does in the ZigBee specification. It may be one of:

- gZbAddrModeIndirect_c

- gZbAddrModeGroup_c

- gZbAddrMode16Bit_c

- gZbAddrMode64Bit_c

The gZbAddrModeIndirect_c is used to indicate if the local binding table is used (I'll explain more about binding later). The gZbAddrModeGroup_c mode is used when transmitting to a group, and dstAddr field is then used as a group ID. The mode gZbAddrMode16Bit_c is used to transmit directly to a node, and the dstAddr field is used to indicate the 16-bit node address. The gZbAddrMode64Bit_c is used to transmit to a node using its IEEE (or 64-bit) address, and dstAddr will reflect that IEEE address.

The second and third parameters to AF_DataRequest() define the application's data (payload). The payload can be any length up to 80 bytes, and may contain any content.

The last parameter, pConfirmId, may take more explanation. Network traffic can be difficult to predict, so care must be taken in applications to either synchronize the data or to allow the data to be received by a node asynchronously. For example, say an application sends two data requests: one to node A, one to node B, in that order, as shown in Figure 4.4.

Due to multi-hops, retries, or perhaps the necessity of route discovery, the confirm from the data request to B may actually arrive before the confirm from the data request to A.

Freescale BeeStack uses a pointer to the confirm ID, a rolling 8-bit number, to allow the application to keep track of which data request is which. For those of you ZigBee experts, this confirm ID is just the APS counter, an over-the-air field used for the same purpose. Some ZigBee stacks solve the problem by allowing only one message to be in flight at a time, and hence, no need for a confirm ID.

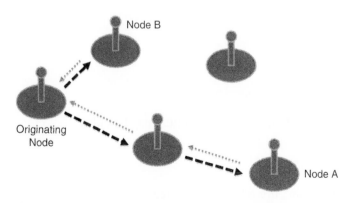

Figure 4.4: ZigBee Data Confirms

Data confirms come into a callback function named `BeeAppDataConfirm()`. The data type for that function is:

```
typedef struct zbApsdeDataConfirm_tag
{
  zbAddrMode_t  dstAddrMode;
  zbApsAddr_t  dstAddr;
  zbEndPoint_t  dstEndPoint;
  zbEndPoint_t  srcEndPoint;
  zbStatus_t  status;
  zbApsCounter_t  confirmId;
} zbApsdeDataConfirm_t;
```

Aside from the `status` and the `confirmId`, the rest of the source and destination information is included in the structure, simplifying code logic in applications. And, as is typical with the Freescale solution, it mirrors the ZigBee specification for APSDE-DATA.confirm.

A data indication means a node is receiving data from another node, whether the data request was unicast, broadcast, or groupcast. The data structure in BeeStack is shown next. As usual, the fields in the Freescale implementation mirror the ZigBee specification's fields APSDE-DATA.indication:

```
typedef struct zbApsdeDataIndication_tag
{
  zbAddrMode_t  dstAddrMode;
  zbNwkAddr_t  aDstAddr;
  zbEndPoint_t  dstEndPoint;
  zbAddrMode_t  srcAddrMode;
  zbNwkAddr_t  aSrcAddr;
  zbEndPoint_t  srcEndPoint;
  zbProfileId_t  aProfileId;
  zbClusterId_t  aClusterId;
  uint8_t    asduLength;
  uint8_t    *pAsdu;
  bool_t  fWasBroadcast;
  zbApsSecurityStatus_t fSecurityStatus;
  uint8_t linkQuality;
  } zbApsdeDataIndication_t;
```

The `dstAddrMode` field on the data indication may be either:

```
gZbAddrModeGroup_c
gZbAddrMode16Bit_c
```

So why not also use indirect or 64-bit mode? Because ZigBee resolves both indirect and 64-bit modes to one of the above, before sending the data request. The over-the-air packet will always be either group or 16-bit. The `srcAddrMode` field above is always `gZbAddrMode16Bit_c`.

I realized I haven't yet explained endpoints, profile IDs, cluster IDs, or security. These fields will become clear before the end of the chapter.

The `asduLength` and `pAsdu` is the length and pointer to the payload, the data that was transmitted by the data request from the other node. By the way, ASDU stands for APS Service Data Unit, the ZigBee way of referring to the packet's data payload.

You'll also note that an application can tell if a data indication was broadcast or unicast with the `fWasBroadcast` field. This field, combined with the `aSrcAddr` field can be quite useful to determine what type of response (if any) to give to the data request, and where the response should go. The source of any message is always known in ZigBee. There is no anonymity.

The last field, `linkQuality`, is not in the ZigBee specification for data indications, but is an extra for Freescale BeeStack users. The link quality indicator (or LQI) can be used to estimate how close the nodes are: the stronger the LQI, the closer the nodes (well, approximately). This linkQuality field will be used in Chapter 8, "Commissioning ZigBee Networks," to ease deployment of a network. LQI can also enable some location-based applications.

The example in this section, *Example 1-1 ZigBee Data Requests*, shows how unicasts, groupcasts, and broadcasts are used within an application. The network is composed of three nodes. The ZigBee Coordinator (ZC) and a ZigBee Router (ZR) are lights. A ZigBee End-Device (ZED) is a switch. The lights are controlled via various ZigBee data request methods, to illustrate the use of each see Figure 4.5.

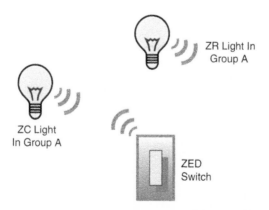

Figure 4.5: ZigBee Data Request Example

Table 4.1: ZigBee Data Request Example UI

Switch	Description
SW1	Uses address mode gZbAddrMode16Bit_c to toggle the ZC light directly.
SW2	Uses address mode gZbAddrModeIndirect_c (the local binding table) to toggle ZC light.
SW3	Uses gZbAddrModeGroup_c to send to group A. Only the ZR light belongs to group A.
SW4	Uses address mode gZbAddrMode16Bit_c to turn all lights on or off via broadcast (0xffff).
LSW2	Uses address mode gZbAddrModeIndirect_c to turn the ZC light on with acknowledgment.
LSW3	Uses address mode gZbAddrModeIndirect_c to turn the ZC light off with acknowledgment.
LSW4	Uses address mode gZbAddrMode64Bit_c to toggle the ZC light.

The user interface on the light switch is as follows. SW1 through SW4 are labeled on the Freescale development boards. The term LSW2, or long switch 2, means to press and hold SW2 for about one second or so (see Table 4.1).

I ran this example and captured the results. I pressed SW1, SW2, SW3, SW4, LSW2, LSW3, and LSW4 in that order. The full example and capture can be found at http://www.zigbookexamples.com:

```
Frame 19 and 21 (Length=30 bytes)
IEEE 802.15.4
   Frame Control: 0x8861
   Sequence Number: 61
   Destination PAN Identifier: 0x0f00
   Destination Address: 0x0000
   Source Address: 0x796f
ZigBee NWK
   Frame Control: 0x0048
   Destination Address: 0x0000
   Source Address: 0x796f
   Radius=10
   Sequence Number=13
ZigBee APS
   Frame Control: 0x00
   Destination Endpoint: 0x08
   Cluster Identifier: On/off (0x0006)
```

```
   Profile Identifier: HA (0x0104)
   Source Endpoint: 0x08
   Counter: 0x3f
ZigBee ZCL
   Frame Control: 0x01
   Transaction Sequence Number: 0x42
   Command Identifier: Toggle (0x02)
```

W1 generated frame 19, and SW2 generated frame 22. Both look the same, as shown in the capture, even though one is gZbAddrMode16Bit_c and the other is indirect through the local binding table. Why is this? Entries in the binding table always resolve to either gZbAddrModeGroup_c or gZbAddrMode16Bit_c before being sent over-the-air.

Notice in the decode, both SW3 and SW4 cause broadcasts (0xffff for the destination node at the NWK layer). SW3 causes a groupcast, so while it looks like a broadcast from the network layer, the APS layer filters the packet by group. Only an endpoint (an application within a node) which is a member of the group will receive the data indication and so, in this case, toggle:

```
Frame 23 (Length = 31 bytes)
IEEE 802.15.4
   Frame Control: 0x8841
   Sequence Number: 63
   Destination PAN Identifier: 0x0f00
   Destination Address: 0xffff
   Source Address: 0x796f
   Frame Check Sequence: Correct
ZigBee NWK
   Frame Control: 0x0048
   Destination Address: 0xffff
   Source Address: 0x796f
   Radius = 10
   Sequence Number = 15
ZigBee APS
   Frame Control: 0x0c
   Group Address: 0x000a
   Cluster Identifier: On/off (0x0006)
   Profile Identifier: HA (0x0104)
   Source Endpoint: 0x08
   Counter: 0x41
ZigBee ZCL
   Frame Control: 0x01
   Transaction Sequence Number: 0x44
   Command Identifier: Toggle (0x02)
```

LSW2 and LSW3 are acknowledged data requests, and you can see the ZigBee APS ACKs in frames 36 and 40. Notice also the IEEE 802.15.4 (MAC) acknowledgments below in frames 35, 37, 39, and 41. With unicasts, ZigBee always acknowledges each hop at the MAC level, regardless of whether APS (end-to-end) acknowledgments are enabled on that data request:

Frame No	MAC Src	MAC Dest	NWK Src	NWK Dest	Protocol	Packet Type
34	0x796f	0x0000	0x796f	0x0000	Zigbee APS Data	HA:On/off
35					IEEE 802.15.4	Acknowledgment
36	0x0000	0x796f	0x0000	0x796f	Zigbee APS	APS Ack
37					IEEE 802.15.4	Acknowledgment
38	0x796f	0x0000	0x796f	0x0000	Zigbee APS Data	HA:On/off
39					IEEE 802.15.4	Acknowledgment
40	0x0000	0x796f	0x0000	0x796f	Zigbee APS	APS Ack
41					IEEE 802.15.4	Acknowledgment

LSW4 uses the 64-bit IEEE address to communicate, in this case, to its parent. But the over-the-air frame uses only the short 16-bit NwkAddr for the destination node. All 64-bit addresses are actually resolved internally to a 16-bit address before transmission. ZigBee does not send data requests to 64-bit destinations. The 64-bit address mode is there for the sake of convenience sake.

In Freescale BeeStack, a ZDP.IEEE_addr_req or ZDP.NWK_addr_req must be initiated prior to using the 64-bit address mode (unless the node is a neighbor).

There are a few interesting things to notice in this data request example. The destination address, group ID, and any other multi-byte fields are little endian. All multi-byte things in ZigBee are little Endian over-the-air. That is, group 0x000a is byte order 0x0a 0x00. The Freescale HCS08 processor, used in these examples, is big Endian. Freescale solves this problem by insisting that all things over-the-air are stored little Endian, even in memory. For example, setting up group A below uses Set2Bytes(aDestAddress, 0x0a00).

Another interesting thing to notice is that when using gZbAddrModeIndirect_c, no destination endpoint or NwkAddr is needed. The destination will be resolved from the binding table entry, based on the source endpoint:

```
void BeeAppHandleKeys(key_event_t events)
{
zbNwkAddr_t aDestAddress;
zbEndPoint_t endPoint=appEndPoint;
```

```
/* Application-mode keys */
if(gmUserInterfaceMode == gApplicationMode_c) {
switch (events) {
  /* SW1-Uses address mode gZbAddrMode16Bit_c to toggle
    the ZC light directly. */
  case gKBD_EventSW1_c:
    Set2Bytes(aDestAddress, 0x0000);
    OnOffSwitch_SetLightState(gZbAddrMode16Bit_c,
      aDestAddress,endPoint,gZclCmdOnOff_Toggle_c, 0);
    break;
  /* SW2-Uses address mode gZbAddrModeIndirect_c
    (the local binding table) to toggle ZC light. */
  case gKBD_EventSW2_c:
    OnOffSwitch_SetLightState(gZbAddrModeIndirect_c,
      aDestAddress,endPoint,gZclCmdOnOff_Toggle_c, 0);
    break;
  /* SW3-Uses gZbAddrModeGroup_c to send to group A.
    Only the ZR light belongs to group A. */
  case gKBD_EventSW3_c:
    Set2Bytes(aDestAddress, 0x0a00);
    OnOffSwitch_SetLightState(gZbAddrModeGroup_c,
      aDestAddress,endPoint,gZclCmdOnOff_Toggle_c, 0);
    break;
  /* SW4-Uses address mode gZbAddrMode16Bit_c to turn
    all lights on or off via broadcast (0xffff). */
  case gKBD_EventSW4_c:
    Set2Bytes(aDestAddress, 0xffff);
    OnOffSwitch_SetLightState(gZbAddrMode16Bit_c,
      aDestAddress,endPoint,gZclCmdOnOff_Toggle_c, 0);
    break;
  /* LSW2-Uses address mode gZbAddrModeIndirect_c to turn
    the ZC light on with acknowledgement. */
  case gKBD_EventLongSW2_c:
    OnOffSwitch_SetLightState(gZbAddrModeIndirect_c,
      aDestAddress,endPoint,gZclCmdOnOff_On_c,
      gApsTxOptionAckTx_c);
    break;
  /* LSW3-Uses address mode gZbAddrModeIndirect_c to turn
    the ZC light off with acknowledgement. */
  case gKBD_EventLongSW3_c:
    OnOffSwitch_SetLightState(gZbAddrModeIndirect_c,
      aDestAddress,endPoint,gZclCmdOnOff_Off_c,
      gApsTxOptionAckTx_c);
    break;
  /* LSW4-Uses address mode gZbAddrMode64Bit_c to toggle
    the ZC light. */
```

```
   case gKBD_EventLongSW4_c:
     /* IEEE address alrady in gaAppIeeeAddr */
     OnOffSwitch_SetLightState(gZbAddrMode64Bit_c,
       aDestAddress,endPoint,gZclCmdOnOff_Toggle_c, 0);
     break;
   /* let ASL handle LSW1 */
   default:
     ASL_HandleKeys(events);
     break;
   }
}
else {
  /* If you are not on Appication Mode, then call
     ASL_HandleKeys, to be eable to use the Configure Mode */
  ASL_HandleKeys(events);
  }
}
```

The `OnOffSwitch_SetLightState()` function simply ends up calling `AF_DataRequest()`, after setting up the proper ZigBee Cluster Library frame as the payload.

The `ASL_HandleKeys()` function is not directly related to ZigBee. This is the default key handler which provides a common user interface for all of the Freescale applications. Chapter 3, "The ZigBee Development Environment," describes this common UI pretty well.

Source full code is available at http://www.zigbookexamples.com.

So this example showed pretty much every type of data request an application could desire. So for unicasts, broadcasts, groupcasts, and indirect data requests through the local binding table, why use any particular mechanism? This will help you decide:

- Use unicasts when a node will be communicating with another node fairly frequently. Unicasts conserve precious bandwidth if used in this way.

- Use acknowledged unicasts if the sending application must absolutely know the message got through.

- Use broadcasts sparingly (that is, not more than once per minute or so), unless radius-limited. Sometimes a broadcast with radius 1 or 2 (neighbors-only) can be very useful. Broadcasts are not acknowledged.

- Use groupcasts, if controlling a group of five or more nodes. Groupcasts can be delivered faster to a large group of nodes than can unicasts. Groupcasts, like broadcasts, should be used sparingly if not radius-limited. Groupcasts are not acknowledged.

Always use the local binding table (gZbAddrModeIndirect_c), unless code space is too tight. Local binding tables allow other external nodes to connect applications. For example, a PC equipped with a ZigBee dongle could connect a switch and a light through a pleasant drag-and-drop interface.

> Use the local binding table (indirect messages) if code space allows.
>
> Use broadcasts and groupcasts sparingly, unless radius limited.

4.2 No Common C API

Something that surprises many people new to ZigBee is the fact that there is no standard C API for it. What? How can ZigBee be a standard, if there is no standard for the API?

In the ANSI C standard, `strlen()` computes the length of a string in characters, and developers can count on this function being available in the C library. In the POSIX standard, the function `open()` opens a device or file and returns a handle, and its counterpart, `close()`, closes that device or file based on the handle. The calls work the same (or nearly so) regardless of the platform.

Not so with ZigBee. The *only* thing that the ZigBee standard requires is correct over-the-air behavior. To the ZigBee Alliance testing houses, a platform or product is a black box, to be tested by octet sequences over-the-air. This is not unique with ZigBee. The practice is actually pretty common with networking protocols: They concentrate on the over-the-air or over-the-wire protocol, and leave improvements in the API up to the vendor.

So what does this mean to you as a developer? Once you pick a ZigBee vendor and write code for that platform, at least some code must be rewritten if you decide to switch stack vendors. To minimize that effort, be sure to separate the application code that interacts directly with the ZigBee stack (and MCU for that matter) from the rest of the program logic. If you are hiring a consulting company to do the software and firmware for you, make sure to specify this as part of the requirements.

All ZigBee stack vendors provide a way to access the standard ZigBee features. And regardless of how the feature is accessed, it will generate the same data, over-the-air.

For example, the ZigBee command APSDE-DATA.request transmits data over-the-air. This command is generated with a call to the C function AF_DataRequest() in the Freescale platform and emberSendDatagram() in the Ember platform. Both functions, although they have different parameters and names, produce exactly the same set of octets over-the-air, as shown in Figure 4.6.

Unfortunately, there is no standard way in ZigBee to detect, over-the-air, which brand of ZigBee is implemented on a remote node. You might be able to guess, based on how

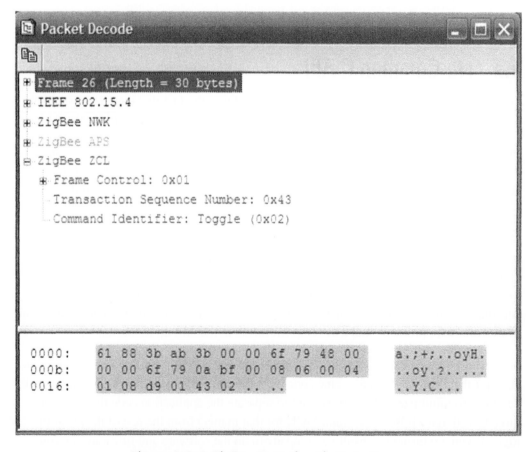

Figure 4.6: A ZigBee Over-the-Air Data Frame

the node performs optional behavior. For example, some stack vendors rebroadcast all three times, regardless of what the neighbors do. Others rebroadcast only if the neighbors have not. Anything ZigBee-certified is standard, and works in an interoperable way with any other ZigBee node. If you need interoperability, make sure the stack vendor you've chosen is a ZigBee Certified Platform.

In the following few sections, I'll show you an API sampling from the four "Golden Unit" vendors: Texas Instruments, Ember, Integration Associates, and Freescale. Golden Units are the official ZigBee boards which make up the ZigBee Compliant Platform test suite, and are the nodes that every certified ZigBee stack must be tested against.

Now, let me introduce you to application development on the four Golden Unit platforms.

> ZigBee contains no standard C API. The API is vendor-specific.
>
> ZigBee does specify precise over-the-air behavior for compatibility among vendors.

4.2.1 Texas Instruments API

Texas Instruments, a ZigBee promoter, makes many worldwide recognized products (see Figure 4.7). For the developer, TI's microcontrollers and RF products are well

Figure 4.7: Texas Instruments SmartRF04EB Evaluation Board

known, including the ultralow power MSP430 MCU. In early 2006, TI purchased ChipCon, a small RF and ZigBee company located in Oslo, Norway. TI still uses the ChipCon brand today on their ZigBee products, and in fact calls their ZigBee radios by the original ChipCon names (for example, the CC2430).

TI is primarily a silicon company, and you can download the TI ZigBee Solution, Z-Stack, directly from the Web for free. Simply go to http://www.ti.com/zigbee.

Z-Stack provides functions that look very similar to the ZigBee specification. The function for transmitting data is `AF_DataRequest()`:

```
afStatus_t AF_DataRequest(afAddrType_t *dstAddr,
   endPointDesc_t *srcEP,
   uint16 cID,
   uint16 len, uint8 *buf, uint8 *transID,
   uint8 options, uint8 radius);
```

The destination endpoint and node address (or group) is contained in the `afAddrType_t` structure. The source endpoint and cluster ID are direct parameters. The `len` and `buf` parameters are the application payload and the pointer to the transaction ID allows the application to control what the transaction ID will be for the APS layer, over-the-air. The transaction ID is automatically incremented by the `AF_DataRequest()`, if the message is buffered.

TI actually offers two interfaces to Z-Stack, the full interface mentioned previously, and the Simple API for Z-Stack. The Simple API offers fewer commands and options, but makes applications considerably easier for the first-time developer. The prototype for the simple data request is:

```
void zb_SendDataRequest (uint16 destination,
   uint16 commandId, uint8 len, uint8 *pData,
   uint8 handle, uint8 ack, uint8 radius);
```

Notice there are no endpoints, and no discussion of clusters or profile IDs; all of these are fixed to a private profile in the Simple API for Z-Stack.

The data indications in the Simple API for Z-Stack come in through:

```
void zb_ReceiveDataIndication(uint16 source,
   uint16 command, uint8 len, uint8 *pData);
```

It can't get much simpler than that! For a private profile, the TI Simple API for Z-Stack is probably easiest to use for the application developer of all the stack vendors.

Texas Instruments offers two ZigBee interfaces: a full one and a simple one.

4.2.2   Ember ZigBee API

Ember is a small start-up company out of Boston, Massachusetts. They are funded by some heavyweight venture capitalists, such as Vulcan. Bob Metcalf, the inventor of Ethernet, is one of the principal investors. Ember is one of 16 ZigBee promoter companies.

Ember makes ZigBee radios, integrated chips, and a ZigBee software stack called EmberZNet. Documentation for the stack can be downloaded from http://www.ember.com. The software is only available in the development kits (see Figure 4.8).

Figure 4.8: Ember EM250 Development Kit

Ember takes a slightly different approach to ZigBee than either TI or Freescale. Ember uses the concept of a transport layer, and enhances the functionality found in ZigBee with a series of data request functions:

- emberSendDatagram()
- emberSendSequenced()
- emberSendMulticast()
- emberSendLimitedMulticast()
- emberSendUnicast() sends APS unicast messages
- emberSendBroadcast() sends APS broadcast messages

Here is an example of using emberSendDatagram():

```
EmberStatus AppSendDatagram(int8u clusterId, int8u *contents,
  int8u length)
{
  EmberMessageBuffer message=EMBER_NULL_MESSAGE_BUFFER;
  EmberStatus status;
  if (length != 0) {
    message=emberFillLinkedBuffers(contents, length);
    if (message == EMBER_NULL_MESSAGE_BUFFER)
      return EMBER_NO_BUFFERS;
  }
  status=emberSendDatagram(0, clusterId, message);
  if (message != EMBER_NULL_MESSAGE_BUFFER)
    emberReleaseMessageBuffer(message);
  return status;
}
```

Ember uses the concept of linked buffers: a set of 32-byte buffers concatenated to form a larger buffer, for the use of over-the-air message functions.

EmberZNet also offers a variety of interesting features, such as over-the-air (or over Ethernet) updates of the software stack. Of course, updating over-the-air is a two-edged sword: On the one hand, it allows a vendor to provide feature enhancements or bug fixes across the ZigBee network. On the other hand, it opens a potential security hole, one that attackers could use to change the behavior of the network through the update mechanism.

Also differently than TI or Freescale, Ember uses Ethernet, not USB, to connect the development PC to the ZigBee boards (not surprising, considering Bob Metcalf's involvement with Ember). An advantage of the Ethernet approach is that a corporation can provide access to the ZigBee devices from anywhere inside the company, if set up correctly by the IT department. A disadvantage lies in price: Ember's kits tend to be more expensive than their competition.

Ember provides the InSight development environment with their kits, which can debug both single nodes, and the entire ZigBee network, over-the-air. InSight allows a developer to:

- Debug hardware
- Monitor application or debug data
- Monitor radio data packets

Most other vendors use the Daintree Sensor Network Analyzer I introduced in Chapter 3, "The ZigBee Development Environment."

> Ember uses the concept of a transport layer and includes over-the-air updates.

4.2.3 Integration Associates ZigBee API

Integration Associates is a fab-less semiconductor company offering high performance analog and mixed-signal semiconductor solutions for wireless and wireline communications. Integration also offers turnkey ASIC development for customers, from initial concept development to high volume production.

You can download information about EZLink, the Integration ZigBee solution, at http://www.integration.com. The documentation for the ZigBee API can be found in the document "ZigBee Common Interfaces: Revision 06 Version User Guide."

Integration Associates integrates the 802.15.4 MAC directly into hardware. This approach works well because 802.15.4 has been stable since 2003, when the IEEE specification was first released. Placing the MAC in hardware has the advantage that certain activities can be accelerated, like encryption and decryption.

Integration offers a popular USB dongle, the IA OEM-DAUB1 2400 (see Figure 4.9) that allows a developer to quickly connect to a ZigBee network via USB port. The dongle can be ordered directly from their Web site.

Integration includes a ready-made set of drivers for the USB dongle that allows desktop applications to take full advantage of ZigBee, monitoring or controlling the network as appropriate from a PC or laptop.

Figure 4.9: Integration Associates ZigBee Dongle

Integration also offers a ZigBee module, the IA-OEM DAMD1 2400. Like the USB dongle, it comes precertified for CE, FCC, IC, and ETSI, in addition to the ZigBee Compliant Platform certification from the ZigBee Alliance. Integration also preloads the modules with a unique MAC address, a step that is often left to the OEM with other solutions.

Integration uses the following functions for sending data:

- AF_INDIRECT_request()

- AF_DIRECT_request()

```
void AF_DIRECT_request
(
   ushort DstShort,
   uchar DstEndpoint,
   uchar SrcEndpoint,
   uchar ClusterId,
   uchar afduLength,
   uchar *afdu,
   uchar TxOptions,
   uchar DiscoverRoute,
   uchar RadiusCounter,
   uchar afduHandle
);
```

You'll notice that the parameters also look an awful lot like the ZigBee APSDE-DATA. request.

The confirm comes in as AF_INDIRECT_confirm() or AF_DIRECT_confirm(), with the usual suspects for parameters:

```
void AF_DIRECT_confirm
(
  ushort DstShort,
  uchar DstEndpoint,
  uchar SrcEndpoint,
  uchar Status,
  uchar afduHandle
);
```

Data indications come in through AF_DIRECT_indication():

```
void AF_DIRECT_indication
(
  uchar DstEndpoint,
  ushort SrcShort,
  uchar SrcEndpoint,
  uchar ClusterId,
  uchar afduLength,
  uchar *afdu,
  uchar WasBroadcast,
  uchar SecurityStatus
);
```

> Integration Associates offers pre-certified modules.
>
> ZigBee networks may be monitored and controlled from a PC through a USB dongle.

4.2.4   Freescale ZigBee API

Freescale is a large corporation with sales and technical offices worldwide. They are headquartered in Austin, Texas, and until recently, they were a publicly traded company. They are now owned by private investors, including BlackStone, and are the world's 10th largest chip maker, according to Forbes. Originally, Freescale was Motorola Semi-Conductor, before it spun off into its own company.

Freescale is a promoter within the ZigBee Alliance and (as their Web site proclaims) sees the potential for ZigBee applications everywhere. Freescale is helping to build that world with a comprehensive ZigBee solution that includes RF ICs, MCUs, sensors, reference designs, protocol stack software, sample application software, and development tools.

Freescale provides a desktop stack configuration tool, called BeeKit, which makes building ZigBee (and other Freescale networking) systems much easier. Freescale offers

Figure 4.10: Freescale MC13213

a range of networking protocols to serve not only the ZigBee markets, but other sensor and control markets. For example, Freescale recently announced EC-Net, a protocol stack running on 802.15.4 designed for television control, which may be connected to ZigBee in the home. Freescale also provides a full 802.15.4 MAC solution as a development environment of its own, without ZigBee (see Figure 4.10).

The Freescale ZigBee solution, BeeStack, has an architecture which looks just like the ZigBee architecture, and is very true to the specification and includes all of the same service access points (SAPs).

I won't go into detail of the API in this section, because the Freescale solution is used throughout this book.

I have to say, I'm a little biased toward the Freescale solution. They make excellent products, have always met the production requirements of my customers, and do a great job supporting San Juan Software as a third-party developer for their platform.

Freescale provides a comprehensive ZigBee solution, with RF ICs, MCUs, sensors, reference designs, protocol stack software, sample application software, and development tools.

4.2.5 Many Stack Vendors

While there is no common C API for ZigBee, the protocol standard makes for a very wide ecosystem. I've just shown you four stack and silicon vendors, but there are many others. Nearly every producer of 8-bit silicon makes, or has available, an 802.15.4 solution with ZigBee. There are even ZigBee stack vendors that span silicon solutions, such as Airbee.

The important thing is that you, as an OEM, have many, many choices. Before selecting the right solution for your product, take the time to examine the choices out there. And don't forget to think about the entire solution: silicon, stack, tools, sensors and peripherals, support, price, and many other factors.

> Consider all the factors when selecting the right ZigBee solution for any given project.

4.3 ZigBee PANs

ZigBee nodes can only send data requests to other nodes on the same network. A single ZigBee network is called a Personal Area Network (PAN).

This section describes PAN IDs, extended PAN IDs, and channels, all concepts which define a single ZigBee network. Application examples are given which show how to adjust these settings in a ZigBee environment.

ZigBee PANs are formed by ZigBee Coordinators. Only ZigBee Coordinators (ZCs) may form a PAN. The other ZigBee node types, ZigBee Routers (ZRs) and ZigBee End-Devices (ZEDs) may join a network, but do not form one themselves.

There is no reason that a given application can't choose at run-time whether to be a ZC, ZR, or ZED, but that would be a ZigBee platform choice. At the time of this writing, the Freescale platform requires the developer to choose at compile-time what ZigBee node-type a given device will be.

4.3.1 Channels

For those of you who haven't heard it before, the 2.4 GHz band is a worldwide unlicensed portion of the RF spectrum for use by many radio products. This band is also known as the ISM (Industrial, Scientific, and Medical) band. WiFi™ exists in this band. So does Bluetooth™. So do cordless phones and microwave ovens. And so does ZigBee.

ZigBee uses the same channel set as specified in 802.15.4. In the 2.4 GHz band, these channels are numbered 11 through 26. Channel numbers 0 through 10 are defined by the sub-1 GHz 802.15.4 radios, but ZigBee (at least to date), doesn't run on the sub-1 GHz radios.

Note that if you use the Freescale SMAC networking stack (which is not ZigBee, but does use the same Freescale 802.15.4 radios), the channels are numbered 0 through 15. You can translate these numbers into MAC/ZigBee channels 11 through 26.

Channels are really just a portion of the RF spectrum. In the case of 802.15.4, each of these channels is separated by 5 MHz in the 2.4 GHz band, as shown in Figure 4.11. 802.15.4 uses Direct Sequence Spread Spectrum (DSSS) to spread the packets into symbols and reassemble them on the other end, verifying that the data was decoded correctly through use of a 16-bit CRC. In Chapter 7, "The ZigBee Networking Layer," I describe the 802.15.4 PHY in a little more detail, but other books describe this PHY standard better than I do. This book is about the ZigBee protocol. If you'd like to learn more about 802.15.4, try *IEEE 802.15.4 Low-Rate Wireless Personal Area Networks: Enabling Wireless Sensor Networks*, by José A. Gutierrez.

The 802.15.4 radio forms the foundation for ZigBee. Two interesting points about this radio is that it is half-duplex (it can listen or talk, but not both at the same time) and accesses one channel at a time. So, a device listening on channel 15 won't hear anything on channels 11 through 14 or 16 through 26.

ZigBee as a protocol does not typically change channels. Bluetooth, like some other wireless technologies, is a channel-hopping protocol and some believe that channel-hopping is required for reliability. Not so. Due to the robust nature of O-QPSK and DSSS, 802.15.4 radios are very robust, even in a noisy RF environment. In fact, tests performed at the ZigBee Alliance with WiFi turned up to maximum on all channels, did

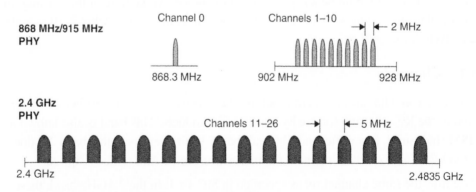

Figure 4.11: IEEE 802.15.4 Channels

not cause ZigBee to lose even one packet, over a series of thousands of data requests. Occasionally there were some retries, but not a single packet was dropped.

So how does a ZigBee device decide what channel to use? It is defined by the Application Profile. In Home Automation, for example, any ZigBee device wishing to join a network must scan all channels, and be able to join any network on any channel. If the profile is a private profile, it may choose to limit the device to one or any set of the 16 available channels.

But scanning channels does take time. When forming a network, ZigBee performs two scans:

- An energy detect scan
- An active scan

The energy detect scan is used to determine which channels are the quietest. The active scan sends out a Beacon Request, and is used to determine what other ZigBee or 802.15.4 PAN IDs are currently in use on that channel within hearing range of the radio. By default, ZigBee chooses the channel with the fewest networks, and which is the quietest (in that order).

Only the active scan (detecting other networks) is used when joining a network. If a network is already formed, and the Beacon Response can be heard by the joining node, it is assumed that the channel is quiet enough for communication.

The scan duration is defined by 802.15.4 specification, and is an integer between 0 and 14. See Table 53—MLME-SCAN.request parameters in the 802.15.4 specification for details. Since the value is used in a formula involving superframe duration, I've converted those times to milliseconds for your convenience (see Table 4.2).

Table 4.2: MAC Scan Durations

Value	Scan Duration (ms)	Value	Scan Duration (ms)
0	31	8	3,948
1	46	9	7,880
2	77	10	15,744
3	138	11	31,473
4	261	12	62,930
5	507	13	125,844
6	998	14	251,674
7	1,981		

In the Freescale platform, both channel selection and scan duration are fully under application control. The defaults for these are properties that can be set in BeeKit. The channels are defined by mDefaultValueOfChannel_c in ApplicationConf.h, and the scan duration by gScanDuration_c in BeeStackConfiguration.h. The defaults are channel 25 and scan duration 3. Each platform will have its own way of defining the channel set and scan duration.

In the example, *Example 4-2 ZigBee Channels*, I demonstrate scanning all the channels and then scanning a single channel for comparison. Notice how much longer it takes to scan all of the channels versus only one. (It's actually 16 to 1 in this case, since the example scans all 16 channels, then only 1 channel).

This example will use two ZigBee nodes: one to form the network (the ZigBee Coordinator), and one to join the network (a ZigBee End-Device). They happen to be a Home Automation light and switch, but they could be any application for the purposes of channel selection.

To follow the example with actual hardware, open the following two projects inside CodeWarrior:

- Chapter04\Example4-2 ZigBee Channels\NcbZcHaOnOffLight.mcp
- Chapter04\Example4-2 ZigBee Channels\SrbZedHaOnOffLight.mcp

If this process is still new to you, read Chapter 3, "The ZigBee Development Environment," which describes how to download images to the Freescale boards, step-by-step.

Load the two programs into the NCB and SRB boards, respectively. If this process is not familiar to you, go back to Chapter 3 and read about developing for the Freescale platform.

Then reset both boards. Press SW1 on both boards to form/join the network. Notice how long the running lights on the LEDs chase each other. This is the amount of time it takes to scan all 16 channels. The actual channel chosen is random (the channel is the third field, line 2 on the NCB board's LCD display), depending on how noisy each channel is. But, usually, channel 11 (the first one tried) is quiet enough to form the network on that channel.

The NCB display may look something like this:

```
Ha OnOffLight
Cfg 3bab 11 000
```

Next, reset both boards. This time, press SW4 on the NCB board to select a channel. Select channel 15. The channel number will change on the LCD display. Note also the hex pattern on the LEDs. Press SW4 on the other (SRB) board to select the same channel. Since the SRB doesn't have a display, make sure the channel pattern on the LEDs match what is on the NCB. Only then, press SW1 on both boards to form/join the network. Notice how much more quickly the network forms?

In the code, channel selection is made through a bit mask set of channels, decided initially at compile time. This channel mask can also be adjusted at run-time like it was using SW4 in the example. The bit mask uses a set of 32 bits to represent channels. Recall that the channels are numbered 11 through 26 by 802.15.4, so bits 11 through 26 are used to represent the channel mask. The comment shows how to set the bits properly, but it's much easier in BeeKit where it provides a check-box for each channel:

```
/*
   The default channel list defines which channels to scan when forming or
   joining a network. Default=0x02000000=channel 25. The channel list is a
   bitmap, where each bit describes a channel (for example bit 12
corresponds to
   channel 12). Any combination of channels can be included. ZigBee
supports
   channels 11-26.
   3 2  2  2  1  1  0  0  0
   1 8  4  0  6  2  8  4  0
   0000 0000 0000 0000 0000 1000 0000 0000 = 0x00000800 = channel 11
   0000 0100 0000 0000 0000 0000 0000 0000 = 0x04000000 = channel 26
   0000 0010 0000 0000 0000 0000 0000 0000 = 0x02000000 = channel 25 (default)
   0000 0111 1111 1111 1111 1000 0000 0000 = 0x07fff800 = all channels 11-26
   0000 0000 1000 0000 0001 0000 0000 0000 = 0x00801000 = channels 23 and 12
*/
#ifndef mDefaultValueOfChannel_c
#define mDefaultValueOfChannel_c    0x06000800
#endif
```

If your application wishes to set this at run-time, and you are not using the common user ASL interface used in all of the Freescale applications, simply set the global variable gSelectedChannel to the desired set of channels prior to calling ZDO_Start().

The scan duration is found in BeeStackConfiguration.h and is just a number, 0 through 14. See Table 4.2, "MAC Scan Durations," for how this number translates into actual time. Scan duration is set at compile-time in the Freescale solution. ZigBee does not mandate a particular scan time for nodes:

```
/*Scan Duration*/
#ifndef gScanDuration_c
```

```
#define gScanDuration_c 3
#endif
```

Often, when defining a private profile application, I'll select channels 15, 20, 25, and 26, as these channels aren't the same as on common WiFi installations. I do this not for ZigBee's sake, but for WiFi's. If the application gets too chatty, it could cause WiFi to miss a packet or two, and degrade the performance of WiFi.

As mentioned before, ZigBee 2007 includes the ability to switch channels. Like forming or joining a network, this is designed to be a rare event for most ZigBee applications. I'll explain that feature in a bit more detail in Appendix A, "ZigBee 2007 and ZigBee Pro."

> Channels are a portion of the RF spectrum.
>
> Of the 16 channels, ZigBee selects the channel with the fewest networks by default.

4.3.2   PAN IDs

ZigBee Personal Area Network identifiers (or PAN IDs) are used to logically separate a collection of ZigBee nodes from other ZigBee nodes in the same vicinity or on the same physical channel. This allows network A and network B to exist in close proximity without interfering with each other, other than consuming over-the-air bandwidth that they both share. For example, take a look at Figure 4.12. The light and dark gray PANs share the same physical space, and in this case the same channel.

ZigBee PAN IDs are 16-bit numbers that range from 0x0000 to 0x3fff. Note this is different than the 802.15.4 specification, which allows PAN IDs from 0x0000 to 0xfffe.

Generally, ZigBee PAN IDs won't conflict (there are 16 K of them, and stacks are encouraged to choose random PAN IDs). ZigBee does not expect to have anything near 16,000 different networks, all within radio range. In our office, we may perhaps have 10 to 20 going at any given time on various channels, and we are developing ZigBee projects. That's a 1 in 13,107 (16 × 16 K/20) chance of a conflict in our case. Not very likely.

In ZigBee 2006, PAN IDs are unique on a given channel: that is, PAN ID 0x1234 could exist on channel 11 and a separate pan 0x1234 could exist on channel 12. In ZigBee 2007, that is disallowed because of the Frequency Agility feature, which allows a network to switch channels if it determines that the current channel is no longer appropriate.

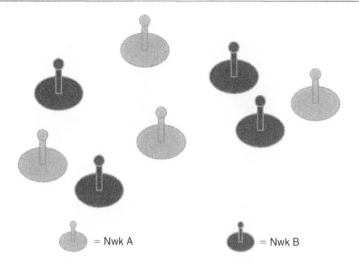

= Nwk A   = Nwk B

Figure 4.12: Separate ZigBee PANs May Coexist on the Same Channel

When forming a network, ZigBee specifies that PAN ID 0xffff means the application doesn't care what the PAN ID is, and is requesting the ZigBee stack to pick one at random that does not conflict with any existing PAN IDs in the vicinity. When joining a network, PAN ID 0xffff means the node doesn't care which network it joins. Any PAN ID will do.

So how does an application decide which PAN ID to use, either for forming or joining? The simple answer is that the Application Profile decides. In Home Automation, the default PAN ID is 0xffff for all devices. For a private profile, any PAN ID may be used.

The following example, *Example 4-3 ZigBee PAN IDs*, shows how to pick which PAN ID a node will join. In the example, two PANs are formed on the same channel, PAN IDs 0x000a and 0x000b. I'll use channel 11 for the demo. Then a switch is started, which joins a PAN. The switch is initially compiled to search for PAN ID 0x000b, so it will always join that PAN (the Ncb Light). Notice the switch can control the NcbLight (see Figure 4.13).

Next, the switch is recompiled to search for PAN ID 0x000a (or, if another SRB board is available, a second switch is started). Reset the switch and notice how it now joins PAN 0x000b this time, and controls the SrbLight, as opposed to the NcbLight (see Figure 4.14).

Like most examples in the book, the example can be followed using the three-node Freescale Network Starter Kit. The BeeKit Solution file, capture file, and source code to the application can be found at http://www.zigbookexamples.com. See Chapter 3, "The

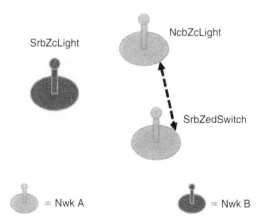

Figure 4.13: Switch Joins PAN A

ZigBee Development Environment," for a full explanation of how to download examples into the hardware. Download the following three projects to the respective boards:

- Chapter04\Example4-3 ZigBee PAN IDs\NcbZcLightPanA.mcp
- Chapter04\Example4-3 ZigBee PAN IDs\SrbZcLightPanB.mcp
- Chapter04\Example4-3 ZigBee PAN IDs\SrbZedSwitchPanA.mcp

Boot all three boards. Once they come up, press SW1 on the switch, and notice that it controls the NCB light, and not the SRB light. If you are sniffing over-the-air, you should see something like this:

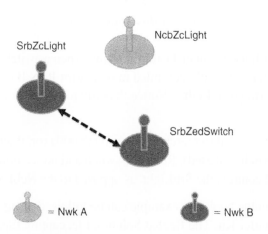

Figure 4.14: Switch Joins PAN B

Seq No	MAC Src	MAC Dest	NWK Src	NWK Dest	Protocol	Packet Type
1		0xffff			IEEE 802.15.4	Command: Beacon Request
2		0xffff			IEEE 802.15.4	Command: Beacon Request
3	0x0000				IEEE 802.15.4	Beacon: BO: 15, SO: 15, PC: 1, AP: 1, Nwk RC: 1, Nwk EDC: 1
4		0xffff			IEEE 802.15.4	Command: Beacon Request
5	0x0000				IEEE 802.15.4	Beacon: BO: 15, SO: 15, PC: 1, AP: 1, Nwk RC: 1, Nwk EDC: 1
6	0x0050c209cc066801	0x0000			IEEE 802.15.4	Command: Association Request
7					IEEE 802.15.4	Acknowledgment
8	0x0050c209cc066801	0x0000			IEEE 802.15.4	Command: Data Request
9					IEEE 802.15.4	Acknowledgment
10	0x0050c20d10046400	0x0050c209cc066801			IEEE 802.15.4	Command: Association Response
11					IEEE 802.15.4	Acknowledgment
12	0x796f	0x0000	0x796f	0xfffd	Zigbee APS Data	ZDP:MatchDescReq
13					IEEE 802.15.4	Acknowledgment
14	0x0000	0xffff	0x796f	0xfffd	Zigbee APS Data	ZDP:MatchDescReq
15	0x0000	0xffff	0x796f	0xfffd	Zigbee APS Data	ZDP:MatchDescReq
16	0x796f	0x0000			IEEE 802.15.4	Command: Data Request
17					IEEE 802.15.4	Acknowledgment
18	0x0000	0x796f	0x0000	0x796f	Zigbee APS Data	ZDP:MatchDescRsp
19					IEEE 802.15.4	Acknowledgment
20	0x0000	0xffff	0x796f	0xfffd	Zigbee APS Data	ZDP:MatchDescReq
21	0x796f	0x0000	0x796f	0x0000	Zigbee APS Data	HA:On/off
22					IEEE 802.15.4	Acknowledgment

Take a closer look at packet 6. Notice that the switch is attempting to associate with PAN A, not PAN B (see Destination PAN Identifier below):

```
Frame 6 (Length = 21 bytes)
IEEE 802.15.4
  Frame Control: 0xc823
  Sequence Number: 59
  Destination PAN Identifier: 0x000a
  Destination Address: 0x0000
  Source PAN Identifier: 0xffff
  Source Address: 0x0050c209cc066801
  MAC Payload
    Command Frame Identifier = Association Request: (0x01)
    Capability Information: 0x84
      ... ...0. = Alternate PAN Coordinator: Not capable of becoming PAN
Coordinator
      .... ..0. = Device Type: RFD
      .... .1.. = Power Source: Receiving power from alternating current
mains
      .... 0... = Receiver on when idle: Disables receiver when idle
      ..00 .... = Reserved
      .0.. .... = Security Capability: Not capable of using security
suite
      1... .... = Allocate Address: Coordinator should allocate short
address
  Frame Check Sequence: Correct
```

Now, have a switch join PAN B. Leave both PAN coordinators running (both the NCB light and SRB light). If you have another SRB board like I do, then download the following project into a second board. Otherwise, reuse the SrbZedSwitchPanA from above and place the following project into that board:

Chapter04\Example4-3 ZigBee PAN IDs\SrbZedSwitchPanB.mcp

Boot the switch for PAN B. Notice that it associates with PAN B and controls the SRB light which is on PAN B, not the NCB light on PAN A as the first switch did.

The default PAN ID is simply a property in BeeKit. Set it as you would other properties, or examine the constant directly in ApplicationConf.h:

```
#ifndef mDefaultValueOfPanId_c
#define mDefaultValueOfPanId_c              0x0B,0x00
#endif
```

Notice the PAN ID (in this case for network B) is a sequence of bytes, little Endian. Everything in ZigBee is little Endian over-the-air. Little Endian (as opposed to big Endian) simply means the lowest byte(s) come first. This may or may not be true of your application data (it is up to the application), but for everything the ZigBee generates, it's little Endian.

It's also easy enough to select the PAN ID under application control at run-time by modifying the variable `gNwkData.aPanId`. Do *not* adjust the PAN ID if already on a network—the NWK layer and MAC have some communication which occurs only at form/join time.

It's interesting to note that the ZigBee specification is rather vague on the topic of which channel on which to form or join a network. It says the node will scan the given set of channels and "choose the best one." This allows each OEM to have some flexibility, but still form and join in a standard way. On the Freescale platform, this process to "choose the best one" is exposed to the application in the file `AppStackImpl.c`. An algorithm could be made, for example, to only look for PAN IDs in the range 0x2000 through 0x20ff. See the function `SearchForSuitableParentToJoin()`.

Sometime a PAN ID is just not enough. A particular application may want to only join a network that has a particular (probably private profile) application on it. One way is to use a special PAN ID and hope, but that's not a very sure way to join the right network. Another way is to join the network and see if the application is on it, but this takes time to join the other network, and may require special code to be implemented.

ZigBee thought of this and provided a mechanism to join only those PANs which are released by your corporation, which leads us to extended PAN IDs.

> The 16-bit PAN IDs allow separate ZigBee networks to share a channel.
>
> The same PAN ID may be used in a different physical locality or on a different channel.

4.3.3   Extended PAN IDs

Extended PAN IDs (EPIDs) are 64-bit numbers that uniquely identify a PAN. ZigBee communicates using the shorter 16-bit PAN ID for all communication except one. The beacon response issued as the result of a beacon request contains an Extended PAN ID to allow a node that wishes to join a network to pick exactly the right one.

Every time a ZigBee node wishes to join a network, it sends out a beacon request. It then pays attention to all of the beacon responses, and picks the "best" network (if any) out of these responses. Take a look at the beacon response (also just called a beacon) from the switch in the last example (packet # 3 from *Example 4-1 ZigBee Extended PAN IDs*, available on-line):

```
Frame 3 (Length=24 bytes)
   Time Stamp: 14:10:23.413
   Frame Length: 24 bytes
   Capture Length: 24 bytes
   Link Quality Indication: 115
   Receive Power: -60 dBm
IEEE 802.15.4
   Frame Control: 0x8000
       .... .... .... .000 = Frame Type: Beacon (0x0000)
       .... .... .... 0... = Security Enabled: Disabled
       .... .... ...0 .... = Frame Pending: No more data
       .... .... ..0. .... = Acknowledgment Request: Acknowledgement not
required
       .... .... .0.. .... = Intra PAN: Not within the PAN
       .... ..00 0... .... = Reserved
       .... 00.. .... .... = Destination Addressing Mode: PAN identifier
and address field are not present (0x0000)
       ..00 .... .... .... = Reserved
       10.. .... .... .... = Source Addressing Mode: Address field contains
a 16-bit short address (0x0002)
   Sequence Number: 221
   Source PAN Identifier: 0x000a
   Source Address: 0x0000
   MAC Payload
      Superframe Specification: 0xcfff
          .... .... .... 1111 = Beacon Order (0x000f)
          .... .... 1111 .... = Superframe Order (0x000f)
          .... 1111 .... .... = Final CAP Slot (0x000f)
          ...0 .... .... .... = Battery Life Extension: Disabled
          ..0. .... .... .... = Reserved
          .1.. .... .... .... = PAN Coordinator: Transmitter is a PAN
Coordinator
          1... .... .... .... = Association Permit: Coordinator accepting
Association Requests
      GTS Specification: 0x00
          .... .000 = GTS Descriptor Count (0x00)
```

```
      .000 0... = Reserved
      0... .... = GTS Permit: Coordinator not accepting GTS Requests
Pending Address Specification: 0x00
      .... .000 = Number of short Addresses pending: 0
      .... 0... = Reserved
      .000 .... = Number of extended Addresses pending: 0
      0... .... = Reserved
   Beacon Payload
      Protocol ID: ZigBee NWK (0x00)
  Frame Check Sequence: Correct
NWK Layer Information: 0x8421
   .... .... .... 0001 = Stack Profile (0x1)
   .... .... 0010 .... = nwkcProtocolVersion (0x2)
   .... ..00 .... .... = Reserved (0x0)
   .... .1.. .... .... = Router Capacity: True
   .000 0... .... .... = Device Depth (0x0)
   1... .... .... .... = End Device Capacity: True
   Extended PAN ID: 0x0050c20d10046400
```

The Source PAN Identifier is 0x000a, but notice the contents of the final field above: the Extended PAN ID (0x0050c20d10046400). This field by default uses the Freescale OUI (0x0050c2) for the top 24 bits, and adds some random numbers for the rest.

EPIDs are not managed by any standards body. I recommend using your OUI, a unique 24-bit number issued by IEEE to companies making networking products, for the top 24 bits of the EPID. See MAC Addresses in Chapter 7, "The ZigBee Networking Layer," for more details on the OUI.

It's also interesting to note that Extended PAN IDs are completely unrelated to PAN IDs, and also to MAC addresses. They are a unique 64-bit number, used solely for helping nodes find the right network.

In this next example, both the NcbZcLightExtPan and the SrbZcSwitchExtPan projects will use the extended PAN ID of 0x1122334455667788. The PAN ID will be set to 0xffff so that the NcbZcLightExtPan will form a random PAN ID and the SrbZcSwitchExtPan will find it. The example will look very similar from a user perspective to the previous one, but the over-the-air mechanism used is quite different (see Figure 4.15).

As in the previous example, this one can be followed using the three-node Freescale Network Starter Kit. The BeeKit Solution file, capture file, and source code to the application can be found in the usual place: http://www.zigbookexamples.com. Download

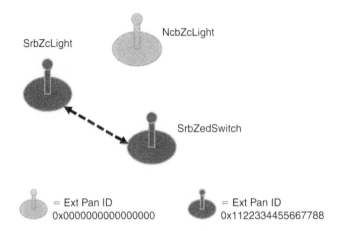

Figure 4.15: Switch Finds the Right Extended PAN ID

the following three projects to the respective boards (you can save yourself a step if you still have the SrbZcLightPanB from the previous example):

- Chapter04\Example4-4 Extended PAN IDs\NcbZcLightExtPan.mcp

- Chapter04\Example4-4 Extended PAN IDs\SrbZcLightPanB.mcp

- Chapter04\Example4-4 Extended PAN IDs\SrbZedSwitchExtPan.mcp

Boot all the boards. Notice the switch always joins the network based on Extended PAN ID, and not PAN B, which has the wrong Extended PAN ID.

An application could look for a range of extended PAN IDs to join. For example, it might look for all EPIDs with the mask of 0x0050c237b0xxxx to ensure that it's any one of a set of EPIDs from a particular manufacturer. To modify Freescale BeeStack to look for a range of EPIDs, modify the routine `SearchForSuitableParentToJoin()` found in `AppStackImpl.c`:

Use Extended PAN IDs (EPIDs) to force nodes to join a specific network.

Table 4.3: ZigBee Addressing Ranges

Name	Range	Description
Channel	11–26	A physical portion of the RF spectrum
PAN ID	0x0000–0x3fff	The address of a network within a channel
NwkAddr	0x0000–0xfff7	The address of a node within a network
Endpoint	1–240	The address of an application within a node
Cluster	0x0000–0xffff	The object within the application
Command	0x00–0xff	An action to take within the cluster
Attribute	0x0000–0xffff	A data item within the cluster

4.4   ZigBee Addressing

Addressing is the way in which a message gets from one place to another in a network. Postal (snail) mail uses one type of address that many are familiar with:

San Juan Software

221 Wood Duck Lane

Friday Harbor, WA 98250

USA

Internet Protocol (IP) addressing is another addressing method:

155.213.123.01

ZigBee addressing takes into account all the information shown in Table 4.3.

There are other ZigBee addressing features that simply translate into one or more of the items shown in the table. For example, Extended PAN IDs really merely determine which PAN ID to join. Another is groups, described later in this section, which defines a set of nodes in the network to address (see Table 4.4).

4.4.1   ZigBee Node Types

Before going into more ZigBee addressing, I'd like to address a concept I've been skirting but need to define formally: ZigBee node types.

Table 4.4: More ZigBee Addressing Features

Name	Range	Description
Extended PAN ID	0x0000000000000000–0xffffffffffffffff	A unique 64-bit identifier for the PAN
Group	0x0000–0xffff	A collection of nodes in a network
Application Profile ID	0x0000–0xffff	Defines a particular application domain

Every node in a ZigBee network is one of three types: a ZigBee Coordinator (ZC), a ZigBee Router (ZR), or a ZigBee End-Device (ZED). In IEEE 802.15.4-speak, both ZCs and ZRs are full-function devices, and a ZED is a reduced-function device (see Table 4.5).

Use a ZigBee Coordinator to form a network. This is the only device type that can. If the network is secure, the ZC must be present to add nodes to the network. Otherwise, once the network starts, it's just a router. The ZC is also the root of the tree when using tree routing in stack profile 0x01, so if the ZC is gone, only mesh routing is available across branches of the tree. Tree routing is explained in-depth in Chapter 7, "The ZigBee Networking Layer."

The ZigBee Coordinator is not needed for normal operation of the network, but is required to allow nodes to join or leave the network, as it contains the Trust Center. Only the Trust Center can decide whether to allow a node on a ZigBee network, or to deny it access.

Table 4.5: ZigBee Node Types

Node Type	Features
ZC	Forms network Routes packets Security Trust Center Allows nodes to join network
ZED	Joins network Routes Packets Allows nodes to join network
ZR	Joins network Battery operated and may sleep Smallest code size Can be RxOnIdle = true (no polling) or RxOnIdle = false (polling)

Use a ZigBee Router to enhance the mesh in the network. ZigBee Routers can extend the range of the network and increase its reliability. ZRs like the ZigBee Coordinator route packets, and also allow other nodes to join the network.

Use a ZigBee End-Device if the node must be battery-operated and sleep during network inactivity. A ZED may be RxOnIde or not. When RxOnIdle is false, ZEDs may sleep for long periods of time. There is no ZigBee-imposed limit on sleeping, but some Application Profiles define a maximum, such as one hour in the Home Automation profile. When an RxOnIdle is false, ZED wakes up. It may transmit immediately, poll its parent to see if any messages are waiting for it, then go back to sleep. When RxOnIdle is true, ZEDs receive messages immediately. In either case, a ZED may transmit any time it wishes. Use an RxOnIdle ZED when an application needs more RAM or flash memory because ZEDs make the smallest code image.

Any application can reside in any ZigBee node type. For example, a ZC, ZR, or ZED could contain a light, switch, temperature sensor, thermostat, gateway, or whatever is appropriate for the physical device.

The following example, *Example 4-5 ZigBee Node Types*, contains one ZigBee node of each type: a ZC, a ZR, and a ZED. In this case, the ZC will be the switch, the ZR and ZED will be lights.

This example can be followed using the three-node Freescale Network Starter Kit. The BeeKit Solution file, capture file, and source code to the application can be found in the usual place: http://www.zigbookexamples.com. To follow the example, download the following three projects to the respective boards:

- Chapter04\Example4-5 ZigBee Node Types\ZcNcbSwitch.mcp
- Chapter04\Example4-5 ZigBee Node Types\ZrSrbLight.mcp
- Chapter04\Example4-5 ZigBee Node Types\ZedSrbLight.mcp

After they are all programmed, turn the nodes on, in any order. Once the ZigBee Coordinator forms the network the other devices will join, in this case on channel 25, PAN ID 0x0f00. Now press SW1 on the ZcNcbOnOffSwitch to toggle the lights.

Notice the delay on the ZED! Over-the-air, you can see the reason for the delay: the ZED doesn't receive the message until it polls its parent for the message. By default, this occurs once every three seconds. I've included time in the following capture so you can see how often the ZED polls its parent for messages. The ZigBee polling command looks like an 802.15.4 Command: Data Request in the captures:

Seq No	Time	MAC Src	MAC Dest	NWK Src	NWK Dest	Protocol	Packet Type
1	11:42:16.000		0xffff			IEEE 802.15.4	Command: Beacon Request
2	11:42:17.604		0xffff			IEEE 802.15.4	Command: Beacon Request
3	11:42:17.609	0x0000				IEEE 802.15.4	Beacon: BO: 15, SO: 15, PC: 1, AP: 1, Nwk RC: 1, Nwk EDC: 1
4	11:42:19.629	0x0050c237b0040002	0x0000			IEEE 802.15.4	Command: Association Request
5	11:42:19.630					IEEE 802.15.4	Acknowledgment
6	11:42:20.126	0x0050c237b0040002	0x0000			IEEE 802.15.4	Command: Data Request
7	11:42:20.127					IEEE 802.15.4	Acknowledgment
8	11:42:20.130	0x0050c237b0040001	0x0050c237b0040002			IEEE 802.15.4	Command: Association Response
9	11:42:20.131					IEEE 802.15.4	Acknowledgment
10	11:42:21.087	0x0000	0xffff			IEEE 802.15.4	Command: Beacon Request
11	11:42:21.089	0x0000				IEEE 802.15.4	Beacon: BO: 15, SO: 15, PC: 1, AP: 1, Nwk RC: 1, Nwk EDC: 1
12	11:42:21.093	0x0001				IEEE 802.15.4	Beacon: BO: 15, SO: 15, PC: 0, AP: 1, Nwk RC: 1, Nwk EDC: 1
13	11:42:23.112	0x0050c237b0040003	0x0000			IEEE 802.15.4	Command: Association Request
14	11:42:23.113					IEEE 802.15.4	Acknowledgment
15	11:42:23.609	0x0050c237b0040003	0x0000			IEEE 802.15.4	Command: Data Request
16	11:42:23.610					IEEE 802.15.4	Acknowledgment
17	11:42:23.612	0x0050c237b0040001	0x0050c237b0040003			IEEE 802.15.4	Command: Association Response
18	11:42:23.613		0x0000			IEEE 802.15.4	Acknowledgment
19	11:42:26.688	0x796f				IEEE 802.15.4	Command: Data Request
20	11:42:26.689					IEEE 802.15.4	Acknowledgment
21	11:42:26.867	0x0000	0xffff	0x0000	0xffff	Zigbee APS Data	HA:On/off
22	11:42:26.880	0x0001	0xffff	0x0000	0xffff	Zigbee APS Data	HA:On/off
23	11:42:29.758	0x796f	0x0000			IEEE 802.15.4	Command: Data Request
24	11:42:29.759					IEEE 802.15.4	Acknowledgment
25	11:42:29.761	0x0000	0x796f	0x0000	0xffff	Zigbee APS Data	HA:On/off
26	11:42:29.762					IEEE 802.15.4	Acknowledgment
27	11:42:32.831	0x796f	0x0000			IEEE 802.15.4	Command: Data Request
28	11:42:32.832					IEEE 802.15.4	Acknowledgment

Notice that the broadcast toggle command, initiated by the ZC (address 0x0000) in packet 21, is repeated by the router (address 0x0001) in packet 22. Then, three seconds later, the polling ZED asks node 0x0000 (its parent) if there are any messages waiting for it. The parent responds, "Why, yes, there are," and promptly sends down the broadcast to the child:

```
Frame 25 (Length = 30 bytes)
IEEE 802.15.4
   Frame Control: 0x8861
   Sequence Number: 254
   Destination PAN Identifier: 0x0f00
   Destination Address: 0x796f
   Source Address: 0x0000
   Frame Check Sequence: Correct
ZigBee NWK
   Frame Control: 0x0048
   Destination Address: 0xffff
   Source Address: 0x0000
   Radius=10
   Sequence Number = 217
ZigBee APS
   Frame Control: 0x08
   Destination Endpoint: 0x08
   Cluster Identifier: On/off (0x0006)
   Profile Identifier: HA (0x0104)
   Source Endpoint: 0x08
   Counter: 0x13
ZigBee ZCL
   Frame Control: 0x01
   Transaction Sequence Number: 0x42
   Command Identifier: Toggle (0x02)
```

Note how the 802.15.4 (MAC) frame indicates a unicast from the parent to the child, but the NWK frame indicates that it's a broadcast (destination 0xffff) from node 0x0000, the ZcNcbOnOffSwitch.

> Any application may reside in any ZigBee node type: ZC, ZR, or ZED.

4.4.2 The Network Address

The network address, also called NwkAddr, short address, or node address, is a 16-bit number used to uniquely identify a particular node on a ZigBee network. The ZigBee Coordinator is always NwkAddr 0x0000. Yes, two ZigBee coordinators can exist on the same channel with NwkAddr 0x0000, because they are on different PAN IDs.

If using stack profile 0x01, which uses tree routing in addition to mesh, the NwkAddr reflects its position in the network. For example, NwkAddr 0x0001 is the first ZR that joined the network, and NwkAddr 0x796F is the first ZED. In stack profile 0x02 (with ZigBee 2007), the NwkAddr is assigned randomly.

Most of the examples so far have used service discovery and binding to determine which nodes in the network with which to communicate, such as using Match Descriptor or End-Device-Bind command.

But service discovery and binding are not required in ZigBee. Once a node is on the network, it can communicate to any other node in the network. Simply transmit a packet to that node address. It is very common to send something to the ZigBee Coordinator (NwkAddr 0x0000), because that node address is the same in every ZigBee network.

In this three-node example, the switch application already knows the topology of the network and the addresses of nodes in the network. It knows that one light will be at node 0x0001 (the first ZR), and the other light will be at node 0x143e (the second ZR). SW1 will toggle the light at NwkAddr 0x0001, and SW2 will toggle the light at NwkAddr 0x143e. No other binding or instructions are needed. Take a look at the very simple capture:

Seq No	MAC Src	MAC Dest	NWK Src	NWK Dest	Protocol	Packet Type
1		0xffff			IEEE 802.15.4	Command: Beacon Request
2		0xffff			IEEE 802.15.4	Command: Beacon Request
3		0xffff			IEEE 802.15.4	Command: Beacon Request
4	0x0000				IEEE 802.15.4	Beacon: BO: 15, SO: 15, PC: 1, AP: 1, Nwk RC: 1, Nwk EDC: 1
5	0x0000	0xffff			IEEE 802.15.4	Command: Beacon Request
6	0x0000				IEEE 802.15.4	Beacon: BO: 15, SO: 15, PC: 1, AP: 1, Nwk RC: 1, Nwk EDC: 1
7	0x0050c237b0040002	0x0000			IEEE 802.15.4	Command: Association Request
8					IEEE 802.15.4	Acknowledgment
9	0x0050c237b0040002	0x0000			IEEE 802.15.4	Command: Data Request
10					IEEE 802.15.4	Acknowledgment
11	0x0050c237b0040001	0x0050c237b0040002			IEEE 802.15.4	Command: Association Response
12					IEEE 802.15.4	Acknowledgment
13	0x0050c237b0040003	0x0000			IEEE 802.15.4	Command: Association Request
14					IEEE 802.15.4	Acknowledgment
15	0x0050c237b0040003	0x0000			IEEE 802.15.4	Command: Data Request
16					IEEE 802.15.4	Acknowledgment
17	0x0050c237b0040001	0x0050c237b0040003			IEEE 802.15.4	Command: Association Response
18					IEEE 802.15.4	Acknowledgment
19	0x0000	0x0001	0x0000	0x0001	Zigbee APS Data	HA:On/off
20	0x0000				IEEE 802.15.4	Acknowledgment
21	0x0000	0x0001	0x0000	0x0001	Zigbee APS Data	HA:On/off
22	0x0000				IEEE 802.15.4	Acknowledgment
23	0x0000	0x143e	0x0000	0x143e	Zigbee APS Data	HA:On/off
24	0x0000				IEEE 802.15.4	Acknowledgment
25	0x0000	0x143e	0x0000	0x143e	Zigbee APS Data	HA:On/off
26	0x0000				IEEE 802.15.4	Acknowledgment

Packets 1 through 18 are the packets needed for the nodes to join the network. But notice that at packet 19, the switch begins to send data to the first light, 0x0001. No discovery occurred, the switch just "assumed" the light would be there.

This is a perfectly acceptable practice for a private profile, because the topology and relationships between nodes could be set up in advance. Public profiles require more sophisticated discovery techniques because the network topology is never known in advance.

To run this demo, download the following three projects into their respective boards:

- Chapter04\Example4-6 ZigBee NwkAddr\ZcNcbHaOnOffSwitch.mcp

- Chapter04\Example4-6 ZigBee NwkAddr\ZrSrbHaOnOffLight1.mcp

- Chapter04\Example4-6 ZigBee NwkAddr\ZrSrbHaOnOffLight2.mcp

Once the code is downloaded and the boards are booted (any order), press SW1 to toggle the SrbZrLight1 (NwkAddr 0x0001). Then press SW2 to toggle the SrbZrLight2 (NwkAddr 0x143e). It's as simple as that.

Take a look at the source code on the switch. It just makes a data request like you've seen before, setting the address to the desired node. It doesn't matter if the node is many hops away: ZigBee will discover a route automatically and find the node. The NwkAddr is all that is needed:

```
void BeeAppHandleKeys(key_event_t events)
{
  zbNwkAddr_t aDestAddress;
  zbEndPoint_t endPoint = appEndPoint;

  switch(events) {
  /* SW1 - send toggle to node 0x0001 */
  case gKBD_EventSW1_c:
    Set2Bytes(aDestAddress, 0x0100);
    OnOffSwitch_SetLightState(gZbAddrMode16Bit_c,
      aDestAddress,endPoint,gZclCmdOnOff_Toggle_c, 0);
    break;
  /* SW2 - send toggle command to node 0x143e */
  case gKBD_EventSW2_c:
  Set2Bytes(aDestAddress, 0x3e14);
    OnOffSwitch_SetLightState(gZbAddrMode16Bit_c,
      aDestAddress,endPoint,gZclCmdOnOff_Toggle_c, 0);
    break;
  }
}
```

Note that the addresses are little Endian (lowest byte first). So address 0x0001 is represented as 0x0100. All ZigBee data types are little Endian over-the-air. It is up to the stack vendor to choose how to implement this in source code. Freescale uses little Endian for all the ZigBee types, even though the HCS08 processor used in the MC13213 is big Endian.

> The 16-bit NwkAddr uniquely identifies a node in the network.

4.4.3   MAC Addresses

The MAC address, also called IEEE address, long address, or extended address, is a 64-bit number that uniquely identifies this board from all other ZigBee boards in the world. This number is large enough to allow for about 4 billion ZigBee boards for every square meter of land on earth. ZigBee believes that will be a large enough address space for the foreseeable future.

The top 24 bits of this address consist of the Organizational Unique Identifier (OUI). The lower 40 bits are managed by the OEM producing the boards. For example, all Freescale development boards use the OUI 0x0050c2. The rest of the bits 0x37b001xxxx define the specific ZigBee development board (see Figure 4.16).

The 64-bit MAC addresses have no direct relationship to the 16-bit NwkAddr. If a node leaves one ZigBee network and joins another, its MAC address will remain the same, but the NwkAddr will likely change.

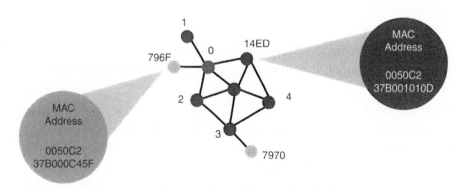

Figure 4.16: MAC Addresses

ZigBee radios do not contain a MAC address. It is up to the OEM producing boards to procure these and place them in flash memory into the board's MCU. The OUI is obtained from IEEE (http://www.IEEE.org). At the time of this writing, an OUI costs a one-time fee of around $1,600 US. If your company already ships networking products (WiFi, Ethernet, etc.) it probably already has an OUI and manages these numbers.

It's interesting to note that no standards body truly polices MAC addresses. If your company places duplicate MAC addresses into nodes that go out in the field, it will confuse the networks and your products won't work, but IEEE won't come after your company. So I guess, in a way, this is self-policing.

In the Freescale solution, for ease of development in the labs, the MAC address is set to a random value by default (address 0x0000000000000000 means to pick a random MAC address). Do not continue to use random MAC addresses in shipping products, or even in field trials.

MAC addresses are used in a number of ZigBee calls, most conspicuously in binding. The reason MAC addresses are used for binding is in case a node is mobile in the network, and it changes its short address. An example of this would be a handheld remote control that moves throughout the house. If the remote moves out of radio range of its parent, it must pick a new parent so that any node which wants to send data to the remote can find where the remote polls for its messages. When a ZED picks a new parent like this, it sends out a device to indicate its new NwkAddr, along with its unique MAC address. Any other node in the network that cares updates its internal tables to reflect this new NwkAddr for the node.

MAC addresses are also used by ZigBee in every secure packet as part of the nonce (the unique number that identifies the packet from all others for authentication reasons).

In the Freescale solution, the MAC address is located in flash memory in the structure gHardwareParametersInit. In the Freescale MC13213 (MCU + radio), this structure is at a fixed location at 0xFF72 which places the MAC address at 0xff81. The function which retrieves the MAC address, BUtl_CreateExtendedAddress(), which is found in BeeUtil.c, can also be overridden to get or to create the MAC address in any way you wish.

Remember that the MAC address, like all over-the-air entities, is little endian. So the MAC address 0x0050c237b0011234 would be represented as the following sequence of bytes in flash memory:

```
0x34 0x12 0x01 0xb0 0x37 0xc2 0x50 0x00
```

At compile-time, the MAC address can be set through Freescale BeeKit or directly in ApplicationConf.h:

```
#ifndef mDefaultValueOfExtendedAddress_c
#define mDefaultValueOfExtendedAddress_c
  0x00,0x00,0x00,0x00,0x00,0x00,0x00,0x00
#endif
```

When going to production, you'll need a way to manage the MAC address numbers if your company does not already have such a system. A simple spreadsheet with production run ranges will do.

The production boards must be programmed with these unique MAC addresses, in addition to any other serial numbers, model identifiers, or anything else that should be in flash memory. If you outsource your production, most production facilities offer this "serialization" service.

> MAC addresses can be obtained from http://www.IEEE.org.
> Every board must have a 64-bit MAC address.

4.4.4 Groups

Groups are a way of collecting a set of nodes into a single addressable entity. A single data request can reach every node in a group. Groups are an optional feature in the ZigBee specification, but are mandatory in some profiles, such as in the Home Automation Profile.

Groups are interesting in that an entire set of devices can perform an action all at once. A ZigBee-enabled home theater system could dim the lights, turn on the DVD player, television, and surround-sound speakers, lower the shades, and turn off the telephone ringer, all with a single over-the-air command.

In another example, all the devices in the entire house could be part of the "I'm going to work" group, which could turn off all lights, and lower the heat in all rooms except the

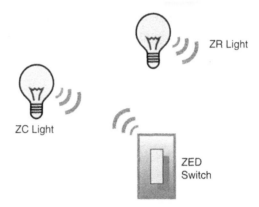

Figure 4.17: ZigBee Groups Example

sun-porch which contains a collection of prized orchids. Over the course of one year, the system pays for itself in reduced energy bills.

Within nodes, applications reside on endpoints. When a groupcast is received, a node checks its endpoints to see if any are members of that group ID. If none match, then the packet is discarded by ZigBee. If there is a match, then the packet is sent to the relevant endpoint(s).

In this three-node example, *Example 4-7 ZigBee Groups*, two lights belong to the same group, but on different endpoints. The switch sends a data request to that group to toggle the light (see Figure 4.17).

To follow the example with hardware, download the source from http://www.zigbook examples.com and then download the following three projects in their relevant boards:

- Chapter04\Example4-7 ZigBee Groups\NcbZcLightGroupA.mcp

- Chapter04\Example4-7 ZigBee Groups\SrbZrLightGroupA.mcp

- Chapter04\Example4-7 ZigBee Groups\SrbZedSwitch.mcp

Only the receiving node(s) need to be part of the group. The sending node needs to know which group to transmit to, but does not need to be a member of that group.

In this example, the lights add themselves to group A with the local function `APSME_AddGroupRequest()`, but this could be done over-the-air using the Groups Cluster, perhaps by a commissioning tool. By default, BeeStack allows up to

16 groups to be defined per node, distributed among any set of endpoints, as this number is what the ZigBee Home Automation Profile requires:

```
typedef struct zbApsmeAddGroupReq_tag {
  zbGroupId_t aGroupId;
  zbEndPoint_t endPoint;
} zbApsmeAddGroupReq_t;

zbStatus_t APSME_AddGroupRequest(zbApsmeAddGroupReq_t *pRequest);
```

The application adds itself to Group A automatically on start-up, then instructs the network to form automatically. This same code resides in both lights:

```
void BeeAppInit (void)
{
  index_t i;
  zbApsmeAddGroupReq_t groupReq;

  …

  /* add the app endpoint to group A */
  groupReq.aGroupId[0] = 0x0A;
  groupReq.aGroupId[1] = 0;
  groupReq.endPoint = appEndPoint;
  APSME_AddGroupRequest(&groupReq);
  /* form the network automatically */
  TS_SendEvent(gAppTaskID, gAppEvtStartNetwork_c);
}
```

On the switch side, it automatically joins the network. When the user (you) presses SW1, the switch sends out a toggle command via Groupcast:

```
void BeeAppHandleKeys(key_event_t events)
{
  zbNwkAddr_t aDestAddress={0x00,0x00};
  zbEndPoint_t EndPoint=0;

  switch (events) {

    /* send toggle command to group A */
    case gKBD_EventSW1_c:
      aDestAddress[0]=0x0A;
      aDestAddress[1]=0;

      OnOffSwitch_SetLightState(gZbAddrModeGroup_c,
        aDestAddress,EndPoint,gZclCmdOnOff_Toggle_c, 0);
      break;
  }
}
```

Notice that when sending to a group, there is no destination endpoint in the APS frame, only a Group Address. ZigBee assumes the broadcast endpoint (0xff) when groupcasts are used. ZigBee, inside the receiving node, then checks all endpoints to see if they are a member of the group. In this case, endpoint 0x08 in the ZC is a member of group A, and endpoint 0x09 in the ZR is also a member of group A. They match, so the light toggles:

```
Frame 24 (Length=31 bytes)
IEEE 802.15.4
      Frame Control: 0x8841
      Sequence Number: 1
      Destination PAN Identifier: 0x1aaa
      Destination Address: 0xffff
      Source Address: 0x0000
      Frame Check Sequence: Correct
ZigBee NWK
      Frame Control: 0x0048
      Destination Address: 0xffff
      Source Address: 0x796f
      Radius=9
      Sequence Number = 245
ZigBee APS
      Frame Control: 0x0c
            .... ..00 = Frame Type: APS Data (0x00)
            .... 11.. = Delivery Mode: Group Addressing (0x03)
            ...0 .... = Indirect Address Mode: Ignored
            ..0. .... = Security: False
            .0.. .... = Ack Request: Acknowledgement not required
            0... .... = Reserved
      Group Address: 0x000a
      Cluster Identifier: On/off (0x0006)
      Profile Identifier: HA (0x0104)
      Source Endpoint: 0x08
      Counter: 0x6b
ZigBee ZCL
      Frame Control: 0x01
      Transaction Sequence Number: 0x42
      Command Identifier: Toggle (0x02)
```

Groupcasts use the underlying ZigBee broadcast mechanism to deliver the packet to all nodes in the network. Notice the Destination Address in the 802.15.4 (MAC) and ZigBee NWK layers: it is 0xffff, the broadcast address.

As with all broadcasts, ZigBee does not send acknowledgments with groupcasts. A confirm does come back, but it simply means that ZigBee *sent* the groupcast. If a sending node needs ACKs, it should either use unicasts (where ACKs can occur), or instruct the receiving node to send a unicast back to the initiator of the groupcast.

By the way, the term groupcast is one in common usage but is not part of the ZigBee specification's official terminology.

> Groups allow a single data request to reach an arbitrary set of nodes.
>
> Groups are added in the receiving nodes only, and are attached to endpoints.

4.4.5 Using Broadcasts

A broadcast is used to send a data request from one node to the entire ZigBee network, at least within a given radius. Broadcasts come in three flavors:

- 0xffff — Broadcast to all nodes
- 0xfffd — Broadcast to all non-sleeping nodes
- 0xfffc — Broadcast to routers only (including the ZigBee Coordinator)

Broadcasts are used by some underlying ZigBee management functions, such as route discovery or NWK_addr_req, and may be also used by applications.

Think of broadcasts as the waves spreading out from a stone dropped in a still pond. The ripples move away from the center. Unlike waves in a pond, broadcasts ripple out with the same strength to a certain number of hops, defined by the radius, and then stop.

Broadcasts should be used with caution, however. A ZigBee network can support only a limited number of them at any given time. Each broadcast is tracked in what's called the Broadcast Transaction Table (BTT). The minimum number of BTT entries is defined by the stack profile. In the case of stack profile 0x01 used for Home Automation, the minimum is nine. ZigBee does not specify maximums for settings, so this number could be larger, but applications can't count on that.

The number of BTT entries can be set through a property in BeeKit, within the Freescale solution. Set `MaxBroadcastTransactionTableEntries_c` found in `BeeStackConfiguration.h` (see Figure 4.18).

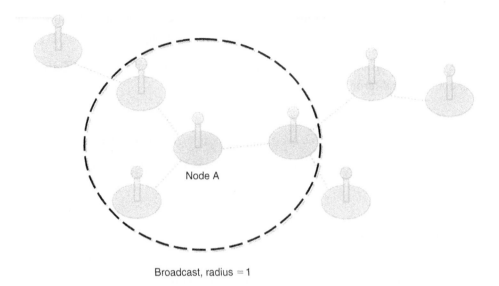

Broadcast, radius = 1

Figure 4.18: Node A Making a Radius 1 Broadcast

The broadcast radius is specified as part of a data request that initiates the broadcast. In the Freescale solution, it is part of the `afAddrInfo_t` type passed as the first parameter to `AF_DataRequest()`:

```
typedef struct afAddrInfo_tag
{
    zbAddrMode_t dstAddrMode;
    zbApsAddr_t dstAddr;
    zbEndPoint_t dstEndPoint;
    zbClusterId_t aClusterId;
    zbEndPoint_t srcEndPoint;
    zbApsTxOption_t txOptions;
    uint8_t radiusCounter;
} afAddrInfo_t;

zbStatus_t AF_DataRequest(afAddrInfo_t *pAddrInfo,
  uint8_t payloadLen, void *pPayload, zbApsCounter_t,
  *pConfirmId);
```

A simple function using broadcast with radius 1 is below. The function assumes the application has already set global variables `appCluster` and `appEndPoint`:

```
void BroadCastRadius1(uint8_t payloadLen, void *pPayload)
{
  afAddrInfo_t addrInfo;
```

```
    addrInfo.dstAddrMode = gZbAddrMode16Bit_c;
    Set2Bytes(addrInfo.dstAddr.aNwkAddr, 0xffff); /* all nodes */
    addrInfo.dstEndPoint = 0xff; /* all endpoints */
    Set2Bytes(addrInfo.aClusterId, appCluster);
    addrInfo.srcEndPoint = appEndPoint;
    addrInfo.txOptions = afTxOptionsDefault_c;
    addrInfo.radiusCounter = 1;

    (void)AF_DataRequest(&addrInfo, payloadLen, pPayload, NULL);
}
```

A radius of 1 or 2 is often useful in communicating to nodes in close proximity. The network might be 20 hops in diameter, but if a broadcast is limited to radius 1 it affects only the neighboring nodes.

One use of this technique I've seen was in hotel rooms. Each room had a small set of nodes that monitored and controlled various devices in the room (air-conditioner, card-key lock, etc.). The room controller had all of the intelligence and would instruct the other devices to go into "active" or "sleep" modes, depending on whether the room was occupied. Since the developer of this application knew all of the nodes would be in close proximity, a single message, broadcast, was able to accomplish the job quickly, and efficiently.

If a message will be sent out repeatedly, broadcasts must be paced within the network. ZigBee nodes that propagate the broadcast may not have enough resources (RAM) or entries in the Broadcast Transaction Table (BTT) to accommodate a large set of broadcasts. This is one of the common mistakes with programmers new to ZigBee. Use broadcasts sparingly, or in a controlled manner, and avoid deployment problems. Remember that route requests use broadcasts.

While highly reliable, broadcasts are not guaranteed to get through. Unlike unicasts, broadcasts have no acknowledgment mechanism. If you need to be absolutely sure that a message got through, use a unicast with acknowledgment, or instruct the nodes receiving the broadcast to reply back to the sender to indicate they received the message.

Broadcasts are also useful for advanced commissioning. In Chapter 8, "Commissioning ZigBee Networks," I'll show you an example of using broadcasts to go beyond the ZigBee mechanisms for finding the correct node and endpoint to bind to.

In Freescale BeeStack, broadcasts are initiated with a call to AF_DataRequest(), just like all other data requests. Use address mode gZbAddrMode16Bit_c, and set the

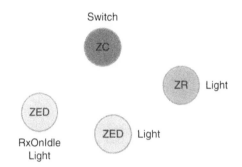

Figure 4.19: ZigBee Broadcast Example

destination address to one of the three broadcast modes: 0xffff, 0xfffd, 0xfffc. Remember, as with all over-the-air fields, to store the field little Endian. For example:

```
Set2Bytes(aBroadCastMode, 0xfdff);
```

This example, *Example 4-8 ZigBee Broadcasts*, again uses lights and a switch. In this case, there will be four nodes: a ZC, a ZR, and two ZEDs, one of which is RxOnIdle = TRUE. What RxOnIdle does is inform the parent that this node won't be sleeping, so to send messages to it at any time. The broadcast address 0xfffd will reach this node, whereas the sleepy ZED will only be reached with a standard 0xffff broadcast address (see Figure 4.19).

The ZC is the switch. The other three nodes are lights. The user interface is simple, as shown in Table 4.6.

Download the following four projects into the respective boards. If using the Freescale NSK, which only has three boards, use just one of the ZEDs.

- Chapter04\Example4-8 ZigBee Broadcasts\ZcNcbHaOnOffSwitch.mcp

- Chapter04\Example4-8 ZigBee Broadcasts\ZrSrbHaOnOffLight.mcp

Table 4.6 ZigBee Broadcast Example UI

Switch	Description
SW1	Sends an on/off toggle command via broadcast address 0xffff (This will toggle all lights.)
SW2	Sends an on/off toggle command via broadcast address 0xfffd (This will toggle ZR and RxOnIdle ZED.)
SW3	Sends an on/off toggle command via broadcast address 0xfffc (This will toggle only the ZR.)

- Chapter04\Example4-8 ZigBee Broadcasts\ZedSrbHaOnOffLight.mcp

- Chapter04\Example4-8 ZigBee Broadcasts\ZedSrbRxOnIdleHaOnOffLight.mcp

Once the boards have booted and joined the network, press SW1 on the ZcNcbHaOnOffSwitch. All the lights in the network toggle!

But something strange happened. The sleepy ZED didn't turn on its light for three seconds. Why? The sleepy (non RxOnIdle) ZED polls its parent once every three seconds, so there will be a delay before the message is received. Its parent has held the message before delivery. This only works if the polling rate is fairly fast (once every six seconds or less), since the MAC will purge the message after this amount of time. I'll describe in-depth sleepy ZigBee devices in the next chapter when I discuss low power.

Now press SW2. Notice that the sleepy ZED doesn't receive this message at all, but the awake (RxOnIdle) ZED does, and so does the ZR.

Now press SW3. Neither of the ZEDs receives this toggle command; it is only for routers, so only the ZR toggles.

Here is the over-the-air capture:

Seq No	MAC Src	MAC Dest	NWK Src	NWK Dest	Protocol	Packet Type
34	0x0000	0xffff	0x0000	0xffff	Zigbee APS Data	HA:On/off
35	0x0001	0xffff	0x0000	0xffff	Zigbee APS Data	HA:On/off
36	0x796f	0x0000			IEEE 802.15.4	Command: Data Request
37					IEEE 802.15.4	Acknowledgment
38	0x0000	0x796f	0x0000	0xffff	Zigbee APS Data	HA:On/off
39					IEEE 802.15.4	Acknowledgment
40	0x0000	0xffff	0x0000	0xfffd	Zigbee APS Data	HA:On/off
41	0x0001	0xffff	0x0000	0xfffd	Zigbee APS Data	HA:On/off
42	0x796f	0x0000			IEEE 802.15.4	Command: Data Request
43					IEEE 802.15.4	Acknowledgment
44	0x0000	0xffff	0x0000	0xfffc	Zigbee APS Data	HA:On/off
45	0x0001	0xffff	0x0000	0xfffc	Zigbee APS Data	HA:On/off
46	0x796f	0x0000			IEEE 802.15.4	Command: Data Request
47					IEEE 802.15.4	Acknowledgment

Notice when the first (0xffff) broadcast is initiated (packet 34), it is repeated by the ZR (packet 35). In fact, broadcasts are repeated by all routers in a ZigBee network, within the given radius. Not until the sleepy ZED polls its parent (packet 36) does the parent give the broadcast to the sleeping light (packet 38). The parent only indicates there is a message waiting on broadcast to all nodes (0xffff). The other two broadcast modes, 0xfffd and 0xfffc do not go to sleeping children.

From the over-the-air capture it's hard to see the RxOnIdle ZED interact with the toggle command, because it does not repeat broadcasts, but if you try this example with real hardware, you'll see the light toggle on the ZedRxOnIdleSrbHaOnOffLight with both broadcast modes 0xffff and 0xfffd, but not with 0xfffc. 0xfffc is only received by ZRs:

```
void BeeAppHandleKeys(key_event_t events)
{
  zbNwkAddr_t aDestAddress;
  zbEndPoint_t endPoint = appEndPoint;

  switch (events) {
    /* SW1-toggle via broadcast 0xffff */
    case gKBD_EventSW1_c:
      Set2Bytes(aDestAddress, 0xffff);
      OnOffSwitch_SetLightState(gZbAddrMode16Bit_c,
        aDestAddress,endPoint,gZclCmdOnOff_Toggle_c, 0);
      break;

    /* SW2-toggle via broadcast 0xfffd */
    case gKBD_EventSW2_c:
      Set2Bytes(aDestAddress, 0xfdff);
      OnOffSwitch_SetLightState(gZbAddrMode16Bit_c,
        aDestAddress,endPoint,gZclCmdOnOff_Toggle_c, 0);
      break;

    /* SW3-toggle via broadcast 0xfffc */
    case gKBD_EventSW3_c:
      Set2Bytes(aDestAddress, 0xfcff);
        OnOffSwitch_SetLightState(gZbAddrMode16Bit_c,
          aDestAddress,endPoint,gZclCmdOnOff_Toggle_c, 0);
      break;
  }
}
```

Broadcasts don't need specific commissioning. There is no "binding" to a broadcast destination address. A node simply broadcasts to a specified radius. The default radius for a broadcast spans the entire network.

Broadcasts can reach many ZigBee nodes with a single data request.

There are three broadcast types: 0xffff (all), 0xfffd (awake), and 0xfffc (routers only).

4.5   Addressing Within the Node

The ZigBee addressing discussed so far, PAN IDs, unicasts, groupcasts, and broadcasts, are all about getting data to a node. But ZigBee goes beyond a networking protocol. ZigBee also offers application-level compatibility. To achieve this, ZigBee contains a variety of concepts I'll explain in detail here: endpoints, clusters, command, attributes, and profiles.

Addressing within ZigBee includes all of the following components:

- PAN ID (MAC)

- NwkAddr (NWK)

- Endpoint (APS)

- Profile ID (APS)

- Cluster (APS)

- Command and/or attribute (ZCL)

Take a look at this packet decode for the toggle command. All of the above components are included. The PAN ID is found in the IEEE 802.15.4 (MAC) portion of the frame, the source, and destination NwkAddr are found in the NWK frame, the source and destination endpoint, profile ID, and cluster ID are found in the APS frame. The command is found in the ZCL frame:

```
Frame 38 (Length = 30 bytes)
IEEE 802.15.4
        Frame Control: 0x8861
        Sequence Number: 67
        Destination PAN Identifier: 0x0f00
        Destination Address: 0x0000
        Source Address: 0x796f
ZigBee NWK
        Frame Control: 0x0048
        Destination Address: 0x0000
        Source Address: 0x796f
```

```
        Radius = 10
        Sequence Number = 19
ZigBee APS
        Frame Control: 0x40
        Destination Endpoint: 0x08
        Cluster Identifier: On/off (0x0006)
        Profile Identifier: HA (0x0104)
        Source Endpoint: 0x08
        Counter: 0x45
ZigBee ZCL
        Frame Control: 0x01
        Transaction Sequence Number: 0x48
        Command Identifier: Off (0x00)
```

4.5.1   Endpoints

Within each node are endpoints. Endpoints, identified by a number between 1 and 240, define each application running in a ZigBee node (yes, a single ZigBee node can run multiple applications).

Imagine a virtual wire connecting an endpoint on one node to an endpoint on another node, from endpoint 1 on node A to endpoint 99 on node B (see Figure 4.20).

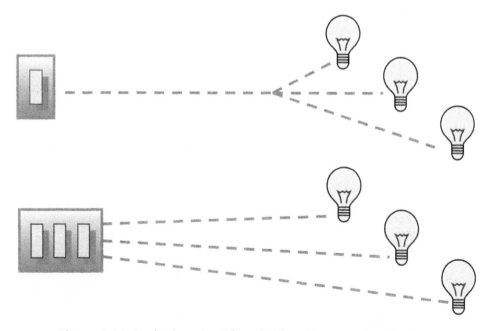

Figure 4.20: Endpoints Are Virtual Wires Between Applications

A node may contain any number of endpoints (up to 240), with any set of endpoint identifiers. For example, a single node might contain just one endpoint, numbered 55. Or a node might contain 20 endpoints, numbered 1 through 10 and 201 through 210.

Endpoints serve three purposes in ZigBee:

- Endpoints allow for different application profiles to exist within each node

- Endpoints allow for separate control points to exist within each node

- Endpoints allow for separate devices to exist within each node

Perhaps a switch may be sold in both commercial and home (or light commercial) markets. This switch, then, might support two Application Profiles: Home Automation and Commercial Building Automation. If so, it would contain two endpoints, one for each profile.

Perhaps a bank of three switches, as depicted in Figure 4.20, contains a single radio. Each of the switches would reside on a separate endpoint within that node, so that the switches operate independently of each other.

Perhaps a single widget contains three separate devices from a ZigBee standpoint. A common example of this is a thermostat. ZigBee defines at least three separate devices to heat or cool a home or office, each of which may be controlled or monitored independently: a thermostat, which includes a human interface, a temperature sensor which informs the thermostat how warm it is, and a heating/cooling unit which physically controls the heater or air-conditioner unit. All three of these devices might be separated, spread out across the office, or they might be together in one widget as is common in homes. In the latter case, each of these devices would reside on a separate endpoint, within a single ZigBee node.

There is a special endpoint called a broadcast endpoint (0xff). Sending a data request to the broadcast endpoint will reach all endpoints within the node that matches the profile ID. The broadcast endpoint is implied when groupcasting (sending a data request with address mode gZbAddrModeGroup_c).

The example in this section, *Example 4-9 ZigBee Endpoints*, shows how a single node can support multiple endpoints. This example only uses two nodes, a ZC light and a ZR switch, but three different endpoints exist within each node. The ZC light has three separately controllable lights in one node, while the ZR switch has three separate

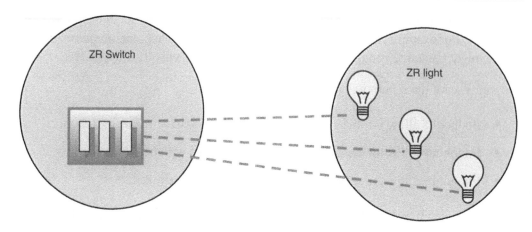

Figure 4.21: ZigBee Endpoints Example

toggle switches in the other (see Figure 4.21). In this case, I've numbered the endpoints 1 through 3 on the switch, and 11, 22, and 33 on the light, just to show exactly which endpoint numbers are used doesn't matter much to ZigBee.

The user interface on this example is pretty simple (see Table 4.7). Each switch, SW1-SW3, toggles a light. Each long switch, LSW1 – LSW3, binds that switch to the light when pressed on both nodes (light and switch). Feel free to bind any light to any switch. If binding a switch to multiple lights, use multiple End-Device-Bind sequences.

Compile and download the following projects into their respective boards:

- Chapter04\Example4-9 ZigBee Endpoints\ZcNcbHaOnOffLight.mcp

- Chapter04\Example4-9 ZigBee Endpoints\ZrSrbHaOnOffSwitch.mcp

Table 4.7: ZigBee Endpoints Example UI

Switch	Use
SW1	Toggles remote light bound on this endpoint
LSW1	Binds EP1
SW2	Toggles remote light bound on this endpoint
LSW2	Binds EP2
SW3	Toggles remote light bound on this endpoint
LSW3	Binds EP3
SW4	Sends to the broadcast endpoint, which will toggle all three remote lights

Next, press LSW1 on both the switch and the light to bind SW1 on the switch to LED1 on the light. Do the same with LSW2 and LSW3, on both boards. Now, press SW1 to toggle LED1, press SW2 to toggle LED2, and press SW3 to toggle LED3. Notice they are all independent of each other.

Feel free to reboot and bind them in any way you choose. These nodes don't keep track of binding after a reset, as a production-ready node might do. I often find it handy to have nodes forget what they know after a reset when making example code. The next chapter will describe how to save all of the information using Non-Volatile Memory.

Notice how the ZDO End-Device-Bind request, shown below and sent when LSW1 was pressed, includes which endpoint will be bound and which clusters the device is looking for:

```
Frame 42 (Length = 53 bytes)
IEEE 802.15.4
      Frame Control: 0x8861
      Sequence Number: 39
      Destination PAN Identifier: 0x0f00
      Destination Address: 0x0000
      Source Address: 0x0001
      Frame Check Sequence: Correct
ZigBee NWK
      Frame Control: 0x0048
      Destination Address: 0x0000
      Source Address: 0x0001
      Radius=10
      Sequence Number = 135
ZigBee APS
      Frame Control: 0x00
      Destination Endpoint: 0x00
      Cluster Identifier: EndDeviceBindReq (0x0020)
      Profile Identifier: ZDP (0x0000)
      Source Endpoint: 0x00
      Counter: 0x1d
ZigBee ZDO
      Transaction Seq Number: 0x04
      End Device Bind Request
            Binding Target: 0x0001
            Source IEEE Address: 00:50:c2:37:b0:04:00:02
            Source Endpoint: 0x03
            Profile ID: HA (0x0104)
            Number Of Input Clusters: 4
            Input Cluster List
                  Cluster 1: Basic (0x0000)
                  Cluster 2: Identify (0x0003)
```

```
                Cluster 3: Groups (0x0004)
                Cluster 4: On/off Switch Configuration (0x0007)
        Number Of Output Clusters: 1
        Output Cluster List
                Cluster 1: On/off (0x0006)
```

In the Freescale solution, the list of endpoints is defined in `EndPointConfig.c` in a structure called, strangely enough, `endPointList`. Most nodes support just one endpoint. Some have a handful. This one has three on/off lights. A node could support a temperature sensor, thermostat, and light switch, or could support a variety of applications on different application profiles, including a private profile. Again, remember the three purposes of endpoints:

- Endpoints allow for different application profiles to exist within each node

- Endpoints allow for separate control points to exist within each node

- Endpoints allow for separate devices to exist within each node

```
const endPointList_t endPointList[3]={
  {&HaOnOffLight1_EndPointDesc, &gHaOnOffLightDeviceDef1},
  {&HaOnOffLight2_EndPointDesc, &gHaOnOffLightDeviceDef2},
  {&HaOnOffLight3_EndPointDesc, &gHaOnOffLightDeviceDef3}
};
```

The `EndPointDesc` field points to an endpoint descript which in turn points to a SimpleDescriptor. Simple Descriptors are a ZigBee construct that describes the endpoint in detail:

```
const zbSimpleDescriptor_t HaOnOffLight3_SimpleDescriptor={
  12,    /* Endpoint number */
  0x4, 0x1, /* Application profile ID */
  0x0, 0x1, /* Application device ID */
  0,     /* Application version ID */
  5,     /* Number of input clusters */
  (uint8_t *)HaOnOffLightInputClusterList,
  0,     /* Number of output clusters */
  (uint8_t*)HaOnOffLightOutputClusterList,
};
```

Freescale also requires a device definition as part of their ZigBee Cluster Library support. If ZCL is not used, a device definition is not required. The device definition, among other things, points to common cluster routines and an instantiation of data for that endpoint. Different instantiations of data allow the three on/off lights to be controlled independently of each other.

One thing that's very important in Freescale's BeeStack: Be sure to register the endpoints. If an endpoint is not registered, then no data can be sent or received on that endpoint. The standard `BeeAppInit()` automatically registers all of the endpoints defined in the endpoint list:

```
void BeeAppInit(void)
{
  index_t i;

  /* register the application endpoint(s) */
  for(i=0; i< gNum_EndPoints_c;++i) {
    (void)AF_RegisterEndPoint(endPointList[i].pEndpointDesc);
  }

  ...

}
```

The on/off light (`ZcNcbHaOnOffLight`) determines which LED to toggle based on the incoming endpoint. The incoming frame comes into `BeeAppDataIndication()`. Then the code is parsed by the ZCL function, `ZCL_InterpretFrame()`, which in turn calls on `BeeAppUpdateDevice()`. The application doesn't have to worry about data coming in on an unregistered endpoint: If the endpoint is not registered, the packet is thrown away. `BeeAppUpdateDevice()` is where the real work happens for turning the light on and off. It doesn't matter to the code in this example whether the endpoint starts at 1, 10, or 193, but it does assume the endpoints are contiguous:

```
void BeeAppUpdateDevice
{
  (
  zbEndPoint_t endPoint,    /* IN: endpoint update happend on */
  zclUIEvent_t event        /* IN: state to update */
  )
{
  static const uint8_t aiLeds[]={LED1, LED2, LED3};
  static const char acLeds[]={'1', '2', '3'};
  static char szLightStr[]="Light1 Off";
  uint8_t iLedIndex;

  /* get an index into the endpoints */
  iLedIndex = endPoint-appEndPoint;
  if(iLedIndex >= gNum_EndPoints_c)
    iLedIndex=0;
  szLightStr[5]=acLeds[iLedIndex];

  switch (event) {
    case gZclUI_Off_c:
      ASL_SetLed(aiLeds[iLedIndex],gLedOff_c);
```

```
      FLib_MemCpy(&szLightStr[7], "Off", 3);
      ASL_LCDWriteString(szLightStr);
      ExternalDeviceOff();
      break;

   case gZclUI_On_c:
     ASL_SetLed(aiLeds[iLedIndex],gLedOn_c);
     FLib_MemCpy(&szLightStr[7], "On", 3);
     ASL_LCDWriteString(szLightStr);
     ExternalDeviceOn();
     break;

   default:
       ASL_UpdateDevice(endPoint,event);
  }
}
```

This example showed how to use endpoints for multiple control points within a single node, both on the sending (switch) and receiving (light) side.

So why pick a particular endpoint number for any given application purpose, whether defining a control point, a new device, or a new profile within the node? If using a ZigBee public profile, such as Home Automation or Commercial Building Automation, the choice is completely arbitrary. I suggest starting at endpoint 1 and going up from there. If you are supporting multiple control points, keep them contiguous (e.g., 1, 2, 3) to make the C coding easier based on the destination endpoint number.

If using a private application profile, it is easiest to use a fixed endpoint for all devices in the network. The default endpoint in Freescale demo applications is 8. If using only one endpoint, just leave it there for private profiles. The network can then avoid the initial start-up discovery of endpoints and event save code-size by not including those functions. I'll talk more about endpoint discovery functions in Chapter 5, "ZigBee, ZDO, and ZDP."

> Think of endpoints as virtual wires connecting applications.
>
> Endpoints allow for separate profiles, devices, and control points to exist within a single node.

4.5.2   Clusters

Clusters, defined by a 16-bit identifier, are application objects. Whereas the NwkAddr and endpoint are addressing concepts, the cluster defines application meaning. The cluster that has been used for many examples in this book is the OnOffCluster

Table 4.8: Some ZigBee Clusters

Cluster Name	Cluster ID
Basic Cluster	0x0000
Power Configuration Cluster	0x0001
Temperature Configuration Cluster	0x0002
Identify Cluster	0x0003
Groups Cluster	0x0004
Scenes Cluster	0x0005
OnOff Cluster	0x0006
OnOff Configuration Cluster	0x0007
Level Control Cluster	0x0008
Time Cluster	0x000a
Location Cluster	0x000b

(ID 0x0006). This cluster knows how to turn something on or off. It doesn't matter whether the something is a light, a pump, a doorbell, or is a mechanism to open and close a door. All of these things have a binary state: on or off, open or closed.

Clusters encapsulate both commands and data. A ZigBee application can determine whether the light is on or off by querying the OnOffAttribute within the OnOffCluster, or it can set the state of that light by commanding the cluster to turn the light on, off, or to toggle it. A small subset of the clusters defined by the ZigBee Cluster Library includes the information shown in Table 4.8.

Clusters only have meaning within a particular profile. For example, a private profile may define cluster 0x0000 as the "shut down the network" cluster, whereas cluster 0x0000 is the Basic Cluster in the Home Automation profile. To keep things simple, all profiles which support the ZigBee Cluster Library use the same cluster IDs. That is, any ZigBee public profile which turns something on or off will use cluster ID 0x0006 for that purpose.

Clusters, in addition to the identifier, have direction. In the SimpleDescriptor which describes an endpoint, a cluster is listed as either input or output. This is used only for the purposes of service discovery: a switch, which has cluster 0x0006 as an output, can find a light, which has cluster 0x0006 as an input. But a switch doesn't find another switch, as they are both outputs on cluster 0x0006 and don't match.

Table 4.9: Morse Code Clusters

Profile ID	Cluster Name	Cluster ID	Data
0xc035	MorseBlink	0x0100	short (0) or long (1)
0xc035	MorseString	0x0101	ASCII string

Clusters contain both code (commands) and data (attributes). Commands cause action. Attributes keep track of the current state of that cluster. A good example of this is the HA DimmingLight. It supports both the OnOff Cluster (0x0006) and the LevelControl Cluster (0x0008). There are commands to turn the light on, off, or toggle it on the OnOff Cluster; but to control the dimming function, commands on the LevelControl Cluster are used. The data on the OnOff Cluster is simply a field: Is the light on, or off. The data on the LevelControl Cluster indicates what dimming level the light is (or will be) at when it is in the "on" state.

Clusters in a private profile may communicate any kind of data. ZigBee places no restrictions. Attributes and commands, which are part of the ZigBee Cluster Library specification, may not even be used at all. For example, San Juan Software uses a private profile, 0xc035, for all of our training and demo programs. You've seen one of these demo programs already in "*Example 3-2: Morse Code*," from Chapter 3, "The ZigBee Development Environment." The Morse Code program uses two clusters, shown in Table 4.9

The Morse Code program does not use the ZigBee Cluster Library, so the meanings of the clusters are completely open. San Juan Software maintains a database of clusters for the training applications on the SJS private profile 0xC035, so the applications can be used together on the same network without conflict.

The following source code shows how the data indication is handled with these clusters from the SJS private profile. Notice that the code needs only to check the cluster ID. Endpoint and profile ID are already handled by the ZigBee stack:

```
void BeeAppDataIndication(void)
{
  apsdeToAfMessage_t *pMsg;
  zbApsdeDataIndication_t *pIndication;

  while(MSG_Pending(&gAppDataIndicationQueue))
  {
```

```
    /* Get a message from a queue */
    pMsg=MSG_DeQueue(&gAppDataIndicationQueue);

    /* ask ZCL to handle the frame */
    pIndication=&(pMsg- > msgData.dataIndication);

    /* user wants to blink (short or long) */
    if(pIndication- > aClusterId == gClusterMorseBlink)
      MorseBlink(pIndication- > pAsdu[0]);

    /* user passed a string to */
    else if(pIndication- > aClusterId == gClusterMorseString)
      MorseString(pIndication- > asduLength, pIndication- > pAsdu);

    /* Free memory allocated by data indication */
    MSG_Free(pMsg);
  }
}
```

> Think of clusters as an object, including both code (commands) and data (attributes). Clusters may have any meaning in a private profile.

4.5.3 Commands

Public profiles use the ZigBee Cluster Library. ZCL introduces two new concepts into ZigBee that you won't find in the main ZigBee stack specification: commands and attributes. ZCL makes it easy to get or to set attributes through a common set of commands, and provides a simple mechanism for issuing cluster-specific commands.

Commands are identified by an 8-bit number, and are either cluster-specific, or cross-cluster. Cluster-specific commands depend on the cluster number, and generally start at 0x00. For example, 0x00 is the "off" command in the OnOff Cluster, whereas 0x00 is "move-to-level" command in the LevelControl Cluster.

ZCL is used in (nearly) all of the ZigBee public profiles, but is not required for private profiles. Considerable code space can be saved by leaving out ZCL, but then of course, the features of ZCL are left out, too.

Over-the-air, ZCL frames are the payload of the APS frame. Take a look at the following ZCL:

```
ZigBee APS
      Frame Control: 0x00
      Destination Endpoint: 0x08
```

```
        Cluster Identifier: On/off (0x0006)
        Profile Identifier: HA (0x0104)
        Source Endpoint: 0x08
        Counter: 0x3f
ZigBee ZCL
        Frame Control: 0x01
            .... ..01=Frame Type: Command is specific to a cluster (0x01)
            .... .0..=Not Manufacturer Specific
            .... 0...=Direction: From the client side to the server side
            0000 ....=Reserved: Reserved (0x00)
        Transaction Sequence Number: 0x48
        Command Identifier: Off (0x00)
```

Notice that the command in the ZCL frame is the identifier for "Off." Command 0x00 is only "Off" in this case because it is the On/Off Cluster (0x0006), as identified in the APS frame. If it were a different cluster, such as the Level Control Cluster, command 0x00 would be "Move to level."

Clusters may work together to form a complete application. For example, a dimming light supports both the On/Off Cluster and the Level Control Cluster, and the on/off behavior has specific meaning when related to a dimming light. For example, "on" turns the device on to the last non-zero level.

> Commands cause action on a cluster.

4.5.4   Attributes

You've seen the command "toggle" many times throughout this book. But what if the application wants to know whether the light is on or off, after toggling it? That's where attributes come in. An application simply needs to read the attribute on that cluster.

Attributes store the current "state" of a given cluster. Collectively, a set of attributes on all clusters supported by a device define the state of that device. There are generic ZCL commands to read and write attributes on any given cluster.

Attributes are identified by a 16-bit number, and may be any value between 0x0000 and 0xffff. In the ZigBee Cluster Library, they tend to start at 0x0000 for any given cluster.

Attributes can even be set up to report automatically at regular intervals, if they change, or both. I'll describe the ZigBee Cluster Library in a lot more detail in Chapter 6, "The

ZigBee Cluster Library." It is there I'll include plenty of examples on reading, writing, and reporting attributes.

> Attributes show the state of a given cluster.

4.6 Profiles

Every data request in ZigBee is sent (and received) on an Application Profile. Application Profile IDs are 16-bit numbers and range from 0x0000 to 0x7fff for public profiles and 0xbf00 to 0xffff for manufacturer specific profiles.

Think of a profile as a domain space of related applications and devices. Public profiles are those specified by the ZigBee Alliance (as opposed to private profiles specified by individual OEMs). Home Automation is a public application profile which defines a wide range of ZigBee networked devices intended for use in the home, including lights and switches, wall outlets, remotes, thermostats, air-conditioners, and heaters. Another public profile, Commercial Building Automation, defines ZigBee devices such as advanced lights and switches, and keyless entry and security systems.

Any number of Application Profiles, both public and manufacturer-specific, may exist in a single ZigBee network. In fact, any number of profiles may exist in a single node on the network, separated on different endpoints.

The ZigBee Alliance continues to work on public profiles to ensure they match the needs of OEMs producing products. It is the OEMs in the ZigBee Alliance who define the profile in the first place. Table 4.10 shows a short list of the ZigBee public profiles.

Table 4.10: ZigBee Public Profile IDs

Profile ID	Profile Name
0101	Industrial Plant Monitoring (IPM)
0104	Home Automation (HA)
0105	Commercial Building Automation (CBA)
0107	Telecom Applications (TA)
0108	Personal Home & Hospital Care (PHHC)
0109	Advanced Metering Initiative (AMI)

Public profiles are designed so that products from one manufacturer (X) can work, right out-of-the-box, with products from another manufacturer (Y). For example, a thermostat from Honeywell can work with a variable-airflow-valve from Trane, or a light from Philips can work with a switch from Leviton.

ZigBee members may also apply for what is called a private profile. Private profiles, officially called Manufacturer Specific Profiles (or MSPs), are not defined by the ZigBee Alliance, but instead are defined by the OEMs making the products. Private profiles are used for those applications that do not need to interact with other vendors' products.

4.6.1 Public Profiles

All of the devices pictured in Figure 4.22 are part of the Home Automation profile. Once commissioned, all these devices perform their appropriate actions, simply and effectively.

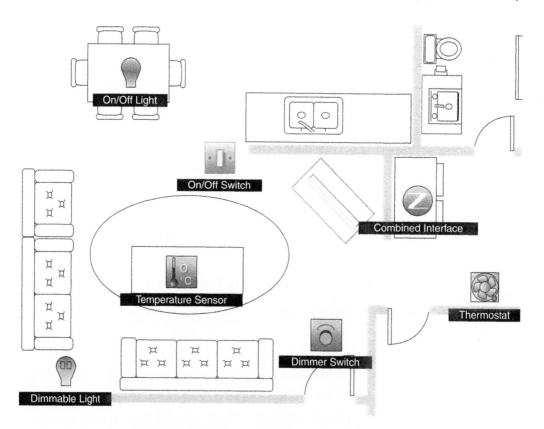

Figure 4.22: Devices in a ZigBee Public Profile

The light switches (both on/off and dimmer) just work as light switches should, and the temperature is sent dutifully to the thermostat once each minute, which controls the heating and/or cooling unit. Lights may be dimmable or not, and may be three-way (toggle) or two-way (on/off).

The Freescale solution supports the Home Automation public profile out-of-the-box, including a large same of ready-made HA devices such as lights, thermostats, and temp sensors. All the devices can be monitored by the PC through the ZigBee Environment Demo (ZeD). I'll not describe the ZigBee Environment Demo more here, since the Freescale documentation already does an adequate job.

Nearly all public profiles use the ZigBee Cluster Library, so I'll leave the example and details to Chapter 6, "The ZigBee Cluster Library."

4.6.2   Manufacturer-Specific Profiles

Manufacturer-specific profiles (MSPs) allow the OEM to define any set of clusters, endpoints, and devices. ZigBee places no restrictions on what type of data is transmitted, other than requesting that the data rate remain reasonable, so the application doesn't flood the channel.

The ZigBee Alliance assigns MSP identifiers upon requests from member companies. As explained before, San Juan Software has been assigned an MSP that we use for training and demo purposes: `0xc035`. Feel free to use this MSP for your own demos, keeping in mind that it's not suitable for shipping products. At the time of this writing, there is no charge for a Manufacturer Specific Profile ID. Simply join the ZigBee Alliance and request one.

To define clusters in a private profile, start by mapping out all of the devices that will exist in the network, and all the information they must communicate. Define the payloads to be as small as possible. It is often just fine to assume some information on both sides. For example, a single command byte might suffice.

The example in this section, *Example 4-10 iPod Controller*, shows how to use a manufacturer-specific profile in an application. It sends a single command byte over the air on a particular cluster. That command byte is translated into a series of bytes that are sent over the serial port to control an iPod.

You've probably noticed most of the examples in this book are purely software examples. That's appropriate because this text is really about ZigBee, which aside from the radio, is all about software. But no embedded book is ever truly complete without at least one hardware project. To make it more fun, this one involves both a ZigBee network and an Apple iPod.

Warning and Disclaimer: This example involves modifying hardware and connecting a Freescale SRB board to an iPod. While this example worked well for the author, no warranty of any kind is implied or expressed that this will actually work in your case. Any damage to your hardware, loss of productivity or loss of life and limb is strictly your responsibility.

Seriously, if you implement this example, do be careful so you don't damage either your Apple iPod or your Freescale SRB board.

The concept is fairly simple. Use ZigBee to control an iPod (see Figure 4.23). One ZigBee node will act as an iPod controller and is wired directly to a connector plugged into the iPod dock connector port. Another ZigBee node will act as the remote control.

The following commands are implemented in this simple remote control:

- Play/Pause
- Skip Forward
- Skip Backward
- Volume Up
- Volume Down

As seen in the figure, the ZED is the remote control. This is appropriate because a real remote control is likely to be battery-operated, and so should be a ZigBee End-Device. The iPod Controller is a ZigBee Coordinator only to keep the example small, with only

iPod Remote
(ZED)

iPod Controller
(ZC)

Figure 4.23: iPod Controller Concept

two nodes. This same example could work across any size ZigBee network, and the iPod controller could be an RxOnIdle ZED, or a ZigBee Router.

To obtain an iPod connector, I used an on-line store headquartered in Sweden (strange how the world is so small these days) at http://home.swipnet.se/ridax/connector.htm. They also have a U.S. office and phone number.

To obtain more information on the iPod connector and the commandset available, take a look at my old favorite, Wikipedia: http://en.wikipedia.org/wiki/Dock_connector, or go to http://ipodlinux.org/Dock_Connector.

To prepare an SRB board for the iPod Controller node, remove the battery pack from the back. There is some double-sided sticky tape, and also the thin wire connection between the battery pack and the board. I simply used my hands and applied pressure. The SRB board doesn't care if the battery pack is gone as it can also receive power through the USB connector.

You will also need a soldering iron. I used my own, but you may have a coworker or friend who can do the soldering.

Assuming you've been following along with examples using the Freescale development kit, the NSK, this example can be completed with the following steps:

1. Buy a connector (this may take a few days to deliver)
2. Buy an iPod (Okay, you probably already have one)
3. Open the case and remove the battery pack from the SRB
4. Solder the two-wire connector to the SRB as pictured
5. Program the modified SRB as the Controller, and another SRB as the remote
6. Enjoy controlling the iPod with the remote SRB board

Pictured in Figure 4.24 are an SRB board on the left, and an iPod on the right, with the dock connector hardwired to the SRB board and plugged into the iPod. The black wire is the ground. Any ground will do, but I used the ground from the BDM (J101, pin 2). Only the Tx wire needs to be connected for this demo, as the SRB will control the iPod, but not receive any information from it. I've included a close-up picture of the soldering in Figure 4.25.

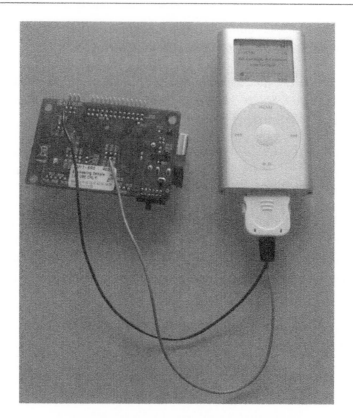

Figure 4.24: iPod Connector

Figure 4.25: Solder Red Wire to TP131 (Tx), Black to Ground

To run the example, compile and download the following two projects into two SRB boards onto the one which you have soldered the iPod connector:

- Chapter04\Example4-10 iPodController\SrbZciPodController

- Chapter04\Example4-10 iPodController\SrbZediPodRemote

These operate on channel 25, PAN ID 0x1aaa, but any set of channels or PAN IDs will do. The application is the same whether it is on the controller or the remote. If the application receives an over-the-air command, that is, a data indication, it sends a corresponding set of bytes to the serial port. If a button is pressed, it sends the command over-the-air via a data request to node 0x0000.

Assuming node 0x0000 for the destination (the ZC) made the application a little simpler, with a bit of work the application could be made more robust complete with binding.

I also show a close-up of the wiring configuration in Figure 4.25. Make sure to solder the red wire to the Tx pin on the UART (TP131 at the back of the SRB board). Make sure to solder the black wire to ground. I used pin 2 of J101, but any ground pin will do.

When the command is sent over-the-air, it looks like the following. This decode depicts command 0x01 which instructs the iPod to skip forward one song:

```
Frame 18 (Length=28 bytes)
IEEE 802.15.4
        Frame Control: 0x8861
        Sequence Number: 205
        Destination PAN Identifier: 0x1aaa
        Destination Address: 0x0000
        Source Address: 0x796f
        Frame Check Sequence: Correct
ZigBee NWK
        Frame Control: 0x0048
        Destination Address: 0x0000
        Source Address: 0x796f
        Radius=10
        Sequence Number=219
ZigBee APS
        Frame Control: 0x00
        Destination Endpoint: 0x01
        Cluster Identifier: (0x0001)
        Profile Identifier: (0xc035)
        Source Endpoint: 0x01
        Counter: 0x4d
        Unknown APS Data: 01
```

Note the profile ID in the ZigBee APS frame (0xc035). This is a private profile assigned to San Juan Software by the ZigBee Alliance. In this profile, cluster 0x0001 means the "iPod" cluster. This cluster is defined to contain a single command byte in the payload, called "APS Data." The commands mirror the features available to the iPod controller.

- Play/Pause   (cmd 0x00)

- Skip Forward   (cmd 0x01)

- Skip Backward   (cmd 0x02)

- Volume Up   (cmd 0x03)

- Volume Down   (cmd 0x04)

The data indication on the iPod Controller simply takes this command byte in and passes it along to the function `SendiPodRemoteCommand()`, which sends a corresponding command string over the serial port at 19,200 baud. The serial port is connected to the iPod via the 2-wire interface:

```
void BeeAppDataIndication(void)
{
  apsdeToAfMessage_t *pMsg;
  zbApsdeDataIndication_t *pIndication;

  while(MSG_Pending(&gAppDataIndicationQueue))
  {
    /* Get a message from a queue */
    pMsg=MSG_DeQueue(&gAppDataIndicationQueue);
    pIndication=&(pMsg- > msgData.dataIndication);

    /* is the cluster for the application */
    if(pIndication- > aClusterId == appDataCluster)
      SendiPodRemoteCommand(pIndication- > pAsdu[0]);

    /* Free memory allocated by data indication */
    MSG_Free(pMsg);
  }
}
void SendiPodRemoteCommand(iPodRemoteCommand_t cmd)
{
  uint8_t *pCmdPtr;

  switch(cmd) {
    case gPlayPause_c:
      pCmdPtr=PlayPause;
      break;
```

```
        case gSkipForward_c:
          pCmdPtr=SkipForward;
          break;
        case gSkipReverse_c:
          pCmdPtr=SkipReverse;
          break;
        case gVolumeDown_c:
          pCmdPtr=VolumeDown;
          break;
        case gVolumeUp_c:
          pCmdPtr=VolumeUp;
          break;
        default:
          pCmdPtr=ButtonRelease;
          break;
      }

      /* send command-all commands are the same size */
      (void)UartX_Transmit(pCmdPtr, sizeof(PlayPause), NULL);
      while (!UartX_TxCompleteFlag);

      /* delay 100 microseconds before sending button release */
      DELAY_100US();

      /* Button release must always be sent after a command */
      UartX_Transmit(ButtonRelease, sizeof(ButtonRelease), NULL);
      while (!UartX_TxCompleteFlag);
}
```

On the sending end, the application simply creates a data request with the command-byte associated with the particular button:

```
void BeeAppHandleKeys(key_event_t keyEvent)
{
   switch(keyEvent) {

     /* play/pause */
     case gKBD_EventSW1_c:
       SendiPodRemoteCommandOTA(gPlayPause_c);
       break;

     /* skip one song forward */
     case gKBD_EventSW2_c:
       SendiPodRemoteCommandOTA(gSkipForward_c);
       break;

     /* skip one song reverse */
     case gKBD_EventLongSW2_c:
       SendiPodRemoteCommandOTA(gSkipReverse_c);
       break;
```

```
    /* volume down */
    case gKBD_EventSW3_c:
      SendiPodRemoteCommandOTA(gVolumeDown_c);
      break;

    /* volume up */
    case gKBD_EventSW4_c:
      SendiPodRemoteCommandOTA(gVolumeUp_c);
      break;
  }
}

void SendiPodRemoteCommandOTA(iPodRemoteCommand_t cmd)
{
  afAddrInfo_t    addrInfo;
  uint8_t         TransmitBuffer[1];

  /* copy iPod command to TransmitBuffer */
  TransmitBuffer[0]=cmd;

  /* set up address information */
  addrInfo.dstAddrMode=gZbAddrMode16Bit_c;
  Set2Bytes(addrInfo.dstAddr.aNwkAddr, 0x0000);
  addrInfo.dstEndPoint=appEndPoint;
  addrInfo.srcEndPoint=appEndPoint;
  Copy2Bytes(addrInfo.aClusterId, appDataCluster);
  addrInfo.txOptions=gApsTxOptionNone_c;
  addrInfo.radiusCounter=afDefaultRadius_c;

  /* send the data request */
  (void)AF_DataRequest(&addrInfo, sizeof(TransmitBuffer),
    TransmitBuffer, NULL);
}
```

Notice the data request is sent to node 0x0000. In a private profile the main application could be on node 0x0000, the ZigBee Coordinator, such as a gateway to the PC, or in this case, the iPod controller, making node discovery unnecessary. The application uses appEndPoint for both the source and destination endpoints. Using a fixed endpoint makes private profiles a little easier, as no endpoint discovery is required. The cluster in this example is appDataCluster, the first cluster in the output cluster list.

> Manufacturer Specific Profile IDs are obtained from the ZigBee Alliance.
>
> MSPs can use fixed endpoints and cluster IDs to make service discovery unnecessary.

4.6.3   Device IDs

Every endpoint includes a profile identifier and a device identifier. Think of a device as a physical thing: a light, temperature sensor, thermostat, or the like. One physical widget might contain multiple devices. For example, a ZigBee thermostat sold at Home Depot may contain a temperature sensor, a thermostat, and a heating/cooling unit controller. From ZigBee's standpoint, these are separate devices. From a packaging standpoint, they are all in one SKU (Shelf-Keeping Unit).

ZigBee Device IDs range from 0x0000 to 0xffff. A short list of device IDs available in the Home Automation Profile are shown in Table 4.11.

Device IDs serve two purposes:

- To allow human-readable displays (your PC, the TV, etc.) to show a proper icon for the device in question.

- To allow ZigBee commissioning tools to be more intelligent.

For example, consider an On/Off Light and a Mains Power Outlet (or switchable outlet, as they are called in some homes). Both devices are identical in the way they are controlled and monitored over-the-air. They both use cluster 0x0006, the OnOff Cluster. But users expect to see an outlet in a commissioning tool when they are connecting a switch to that outlet, and they expect to see a light when they are connecting a switch to that light.

Keep in mind that ZigBee does not specify the human interface to devices. A switch might be a push button, a rocker switch, an e-field sensor plate, accelerometer, or anything else the manufacturer can imagine.

Table 4.11: Device IDs Available in the Home Automation Profile

Name	Identifier	Name	Identifier
Range Extender	0x0008	Light Sensor	0x0106
Mains Power Outlet	0x0009	Shade	0x0200
On/Off Light	0x0100	Shade Controller	0x0201
Dimmable Light	0x0101	Heating/Cooling Unit	0x0300
On/Off Light Switch	0x0103	Thermostat	0x0301
Dimmer Switch	0x0104	Temperature Sensor	0x0302

Each ZigBee Public Profile contains a specific list of devices. Manufacturers are welcome to extend that list with their own devices, provided that they use the "Manufacturer Specific Extension" field provided as part of the ZigBee Cluster Library interface. Or they can make a device on another (MSP) profile that can interact with specific public profile devices, by using the existing device IDs and clusters.

> Device IDs are mainly used for commissioning tools.
>
> ZigBee performs service discovery based on profile IDs and cluster IDs, not device IDs.

4.6.4   The Simple Descriptor

The simple descriptor ties everything together on an endpoint, and an endpoint defines an application. Why the simple descriptor is not simply called an endpoint descriptor, I don't know.

The simple descriptor contains many of the fields I've been describing: an endpoint ID, a profile ID, cluster IDs, and a device ID. Its format is as follows:

```
typedef struct zbSimpleDescriptor_tag
{
  zbEndPoint_t endPoint;
  zbProfileId_t aAppProfId;
  zbDeviceId_t aAppDeviceId;
  uint8_t appDevVerAndFlag;
  zbCounter_t appNumInClusters;
  uint8_t *pAppInClusterList;
  zbCounter_t appNumOutClusters;
  uint8_t *pAppOutClusterList;
}zbSimpleDescriptor_t;
```

The simple descriptor of any endpoint can be retrieved over-the-air with the ZDP command `Simple_Desc_req`, and is used when performing service discovery (which I'll describe in detail in the next chapter).

The simple descriptor does not list the commands and/or attributes available on any given cluster. It is expected that a device "just knows" the capabilities of the cluster, if, say, the OnOff Cluster is supported on this endpoint. As a protocol for 8-bit devices (at least at the low end), ZigBee attempts to balance flexibility with implementation size.

Within any given device, some clusters are mandatory, and some may be optional. By asking for the simple descriptor, any device in the network can determine which clusters are supported by any other device, and can then make an informed decision about optional functionality. The response to a `Simple_Desc_req` is mandatory in all devices.

You'll notice there is both an input and output cluster list in the simple descriptor. This is how ZigBee determines which devices match. If one side has an output cluster (say a dimmer switch) and the other side has the same cluster as an input cluster (say, a dimming light), then they match. Any overlap indicates a match, so if an application is looking for a particular cluster, it can send a "simplified" simple descriptor to match just that one cluster.

> The simple descriptor ties all the endpoint information together in one place.

4.7   ZigBee Application Support Sublayer (APS)

Okay. Don't ask me why the Application Support Sublayer uses the three-letter acronym APS. It's obvious, but don't ask me.

The Application Support Sublayer, or APS, sits above the NWK layer, and is the layer in ZigBee which understands applications. The APS frame over-the-air includes endpoints, clusters, profile IDs, and even groups.

APS is responsible for the following activities:

- Filtering out packets for non-registered endpoints, or profiles that don't match
- Generating end-to-end acknowledgment with retries
- Maintaining the local binding table
- Maintaining the local groups table
- Maintaining the local address map

APS has the job of filtering out packets for endpoints that don't exist in the node. APS filters packets that don't match profile IDs. APS also filters duplicate packets, which can happen in a network that supports automatic retries.

It's the job of APS to perform automatic retries, if acknowledgment is requested by the sender, to maximum the chance of successful transmission and to inform the sender whether or not the packet was delivered.

APS also maintains a variety of application-level tables. Binding is all about connecting an endpoint on this node to one or more endpoints on other nodes. Groups are all about an arbitrary collection of applications residing on an arbitrary set of nodes throughout the network. The address map associates a 64-bit MAC address with a ZigBee 16-bit NwkAddr.

APS and the Application Framework (AF) together form the ZigBee interface used by applications (see Figure 4.26). Lower layers are not called upon directly, but are used by APS and ZDO.

The Application Framework does not have an over-the-airframe of its own, but instead is the set of routines, or API, that the ZigBee stack vendor has chosen for applications to interact with ZigBee. This includes how endpoints are implemented, and how data requests, confirms, and indications are implemented for that particular vendor. In the

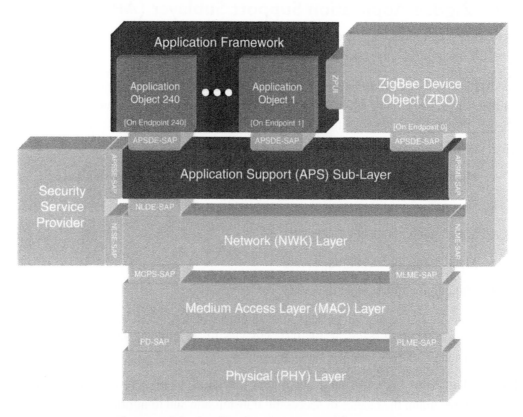

Figure 4.26: Application Support Sublayer (APS)

Freescale solution, you've already seen the Application Framework at work with the functions:

- AF_DataRequest()

- BeeAppDataConfirm()

- BeeAppDataIndication()

And with the endPointList found in EndPointConfig. c.

4.7.1 APS ACKs

While the MAC layer provides per-hop acknowledgments, the APS layer is what provides end-to-end acknowledgments, also called ACKs.

To illustrate, take a look at Figure 4.27. Suppose a switch (the ZED) wants to turn on a light (the ZR), and it wants to verify that the light received the command. The switch uses the optional ACK feature in the AF_DataRequest() txOptions field.

The distance between the nodes is irrelevant. They could be neighbors, or 10 hops away. The effect is the same. Suppose the initial data request got through to the ZR. It then processes that command immediately, perhaps toggling the light. But the APS ACK did not make it back, for some reason. So APS will automatically retry after the time-out period (which defaults to 1.5 seconds). But this retry, labeled (2), doesn't make it through. APS tries again, and this time it succeeds. Only then does APS inform the sender that the results were successful.

By the way, this scenario is extremely unlikely. It is used merely as an illustration. ZigBee uses up to three MAC ACKs per hop, so unless the channel is so noisy that

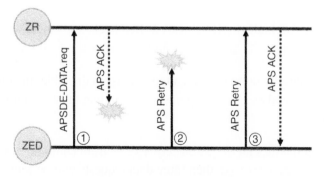

Figure 4.27: APS Retries

communication is impossible, or the path is broken because one node has dropped off the network, or the physical environment has changed (one of the routers along the route can no longer hear its neighbors), APS ACKs are rarely called into play.

APS is smart enough not to send the packet up to the application twice. In Figure 4.27, both (1) and (3) make it through, but because the data request was already heard at (1), the copy at (3) will be dropped by the APS layer after it sends the ACK to the sender. Your application doesn't have to have any special logic to handle duplicates. ZigBee does it for you.

4.7.2   APS Binding

Binding in ZigBee allows an endpoint on one node to be connected, or "bound" to one or more endpoints on another node. Think of a switch controlling a light, or a temperature sensor sending its data to a thermostat. The sender (switch or temperature sensor) is "bound" to the receiving device (light or thermostat).

Binding is unidirectional, in that the switch is bound to the light, but the light isn't bound to the switch.

Bindings are stored locally by the sending node (the switch or temperature sensor). Each binding table entry stores the following information:

- Source endpoint

- Destination NwkAddr and endpoint or destination group

- Cluster ID

The binding table size defaults to five entries in BeeStack, and can be set by changing gMaximumApsBindingTableEntries_c in BeeStackConfiguration.h.

If multiple entries from the same source endpoint exist in the table, then multiple destinations will be sent also. For example, say the binding table contains the entries shown in Table 4.12.

If the application then issues an AF_DataRequest() from endpoint 5 with address mode gZbAddrModeIndirect_c, then three data requests will be sent over the air, one

Table 4.12: ZigBee Binding Table

Src EP	Destination Addr	Addr/Grp	Dst EP	Cluster ID
5	0x1234	A	12	0x0006
6	0x796F	A	240	0x0006
5	0x9999	G	–	0x0006
5	0x5678	A	44	0x0006

to node 0x1234 endpoint 12, one broadcast to group 0x9999, and one to node 0x5678 endpoint 44 in that order:

```
afAddrInfo_t  ad drInfo;
addrInfo.dstAddrMode=gZbAddrModeIndirect_c;
addrInfo.srcEndPoint=5;
addrInfo.txOptions=gApsTxOptionDefault_c;
addrInfo.radiusCounter=afDefaultRadius_c;
Set2Bytes(addrInfo.aClusterId, 0x0006);

  (void)AF_DataRequest(&addrInfo, 10, "ToggleLed2", NULL);
```

Is it starting to become clear just how powerful binding is?

Unless constrained by too little memory, there is no reason not to include binding in all ZigBee devices. Binding, coupled with the over-the-air ZDP binding commands, allows any endpoint on any node to be connected easily with any endpoint on any other node. Binding (or determining which nodes in the network talk together) is one of the critical steps when setting up a ZigBee network. Binding makes it just that much easier.

Local binding commands are supported by the APS layer. Over-the-air binding commands are supported by the ZigBee Device Profile (ZDP).

The local (APSME) binding calls (shown in Table 4.13) return immediately in the Freescale solution. After all, they just manipulate in-memory tables. The over-the-air (ZDP) binding calls issue a callback, but I'll describe that in detail in Chapter 5, "ZigBee, ZDO, and ZDP."

Binding is not required, at least not in Manufacturer Specific Profiles. Feel free to send directly to individual nodes, groups, or to use a broadcast mode, but binding does make the commissioning process easier and more flexible.

> Use binding to simplify the commissioning process.

4.7.3   APS Groups

As explained previously, groups provide a filter for data indications. If an endpoint doesn't belong to the group, it doesn't receive the incoming over-the-air message.

In addition to matching the group ID, APS also matches the profile ID for that endpoint. If both IDs match, the message is sent through to the endpoint. If not, the packet is dropped. When debugging, if you see a packet going over-the-air, but the packet is not received at the data indication, check both these fields first. Having the wrong group or profile ID has caused me debugging grief a number of times.

Table 4.13: ZigBee Binding Commands

ZigBee Name	Freescale Prototype
APSME-BIND.request	zbStatus_t APSME_BindRequest(zbApsmeBind Req_t *pUnbindReq);
APSME-UNBIND.request	zbStatus_t APSME_UnbindRequest(zbApsme UnbindReq_t *pUnbindReq);

ZDP. Bind_req	void
	APP_ZDP_BindUnbindRequest
	(
	    zbCounter_t *pSequenceNumber,
	    zbNwkAddr_t aDestAddress,
	    zbMsgId_t BindUnbind,
	    zbBindUnbindRequest_t
	*pBindUnBindRequest
	    );

ZDP.Unbind_req	void
	APP_ZDP_BindUnbindRequest
	(
	    zbCounter_t *pSequenceNumber,
	    zbNwkAddr_t aDestAddress,
	    zbMsgId_t BindUnbind,
	    zbBindUnbindRequest_t
	*pBindUnBindRequest
	    );

Table 4.14: APS Group Management Commands

ZigBee Name	Freescale Equivalent Function
APSME-ADD-GROUP.request	zbStatus_t ApsmeAddGroup(zbApsmeAddGroupReq_t *pRequest);
APSME-REMOVE-GROUP.request	zbStatus_t ApsmeRemoveGroup(zbApsmeRemoveGroupReq_t *pRequest);
APSME-REMOVE-ALL-GROUPS.request	zbStatus_tApsmeRemoveAllGroups(zbApsmeRemoveAllGroupsReq_t *pRequest);
(none)	bool_tApsGroupIsMemberOfEndpoint(zbGroupId_t aGroupId, zbEndPoint_t endPoint);

BeeStack keeps track of groups in a small table, the size of which is defined by the BeeKit property gApsMaxGroups_c found in BeeStackConfiguration.h.

The local commands for manipulating groups are shown in Table 4.14. These commands are pretty self-explanatory. The ApsmeAddGroup() function adds a group to an endpoint. ApsmeRemoveGroup() function removes a group from an endpoint. ApsmeRemoveAllGroups() function removes all groups from an endpoint. The ApsGroupIsMemberOfEndpoint() checks to see whether an endpoint is a member of a particular group.

In addition to the APS local group management functions, there are over-the-air commands group management functions as part of the ZigBee Cluster Library. These commands include AddGroup, RemoveGroup, RemoveAllGroups, AddGroupIfIdentifying and others, and are described in Chapter 6, "The ZigBee Cluster Library."

> APS provides local group commands. ZCL provides over-the-air group commands.

4.7.4 APS Address Map

The APS layer contains a table called the address map. This table associates the 16-bit ZigBee NwkAddr with the 64-bit IEEE (or MAC) address, as shown in Table 4.15.

Some ZigBee commands, such as binding, use only the IEEE address, but ZigBee needs the 16-bit NwkAddr to communicate, so it must somehow associate these two addresses internally. Also, some nodes may be mobile in a ZigBee network, such as ZigBee End-Devices, and these mobile devices may change their 16-bit NwkAddr. If they do, they

Table 4.15: APS Address Map

NwkAddr	IEEE Addr
0x0000	0x0050c237b0040102
0x0001	0x0050c237b0045ae3
0x796f	0x0050c237b004c290

announce this fact to the network with a Device Announce command. Every node then updates its internal tables and the binding is preserved.

The size of the address map table is set with the `gApsMaxAddrMapEntries_c` property found in `BeeStackConfiguration.h` in the Freescale solution.

In BeeStack, there is no reason to add entries to the address map directly. Instead, use the ZDP commands `ZDP.IEEE_addr_req` or `ZDP.NWK_addr_req`, either of which will add that entry into the address map, if it is not there already. The application may need to clear unused entries from the address map if it will be speaking to many different nodes over time. Usually, though, an application is commissioned to talk to a small set of nodes, and stays that way for years, possibly the lifetime of the network.

In BeeStack, if you'd like to manipulate the address map, use:

```
addrMapIndex_t APS_AddToAddressMap(zbIeeeAddr_t aExtAddr,
                                    zbNwkAddr_t aNwkAddr);
void APS_RemoveFromAddressMap(zbIeeeAddr_t aExtAddr);
addrMapIndex_t APS_FindIeeeInAddressMap(zbIeeeAddr_t aExtAddr);
```

The `zbIeeeAddr_t` is the 64-bit MAC address. The `zbNwkAddr_t` is the 16-bit NwkAddr. The `addrMapIndex_t` returns the index (0-n) in the address map for the entry, or `gNotInAddressMap_c` if not found, or `gAddressMapFull_c` if it couldn't be added.

4.8   ZigBee AES 128-Bit Security

The ZigBee security suite is built on the Advanced Encryption (AES-128 bit) Standard, a well-respected block cipher algorithm published by the National Institute of Standards and Technology (NIST). To read more about AES, go to http://www.nist.gov.

If you read the security section in the ZigBee specification, you might get a headache. The language is pretty thick, and there are a lot of security options, including security

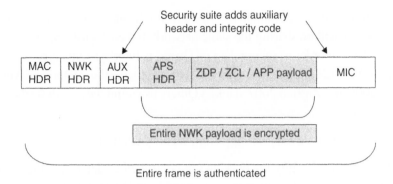

Figure 4.28: ZigBee Secure Data Frame

levels, a variety of key types, CCM*, and so on. Security encompasses 120 pages of the 534-page ZigBee specification. But from an application standpoint, ZigBee security is simple. It is always there.

That's right! There are no code changes or special parameters to set on a data request to include security. It's just there (see Figure 4.28).

ZigBee both encrypts and authenticates packets. The encrypted portion (the NWK payload), cannot be understood by any nodes sniffing the air. This includes sensitive customer data, such as billing or medical records, or any other application payload, including what clusters, profile, and endpoints are used.

ZigBee authenticates the entire frame. Authentication is required in order to prevent replay attacks and to prevent any node from falsely injecting a packet into the network. A replay attack is simple to perform. Set an 802.15.4 device listening on a channel or set of channels. When a packet is heard, replay it byte-for-byte. ZigBee will simply throw away these packets, perhaps after a delay.

Denial of service is another type of wireless attack, and is something that's very difficult to prevent. I can write a small bit of code (no, I'm not providing it in the book!) that constantly transmits. Any radios within hearing range will not be able to transmit because all of the bandwidth is used. This is the equivalent to someone cutting the power to a building, or cutting the office broadband connection to the Internet. With ZigBee, using signal strength (LQI) makes it fairly easy to track down the culprit.

ZigBee uses a 128-bit key for the entire network, called a network key. It is assumed that if a node is allowed on a ZigBee network, it is trusted. This is similar to allowing someone in your home. You at least trust them not to steal the silver.

Some applications require additional security beyond the network key, for example, if multiple customers will be sharing the same network, but each customer may have his own sensitive data. In order to secure data on shared networks of this type, some vendors use a different AES 128-bit key to secure the APS payload. (this method is available to both ZigBee 2006 and ZigBee 2007 stacks.) Others use a link key, as described in ZigBee 2007 specification.

In short, ZigBee provides a very strong, solid, security solution.

Here is a challenge. Using only the over-the-air octets below, determine the AES 128-bit key. Just to make it easier, the following command is an HA OnOff Toggle command, that is, a switch is toggling a Home Automation Light:

```
0000:    61 88 2c aa 1a 00 00 6f    a.,*...o
0008:    79 48 02 00 00 6f 79 0a    yH...oy.
0010:    3c 28 03 00 00 00 00 3c    <(.....<
0018:    03 98 07 c2 50 00 00 8f    ...BP...
0020:    8b 0d f3 67 15 08 5a 11    ..sg..Z.
0028:    da 03 83 09 9c ae .. ..    Z.......
```

The example in this section, *Example 4-11 ZigBee Security*, uses the standard HA on/off light and switch. Notice that there is no special source code for this example on the Web site or in this book, just the BeeKit solution file and the capture file. No special source code is required to demonstrate security, because ZigBee secures packets automatically.

To follow the example with hardware using the Freescale NSK Kit, compile and download the following projects into the respective boards:

- Chapter04\Example4-11 ZigBee Security\NcbZcHaOnOffLight.mcp

- Chapter04\Example4-11 ZigBee Security\SrbZedHaOnOffSwitch.mcp

The example is on channel 25, PAN ID 0x0f00. Follow these steps (this example uses the standard Freescale user interface):

1. After downloading the images, boot both boards

2. Press SW1 on both nodes to form/join the network

3. Once joined, press SW3 on both nodes to bind them

4. Press LSW1 on both nodes to go to application mode

5. Press SW1 on the switch (SRB board) to toggle the remote light, securely

When creating new projects in BeeKit, be sure to enable security in the BeeKit Project Wizard as seen in Figure 4.29.

Due to code-size restrictions on the HCS08, Freescale disables security by default, so it can include more of the other features of ZigBee. If you enable security, especially in the Home Automation applications, you will probably need to disable some other features. Look at the ZDP and HA over-the-air commands for likely candidates. Many of these may not be required by your application. Or you can use the generic application as a starting point.

BeeKit allows you to export the entire project, or just properties. Unlike setting the channel list or PAN ID, which only requires exporting properties (a relatively fast operation), changing the security setting requires exporting the entire project again. In the case of this example, security was enabled from the start.

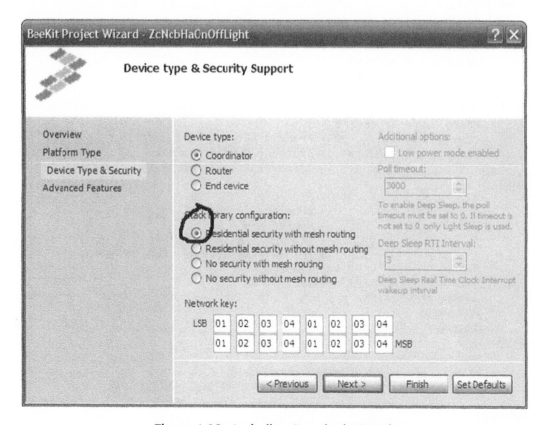

Figure 4.29: Including Security in BeeKit

Some public profiles, like Home Automation, transmit the key in the clear to joining nodes to allow any node to join any network, simply and easily. Another mechanism, called Join Enable, is used to prevent unauthorized nodes from joining the network.

If using a more restrictive public profile, such as Commercial Building Automation or a private profile, ZigBee allows the use of a pre-configured key. Pre-configured keys are never sent over the air. The node must already "know" the key, usually through some configuration tool, or as pre-installed at the factory.

If the protocol analyzer (such as Daintree SNA) doesn't know the key, the decode will look similar to the following code:

```
Frame 40 (Length=48 bytes)
        Frame Length: 48 bytes
        Link Quality Indication: 145
IEEE 802.15.4
        Frame Control: 0x8861
        Sequence Number: 44
        Destination PAN Identifier: 0x1aaa
        Destination Address: 0x0000
        Source Address: 0x796f
        Frame Check Sequence: Correct
ZigBee NWK
        Frame Control: 0x0248
            .... .... .... ..00 = Frame Type: NWK Data (0x00)
            .... .... ..00 10.. = Protocol Version (0x02)
            .... .... 01.. .... = Discover Route: Enable route discovery
(0x01)
            .... ...0 .... .... = Multicast
            .... ..1. .... .... = Security: Enabled
            .... .0.. .... .... = Source Route
            .... 0... .... .... = Destination IEEE Address: Not Included
            ...0 .... .... .... = Source IEEE Address: Not Included
            000. .... .... .... = Reserved
        Destination Address: 0x0000
        Source Address: 0x796f
        Radius=10
        Sequence Number=60
ZigBee AUX
        Security Control: 0x28
            .... .101 = Security Level: 5
            ...0 1... = Key Identifier: Network (0x01)
            ..1. .... = Extended Nonce: Sender Address Field: Present
(0x01)
            00.. .... = Reserved: (0x00)
```

```
        Frame Counter: 0x03
        Source Address: 0x0050c20798033c00
        Key Sequence Number: 0x00
        MIC: ae:9c:09:83
NWK Payload Decryption Failed: 8f:8b:0d:f3:67:15:08:5a:11:da:03
```

Did you notice that decryption failed? MAC, NWK, and AUX headers are not encrypted. Only the payload of the NWK frame is encrypted (the APS and ZCL frames). However, the entire frame is authenticated. Not a single bit can change without re-authenticating using the proper 128-bit key. Note that the security bit is enabled in the NWK Frame Control field.

Once the sniffer knows the key, the packet can be properly decoded. Take a look at the same packet when the sniffer knows the key:

```
Frame 40 (Length=48 bytes)
   Frame Length: 48 bytes
   Link Quality Indication: 145
IEEE 802.15.4
   Frame Control: 0x8861
   Sequence Number: 44
   Destination PAN Identifier: 0x1aaa
   Destination Address: 0x0000
   Source Address: 0x796f
   Frame Check Sequence: Correct
ZigBee NWK
        Frame Control: 0x0248
            .... .... .... ..00 = Frame Type: NWK Data (0x00)
            .... .... ..00 10.. = Protocol Version (0x02)
            .... .... 01.. .... = Discover Route: Enable route discovery
(0x01)
            .... ...0 .... .... = Multicast
            .... ..1. .... .... = Security: Enabled
            .... .0.. .... .... = Source Route
            .... 0... .... .... = Destination IEEE Address: Not Included
            ...0 .... .... .... = Source IEEE Address: Not Included
            000. .... .... .... = Reserved
        Destination Address: 0x0000
        Source Address: 0x796f
        Radius=10
        Sequence Number=60
ZigBee AUX
   Security Control: 0x28
      .... .101 = Security Level: 5
      ...0 1... = Key Identifier: Network (0x01)
      ..1. .... = Extended Nonce: Sender Address Field: Present (0x01)
      00.. .... = Reserved: (0x00)
```

```
        Frame Counter: 0x03
        Source Address: 0x0050c20798033c00
        Key Sequence Number: 0x00
        MIC: ae:9c:09:83
ZigBee APS
        Frame Control: 0x00
        Destination Endpoint: 0x08
        Cluster Identifier: On/off (0x0006)
        Profile Identifier: HA (0x0104)
        Source Endpoint: 0x08
        Counter: 0x22
ZigBee ZCL
        Frame Control: 0x01
                .... ..01 = Frame Type: Command is specific to a cluster
(0x01)
                .... .0.. = Manufacturer Specific=false (0x00)
                .... 0... = Direction: From the client server (0x00)
                0000 .... = Reserved: Reserved (0x00)
        Transaction Sequence Number: 0x42
        Command Identifier: Toggle (0x02)
```

Now the frame looks like the standard toggle command on the HA OnOff Cluster in the decoded APS and ZCL frames above.

To summarize, ZigBee security both encrypts, which prevents rogue nodes from listening to sensitive data, and authenticates, to prevent rogue nodes from injecting false data or commands into the network. If a node is allowed to join the network, it is considered "trusted." However, applications that share a network infrastructure that contain data that should not be seen by other nodes in the network can further encrypt using a link key, or with application-level security.

> ZigBee both authenticates and encrypts packets using the AES 128-bit standard.
>
> ZigBee supports security automatically. No special coding necessary.

CHAPTER 5

ZigBee, ZDO, and ZDP

It's all well and good to know how to transmit data to another node through an APSDE-DATA.request, and what endpoints and groups are all about, but how does a node in a ZigBee network decide which other node(s) in the network to talk to? How is the network set up and maintained?

ZigBee contains two sets of services for network commissioning and maintenance:

- The ZigBee Device Object (together with the ZigBee Device Profile)
- The ZigBee Cluster Library

This chapter describes the ZigBee Device Object (ZDO) and the ZigBee Device Profile (ZDP). The next chapter (Chapter 6) describes the ZigBee Cluster Library (ZCL).

But first, before delving into ZDO, the *real* story behind the ZigBee name.

"Hey, Big Z! Come look at this!" Ford Prefect shouted, staring down at his computer console.

Zaphod Beeblebrox swiveled one of his two heads toward Ford, saying, "Is it about me?"

"Nah. More interesting than that. Take a look at the new Heart of Gold Mark II! Remember the last one with that annoying personality that was always asking you to say 'please' before it would open a door, or giving you extra tidbits of information you didn't ask for every time you queried the computer? Well, in the Mark II they got rid of it. They replaced it with some new wireless technology that automatically handles, well—everything! It opens doors automatically, it makes the lights follow you around the ship, and quiets the music down when you start talking, it says here, almost like it reads your mind."

"Yeah, baby, but I'm of two minds, and I can't seem to get them to agree. For example, my second head is sleeping right now, you see." In fact, Zaphod's other head was snoring, loudly.

"Well, Big Z, I'm going to steal it," said Ford, matter-of-factly.

"What, my head?" asked Zaphod.

"The HoG Mark II."

"Not cool," quipped Zaphod. "Already been done. I stole the first Heart of Gold, remember? Anyway, how would you do it?"

"Toss me another Pan Galactic Gargle Blaster, while you toss yours down. We're hitching a ride." Ford fingered his electronic thumb.

"You can't hitch a ride on the most expensive ship in the Galaxy with just an electronic thumb. It will never work. Impossible!"

"That's exactly why it is going to work," said Ford calmly. "It's just so amazingly improbable that it's nearly impossible. Probability drive. Remember?"

It's a well-known fact to anyone who has ever read The Hitchhiker's Guide to the Galaxy that the only way to handle hitching a ride on a passing space ship and still keep your mind was to be out-of-your-mind drunk when it happened. This fact was almost as well-known as the use of the electronic thumb, the interstellar equivalent of extending your thumb on the side of the road on 1960s Earth (a planet somewhere in the unpopular arm of a small spiral galaxy). It was significantly less well-known that the Heart Of Gold, and subsequently, the Heart Of Gold Mark II, achieved interstellar travel through the use of a probability drive, a drive which ignored very likely and very constant things such as the speed of light, and instead landed you, quite improbably, exactly where you didn't even know you wanted to go, and did it in almost no time at all.

After a few more Pan Galactic Gargle Blasters (Ford stopped counting after three), Ford Prefect said "Big Z, are you ready?" It actually sounded more like, "BigZeeuready," what with the slurring and all.

"And why are we stealing it, exactly?" asked Zaphod, talking mostly to the floor, which wasn't talking back. Ford answered instead.

"We're stealing it to get Trillian back."

"Ah. Trillian?" queried Zaphod.

"Yes, Big Z, Trillian. Remember her? You picked her up from Earth many years ago. She was with us on the last Heart of Gold."

"Ah. And what's the name of that there ol' thingy in the Heart of Gold Mark II that automatically handles, doors an' lights and well, everything?" asked Zaphod.

Ford, who was now also staring mostly at the floor, slurred "Hmm, Zig B? What?"

"Ah. ZigBee. Strange name for a technology, ZigBee."

At that moment, the electronic thumb started beeping and blinking madly. For some reason, engineers love to make gadgets beep and blink. In addition to beeping and blinking, the electronic thumb did what it was actually designed to do and winked them out of existence, to reappear right in the cargo hold somewhere inside the Heart of Gold Mark II.

The ZigBee Device Object (ZDO, shown in Figure 5.1) is simply the application running on endpoint 0 in every ZigBee device. (Remember, application endpoints are numbered 1 through 240.)

This application, ZDO, keeps track of the state of the ZigBee device on and off the network, and provides an interface to the ZigBee Device Profile (ZDP), a specialized Application Profile (with profile ID 0x0000) for discovering, configuring, and maintaining ZigBee devices and services on the network.

As you can see from the figure, ZDO not only interacts with APS, but also interacts directly with the network layer. ZDO controls the network layer, telling it when to form or join a network, and when to leave, and provides the application interface to network layer management services. For example, ZDO can be configured to continue attempting to join a network until it is successful, or until a user-specified number-of-retries has occurred before giving up, and informing the application of the join failure.

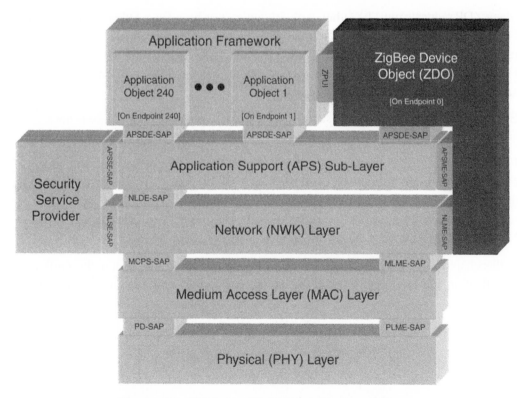

Figure 5.1: ZDO Is a ZigBee Application Object

The over-the-air Application Profile supported by ZDO, called the ZigBee Device Profile (ZDP), is no different than any other, and in most stacks is handled just like any other application object on an endpoint. ZDP services are separated into client and server. Client side services (also called requests), are always optional in ZigBee, but many of the server side ZDP services (also called responses), are mandatory.

Nearly every service follows the same pattern when used. A client device (the node which is doing the asking) first makes a request. The server device then sends the response back to the client device. The cluster number for the response is exactly the same as the cluster number for the request, but with the high bit set. For example, the ZDP command **IEEE_addr_req** is cluster 0x0001, and **IEEE_addr_rsp** is cluster 0x8001.

It doesn't matter how many hops the nodes are from each other. The nodes A and B could be 10 hops away from each other, and the ZDP request/response mechanism will work in exactly the same way, just as it does for applications sending data on an application endpoint (see Figure 5.2).

Many ZDP requests must be either explicitly unicast or broadcast. Others can unicast or broadcast at the client node's discretion (typically with different responses). If a ZDP request is broadcast, only the node that has the requested information returns any data. For example, **NWK_Addr_req** is broadcast, but only the node that matches the IEEE address, provided in the request, responds.

Every ZDP response starts with a status byte. If the particular optional service is not supported by the receiving node, the status returned will be **gZdoNotSupported_c** (0x84).

For sleeping devices, the parents of the device keep track of the IEEE and short address of the child, and will respond for them. However, all other information about the sleeping device, such as the list of active endpoints, are not recorded by the parent and must be retrieved directly from the devices themselves. In Chapter 8, "Commissioning ZigBee Networks," I'll discuss the means of commissioning sleeping devices.

In this chapter, I've organized the ZDP services slightly differently than in the ZigBee specification. For one thing, I've put the request and responses in the same section. The

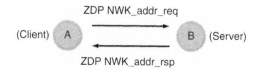

Figure 5.2: ZDP Request and Response

ZigBee specification organizes the services numerically, so the request and responses are many pages apart. Also, I've organized the ZDP services by usage; so, for example, all the node-wide services are together.

ZDP services include the following categories:

- Device discovery services

- Service discovery services

- Binding services

- Management services

After discussing ZDP, I'll discuss how applications interact with ZDO, including:

- Starting and stopping the network through ZDO

- ZDO and low power nodes

5.1 Device Discovery

The ZigBee Device Profile (ZDP) contains a set of commands for discovering various aspects about nodes in the network. The ZigBee specification calls these "device discovery services," which can be confusing because endpoints contain device IDs which really describe individual ZigBee applications running in that node. So, when you see ZDP Device Discovery, think node-wide (not application/endpoint specific) services.

Device discovery services have a few things in common:

- They provide additional information about a node.

- They are all optional from the client side, but some server side processing is mandatory (a common subset among all ZigBee devices).

- They are node-wide, and do not represent any particular application, or Application Profile residing on an endpoint in the node.

The ZDP device discovery services are listed below in Table 5.1. Notice that all the ZDP services on the client side are optional. ZigBee does not require that a node be able to send **NWK_addr_req,** for example. But on the server side of this equation (a node receiving a **NWK_addr_req** and responding to it), the ZDP service is mandatory.

Table 5.1: ZigBee Device Profile Device Discovery Services

Device Discovery Services	Unicast (U), Broadcast (B) or Either (U,B)	Client Transmission (Request)	Server Processing (Response)
NWK_addr_req	U,B	O	M
IEEE_addr_req	U	O	M
Node_Desc_req	U	O	M
Power_Desc_req	U	O	M
Complex_Desc_req	U	O	O
User_Desc_req	U	O	O
User_Desc_set	U	O	O
Device_annce	B	O	M

This makes sense if you think about how the service is used. A tool may want to collect the IEEE (aka MAC) address of every node in the network (using **IEEE_addr_req**, for example) so all nodes in the network must support the server side (**IEEE_addr_rsp**). But only the tool needs to support the client side.

What happens if a given client issues two ZDP requests in a row? How does the client application know which response belongs to which request? Some stack vendors have solved this problem by only allowing a single request to be issued at any one time. Other stack vendors, such as Freescale, provide a *transaction ID* which correlates the request with the response. This rolling 8-bit transaction ID is sent with each request, meaning, in theory, that a single application could have up to 256 requests in flight at once. Normally, however, an application makes one or two requests, and then waits for the response.

In the Freescale ZigBee solution, all ZDP requests begin with the prefix **ASL_** (for example, **ASL_NWK_addr_req**()). Simply look up the particular ZDP request in the table, or the ZigBee specification, and prefix it with **ASL_**. Why ASL, and not ZDP? ASL stands for Application Support Library, which is the prefix used for all optional application-level commands in Freescale BeeStack.

The response to a ZDP request may take some time to come back, because, perhaps, the responding node may be many hops away. In a BeeStack application, this occurs through a C callback function registered with **Zdp_AppRegisterCallBack**().

Each ZDP request in BeeStack requires a destination address, which may be unicast or broadcast, as the ZigBee specification allows.

```
void ASL_NWK_addr_req
(
  zbCounter_t *pSequenceNumber,
  zbNwkAddr_t aDestAddress,
  zbIeeeAddr_t aIeeeAddr,
  uint8_t requestType,
  index_t startIndex
);
```

One thing that is not always obvious with Freescale BeeStack (and this is true of other stack vendors as well) is that optional ZDP services are not enabled by default. In fact, they are compiled-out by default. Often ZigBee stacks run in systems that are very limited by RAM and Flash (ROM), which means every byte can be precious. Services that might not be used by the application are turned off to conserve space.

To enable the optional ZDP services, enable either the client-side service, server-side service, or both. For example, to enable both the server and client for **NWK_addr_req**, enable both **gNWK_addr_req_d** and **gNWK_addr_rsp_d** in BeeStack. All the ZDP services, even the mandatory ones, can be enabled or disabled through Freescale BeeKit, the graphical BeeStack configuration tool.

Although Freescale BeeKit allows it, I don't recommend disabling the mandatory ZDP services unless your company controls all the nodes in the ZigBee network, and you are willing to live with a (slightly) incompatible ZigBee stack. Certainly, the product cannot be certified by ZigBee if the mandatory ZDP services are disabled.

For some application profiles, such as Home Automation, some of the ZDP services listed as optional by the ZigBee specification are mandatory for certain devices in that profile. ZDP binding is a good example of this.

> Use ZDP to discover which nodes to talk to in a ZigBee network.
>
> Optional ZDP services may be mandatory in the application profile.
>
> Remember to enable the optional services if they are needed by a BeeStack application.

5.1.1   NWK_addr_req and IEEE_addr_req

Use ZDP network address request (**NWK_addr_req**) when you already know the MAC address of a node (also called its IEEE or long address), but want to find its short, 16-bit address on the network. This service request can be broadcast or unicast.

Table 5.2: NWK_addr_req/rsp

NWK_addr_req	NWK_addr_rsp

typedef struct zbNwkAddrRequest_tag
 	
{
   zbIeeeAddr_t   aIeeeAddr;
   uint8_t   requestType;
   zbIndex_t   startIndex;
} zbNwkAddrRequest_t;
	typedef struct zbExtendedDevResp_tag
	
	{
	   zbStatus_t iStatus;
	   zbIeeeAddr_t aIeeeAddrRemoteDev;
	   zbNwkAddr_t aNwkAddrRemoteDev;
	   zbCounter_t numAssocDev;
	   zbIndex_t startIndex;
	   zbNwkAddr_t
	aNwkAddrAssocDevList[1];
	} zbExtendedDevResp_t;

For example, say the gateway in a particular ZigBee network (which may or may not be on the ZigBee Coordinator) is known to be IEEE address **0x0050c237b0041234**. Issue a **NWK_addr_req** and the gateway will respond with its short address on the network.

Unfortunately, there is no ZigBee-standard way to find nodes within a range of IEEE addresses.

IEEE_addr_req is the converse of **NWK_addr_req** (see Tables 5.2 and 5.3). It returns the IEEE address of a node, given a 16-bit short address. This command is unicast to the destination. The responses are exactly the same for the two commands, and the requests are quite similar.

Table 5.3: IEEE_addr_req/rsp

IEEE_addr_req	IEEE_addr_rsp

typedef struct zbIeeeAddrRequest_tag
 	
{
   zbNwkAddr_t   aNwkAddrOfInterest;
   uint8_t   requestType;
   zbIndex_t   startIndex;
} zbIeeeAddrRequest_t;
	typedef struct zbExtendedDevResp_tag
	
	{
	   zbStatus_t iStatus;
	   zbIeeeAddr_t aIeeeAddrRemoteDev;
	   zbNwkAddr_t aNwkAddrRemoteDev;
	   zbCounter_t numAssocDev;
	   zbIndex_t startIndex;
	   zbNwkAddr_t aNwkAddrAssocDevList[1];
	} zbExtendedDevResp_t;

Notice that the first byte of the response is a status byte. This will be 0x00 (success) if the response is valid. If this contains an error code, then the rest of the information will not be included in the response. Every ZDP response begins with a status code, so be sure to check it in your applications before assuming that the rest of the information is valid.

Both **NWK_addr_req** and **IEEE_addr_req** contain a **requestType** field. The request type field affects whether the extended information is included in the response. Use **requestType** 0x00 to get only the IEEE and NWK address for one node. Use the extended **requestType** 0x01 to get the information for the node and for all its children as well. Remember, only routers will have children.

This particular request is generally broadcast across the network. If the request is broadcast, and the targeted NWK or IEEE address does not exist on the network, then no over-the-air response is issued. The client application should set up a time-out to let itself know that the node couldn't be found, perhaps to try again at another time.

Why unicast this command? It's a good way to see if a particular device is the child of a given parent. For example, say that you want to ensure that node XYZ is a child of the room controller in a hotel room. Issue a unicast to that room controller (a ZigBee Router) and it will respond either with an error code, or with the short address of the child.

A start index is normally used if the response can't fit in a single over-the-air packet (a payload of about 80 bytes). This field isn't actually needed in **NWK_addr_req** or **IEEE_addr_req** because the response will *always* fit, so always set it to 0.

The example in this section, *Example 5-1—ZDP NWK_addr_req*, uses **NWK_addr_req** to find the short address of a particular node, in this case the node with the IEEE address of 0x0050c237b0040002 (see Figure 5.3). The application on the ZC sends a broadcast across the network, and the node with proper IEEE address responds with its short address.

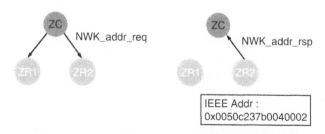

Figure 5.3: Example 5-1—ZDP NWK_addr_req

The BeeKit solution file for this example, found in the directory "Chapter05\Example 5-1-ZDP NWK_addr_req," contains three projects: one for an NCB ZigBee Coordinator (ZC)—the node making the request, and two for ZigBee Routers (ZR)—one of which is the node we're looking for.

To run the example, program the three boards (ZcNcb, Zr1Srb, Zr2Srb, respectively). Next, form the network with the ZcNcb (ZigBee Coordinator) board by pressing SW1. Join the other two nodes in any order, pressing SW1 on each of them. When the LEDs have finished chasing each other, the nodes are on the network. Then, press SW2 on the NCB board. The NCB will send out the **NWK_addr_req** and should display the short address of the node we're looking for. In the figure, this would be **0x143e**.

Try booting all the nodes, joining the routers in the opposite order (so that ZR2 boots first). Notice the NwkAddr returned is now **0x0001**.

> Use NWK_addr_req and IEEE_addr_req to find nodes based on short or long address.
>
> NWK_addr_rsp and IEEE_addr_rsp populate the address map.

5.1.2   ZigBee Descriptors

ZigBee uses *descriptors* to describe a node and its properties, allowing other applications running in the network to discover these properties over-the-air. Node-wide descriptors include the node descriptor, the power descriptor, the complex descriptor, and the user descriptor.

Of these descriptors, I find the **Node_Desc_req** the most useful (see Table 5.4). The results of this include the ZigBee node type (ZR, ZC, or ZED) and the manufacturer ID (a 16-bit ZigBee assigned number that uniquely identifies the manufacturer of the device).

The node descriptor contains a variety of fields, including the node type of the device (whether the node is a ZigBee Coordinator, Router, or End-Device), the manufacturer's code, whether the optional user and complex descriptors are present, and whether the node supports fragmentation.

Use this command when the application needs to know the manufacturer ID, whether the destination node can support the optional fragmentation, or if any other optional service

Table 5.4: Node Descriptor Request and Response

Node_Desc_req	Node_Desc_rsp

typedef struct
zbNodeDescriptorRequest_tag
{
   zbNwkAddr_t   aNwkAddrOfInterest;
} zbNodeDescriptorRequest_t;
	typedef struct
	zbNodeDescriptorResponse_tag
	{
	   zbStatus_t status;
	   zbNwkAddr_t
	aNwkAddrOfInterest;
	   zbNodeDescriptor_t nodeDescriptor;
	} zbNodeDescriptorResponse_t;
	
	   typedef struct zbNodeDescriptor_tag
	{
	
	   uint8_t logicalType;
	   uint8_t apsFlagsAndFreqBand;
	   uint8_t macCapFlags;
	   uint8_t aManfCodeFlags[2];
	   uint8_t maxBufferSize;
	   uint8_t aMaxTransferSize[2];
	   zbServerMask_t aServerMask;
	} zbNodeDescriptor_t;

is present. I rarely use this command in actual applications, except perhaps to find the manufacturer ID. That can be useful if a particular application wants to use extended commands only available from a particular manufacturer.

The other descriptors include the power descriptor, which defines which power modes this node supports, and the user descriptor, which contains a user definable string to identify the location (such as living room or office). These descriptors are all optional in the ZigBee spec. The user descriptor is settable over-the-air, the rest are only gettable.

Tables 5.5, 5.6, and 5.7 describe each of the other descriptors. By and large, these descriptors (with the exception of the Node descriptor) have been supplanted by the ZigBee Cluster Library (ZCL) Basic Cluster. If you are using a profile such as Home Automation (HA), or Automatic Metering (AMI), which use the ZigBee Cluster Library, use the Basic Cluster mechanism instead.

Table 5.5: Power Descriptor Request and Response

Power_Desc_req	Power_Desc_rsp

void ASL_Power_Desc_req
 	
(
  zbCounter_t *pSequenceNumber,
  zbNwkAddr_t aDestAddress
);
	typedef struct zbPowerDescriptor_tag
	
	{
	    uint8_t currModeAndAvailSources;
	    uint8_t currPowerSourceAndLevel;
	} zbPowerDescriptor_t;

Table 5.6: User Descriptor Request and Response

User_Desc_req	User_Desc_rsp

void ASL_User_Desc_req
 	
(
  zbCounter_t *pSequenceNumber,
  zbNwkAddr_t aDestAddress
);
	typedef struct
	
	zbUserDescriptorResponse_tag
	{
	    zbStatus_t status;
	    zbNwkAddr_t aNwkAddrOfInterest;
	    uint8_t aUserDescriptor[16];
	} zbUserDescriptorResponse_t;

Table 5.7: Complex Descriptor Request and Response

Complex_Desc_req	Complex_Desc_rsp

void ASL_User_Desc_req
 	
(
  zbCounter_t *pSequenceNumber,
  zbNwkAddr_t aDestAddress
);
	typedef struct
	
	zbComplexDescriptor_tag
	{
	    uint8_t fieldCount;
	    uint8_t aLanguageAndCharSet[4];
	    uint8_t aManufacturerName[6];
	    uint8_t aModelName[6];
	    uint8_t aSerialNumber[6];
	    uint8_t aDeviceUrl[17];
	    uint8_t aIcon[4];
	    uint8_t aIconUrl[9];
	} zbComplexDescriptor_t;

Descriptors describe the node.

Use the ZigBee Cluster Library (ZCL) basic cluster rather than the power, complex and user descriptors.

Table 5.8: Device Announce Fields

Device_annce

typedef struct zbEndDeviceAnnounce_tag
 	
{
   zbNwkAddr_t   aNwkAddress;
   zbIeeeAddr_t   aIeeeAddress;
   macCapabilityInfo_t capability;
} zbEndDeviceAnnounce_t;

5.1.3   Device Announce

The ZDP **Device_annce** command is issued by the ZigBee stack, not by the applications. Occasionally in a network, a device must change its short address while still on the network. In Stack profile 0x01, this occurs when an end-device loses track of its parent and needs to find a new one. In Stack profile 0x02, this occurs when an address conflict is detected.

Device_annce can also occur if an end-device wants to tell its parent to start buffer packets for it while it sleeps (called an RxOnIdle = FALSE device), or wants its parent to quit buffering packets because the device won't be sleeping anymore. (Perhaps it was plugged into mains power.)

All the **Device_annce** command accomplishes is to instruct any node in the network that cares about this node to update its internal tables, such as the neighbor table, address map, and binding table (see Table 5.8). The over-the-air device announce structure is fairly simple: a short address, IEEE address, and MAC capabilities flags.

The example in this section, *Example 5-2 Device_annce*, demonstrates device announce occurring when a child changes to a new parent (see Figure 5.4). A node is set up to look

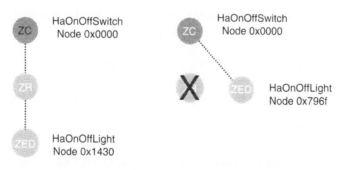

Figure 5.4: Example 5-2—Device_annce

for a new parent after it has lost contact with its original parent for three polling periods in a row. In the left portion of the figure, the ZED has a ZigBee router for a parent. But for some reason (Okay, because I turned it off), the ZED loses contact with its parent. The node then looks for a new parent, and finds the ZigBee Coordinator, shown on the right.

This same thing would happen if the ZED were, perhaps, a roaming remote control device. The same thing could occur if something happened to the link, such as if a large wall of metal or body of water was placed between the ZED and its parent.

To run the example shown above, use the BeeKit solution found in the folder "Chapter05\Example 5-2 Device_annce." This BeeKit solution contains three projects: a ZcNcbSwitch, a ZrSrbRangeExtender, and a ZedSrbLight. Export the solution, and import, compile, and download each project into their respective boards.

The steps to see the demo (and produce a capture) are:

1. Turn on Daintree to record on channel 25.

2. Boot and form a network with the ZcNcbSwitch and ZrSrbRangeExtender boards, by pressing SW1.

3. Turn off joining the ZcNcbSwitch by pressing SW2.

4. Join the network with ZedSrbLight by pressing SW1.

5. Bind the switch and light, by pressing SW3 (in any order) on both ZcNcbSwitch and ZedSrbLight.

6. Go to Application (as opposed to Configuration) Mode on both light and switch, by pressing and holding switch 1 (LSW1).

7. Toggle the light, by pressing SW1 on the ZcNcbSwitch.

8. Force the light to move to a new parent, by turning off ZcSrbRangeExtender.

9. Toggle the light again, by pressing SW1.

Notice the ZcNcbSwitch knows where to find the light (at 0x796f), even though it has moved (from 0x1430).

To see this action over-the-air, take a look at the following excerpts from the Daintree capture. First of all, notice that the switch (node 0x0000) is sending to the light (node 0x1430), which is a child of the range extender (node 0x0001):

```
87   +00:00:00.571  0x0000  0x0001  0x0000  0x1430  0x50  Zigbee APS Data
HA:On/off
88   +00:00:00.001                                         IEEE 802.15.4
Acknowledgment
89   +00:00:00.452  0x1430  0x0001                         IEEE 802.15.4
Command: Data Request
90   +00:00:00.001                                         IEEE 802.15.4
Acknowledgment
91   +00:00:00.003  0x0001  0x1430  0x0000  0x1430  0x75  Zigbee APS Data
HA:On/off
```

Now, the child has lost track of its parent. So, it issues a rejoin request to join a new parent. And then it announces via a broadcast its new short address to the network with **Device_annce**, called ZDP:EndDeviceAnnce:

```
138  +00:00:00.421  0x1430  0x0000  0x1430  0x0000  0x1a  Zigbee NWK
NWK Command: Rejoin Request
139  +00:00:00.001                                         IEEE 802.15.4
Acknowledgment
140  +00:00:00.409  0x1430  0x0000                         IEEE 802.15.4
Command: Data Request
141  +00:00:00.001                                         IEEE 802.15.4
Acknowledgment
142  +00:00:00.005  0x0000  0x1430  0x0000  0x1430  0x51  Zigbee NWK
NWK Command: Rejoin Response
143  +00:00:00.002                                         IEEE 802.15.4
Acknowledgment
144  +00:00:00.005  0x796f  0x0000  0x796f  0xffff  0x1b  Zigbee APS Data
ZDP:EndDeviceAnnce
145  +00:00:00.002                                         IEEE 802.15.4
Acknowledgment
146  +00:00:00.018  0x0000  0xffff  0x796f  0xffff  0x1b  Zigbee APS Data
ZDP:EndDeviceAnnce
```

Finally, notice that the ZcNcbSwitch still knows where to find the light. Instead of sending to address 0x1430, it sends to address 0x796f, the light's new short address:

```
164  +00:00:00.002  0x0000  0x796f  0x0000  0x796f  0x52  Zigbee APS Data
HA:On/off
```

One thing to be aware of: **Device_annce** is a broadcast, and every ZigBee network is limited by the number of broadcasts it can sustain at any given time. Don't design a network where children need to move constantly or the network may be overloaded.

5.2   Service Discovery

In addition to the services related to devices, or nodes, ZDP also contains a variety of standard services for querying the applications within those nodes (see Table 5.9). As with the device discovery services, most of the ZDP service discovery services are optional. Only a few service side responses are required.

5.2.1   Discovering and Matching Endpoints

Discovering application endpoints and the services they support is a common commissioning step in ZigBee. Different manufacturers may choose different endpoints for their applications. For example, a manufacturer of a switch (Leviton, perhaps) may choose endpoint 3 for their switch. Philips may choose endpoint 8 for their light. So how does an application which needs to bind this switch to the light find these endpoints?

Table 5.9: ZDP Service Discovery Services

Service Discovery Services	Client Transmission (Request)	Server Processing (Response)
Simple_Desc_req (unicast)	O	M
Extended_Simple_Desc_req (unicast)	O	O
Active_EP_req (unicast)	O	M
Extended_Active_EP_req (unicast)	O	O
Match_Desc_req (broadcast)	O	M
System_Server_Discover_req	O	O
Find_node_cache_req (broadcast)	O	O
Discovery_Cache_req (unicast)	O	O
Discovery_store_req (unicast)	O	O
Node_Desc_store_req (unicast)	O	O
Power_Desc_store_req (unicast)	O	O
Active_EP_store_req (unicast)	O	O
Simple_Desc_store_req (unicast)	O	O
Remove_node_cache_req (unicast)	O	O

Table 5.10: Active Endpoint Request and Response

Active_EP_req	Active_EP_rsp

void ASL_Active_EP_req
 	
(
 	
   zbCounter_t *pSequenceNumber,
   zbNwkAddr_t aDestAddress
);
	typedef struct zbActiveEpResponse_tag
	
	{
	
	   zbStatus_t status;
	   zbNwkAddr_t aNwkAddrOfInterest;
	   zbCounter_t activeEpCount;
	   zbEndPoint_t pActiveEpList[1];
	} zbActiveEpResponse_t;

ZDP can locate active endpoints through **Active_EP_req** (see Table 5.10). This call returns a list of the active endpoints in a node. The application then calls **Simple_Desc_req**, which returns a description of the endpoint (see Table 5.11). The simple descriptor really should have been called the endpoint descriptor, as that is the object it describes.

Table 5.11: Simple Descriptor Request and Response

Simple_Desc_req	Simple_Desc_rsp

void ASL_Simple_Desc_req
 	
(
 	
   zbCounter_t *pSequenceNumber,
   zbNwkAddr_t aDestAddress,
   zbEndPoint_t endPoint
);
	typedef struct zbSimpleDescriptor_tag
	
	{
	
	   zbEndPoint_t endPoint;
	   zbProfileId_t aAppProfId;
	   zbDeviceId_t aAppDeviceId;
	   uint8_t appDevVerAndFlag;
	   zbCounter_t cNumInClusters;
	   zbClusterId_t *pInClusterList;
	   zbCounter_t cNumOutClusters;
	   zbClusterId_t *pOutClusterList;
	} zbSimpleDescriptor_t;
	
	typedef struct
	zbSimpleDescriptorResponse_tag
	
	{
	
	   zbStatus_t status;
	   zbNwkAddr_t aNwkAddrOfInterest;
	   zbSize_t length;
	   zbSimpleDescriptor_t sSimpleDescriptor;
	} zbSimpleDescriptorResponse_t;

The simple descriptor basically describes everything there is to know about the endpoint: its Application Profile ID, and its list of endpoints, both input and output.

The **Extended_Simple_Desc_req** and **Extended_Active_EP_req** were added in ZigBee 2007 in case the simple descriptor or active endpoint list were too large to fit into a single packet. For example, assume that a node supports all 240 endpoints. Each active endpoint returned in an **Active_EP_req** requires one byte. That's at least 240 bytes, far too large to fit into the 127 byte 802.1.54 PHY. Likewise, if the cluster list is too long, an **Extended_Simple_Desc_req** might be needed. Normally, however, the standard versions are sufficient. It's a rare ZigBee network that contains nodes with that many endpoints or clusters on an endpoint.

Match_Desc_req can be used to find a particular service anywhere across the network (see Table 5.12). As input, it takes a simple descriptor, and as output it provides a matching list of endpoints from any node that matches. It matches both profile ID and input/output cluster lists. The profile ID must be the same, and at least one input must match one output cluster, or vice versa. Any overlap will do. Think of it this way.

Table 5.12: Match Descriptor Request

Match_Desc_req	Match_Desc_rsp

typedef struct
zbSimpleDescriptor_tag
{
  zbEndPoint_t   endPoint;
  zbProfileId_t aAppProfId;
  zbDeviceId_t aAppDeviceId;
  uint8_t   appDevVerAndFlag;
  zbCounter_t   cNumInClusters;
  zbClusterId_t *pInClusterList;
  zbCounter_t   cNumOutClusters;
  zbClusterId_t *pOutClusterList;
} zbSimpleDescriptor_t;
void ASL_MatchDescriptor_req
(
  zbCounter_t *pSequenceNumber,
  zbNwkAddr_t aDestAddress,
  zbSimpleDescriptor_t
*pSimpleDescriptor
);
	typedef struct
	zbMatchDescriptorResponse_tag
	{
	  zbStatus_t status;
	  zbNwkAddr_t aNwkAddrOfInterest;
	  zbSize_t matchLength;
	  zbEndPoint_t matchList[1];
	} **zbMatchDescriptorResponse_t;**

A switch has an On/Off Cluster (0x0006) as an output. A light has an On/Off Cluster as an input. They match. Two lights would not match.

Match_Desc_req may be broadcast (with 0xfffd) or unicast. Here is a simple experiment to cause a flurry of route requests and unicasts in a ZigBee network. Send out a **Match_Desc_req** with the Basic Cluster (0x0000) listed as an output cluster, on profile ID 0x0104. Every node in the network on the Home Automation profile will respond.

5.2.2 Backing Up and Caching Discovery Information

ZigBee utilizes the concepts of backing up and also caching the discovery information. This includes the following ZDP commands:

- System_Server_Discover_req
- Find_node_cache_req
- Discovery_Cache_req
- Discovery_store_req
- Node_Desc_store_req
- Power_Desc_store_req
- Active_EP_store_req
- Simple_Desc_store_req
- Remove_node_cache_req

The concept is fairly simple. The **System_Server_Discovery_req** permits nodes in the network to find the primary cache for everything from endpoints, to simple descriptors, to node descriptors, assuming nodes in the network stored copies of their information on the cache. Then, a commissioning tool or other node can retrieve the information. The trouble with this is that no vendors actually implement the primary discovery cache in a network. In fact, at the time of this writing, Freescale is the only vendor that has actually implemented these optional ZDP commands in their stack.

My advice is not to use them. Get the information directly from the nodes themselves. Or do without.

If you would like to use these commands anyway, here's how to do it. Use a **Discovery_store_req** first to allocate the space on the discovery cache for the various

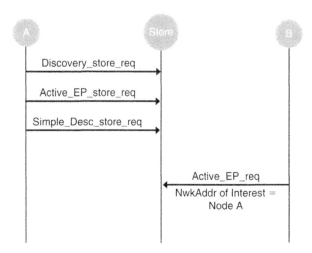

Figure 5.5: Discovery_store_req

items, including endpoints and simple descriptors, as seen in Figure 5.5. Then use the various store commands (e.g., **Simple_Desc_store_req**) to actually store the data on the cache. Using **System_Server_Discovery_req**, other nodes in the network can find the cache and request the sleeping node's information using commands such as **Active_EP_req**.

5.3   Binding

In Chapter 4, "ZigBee Applications," you learned all about APS (local) binding. I'll give a quick refresher here, and then talk about ZDP binding.

Binding provides a mechanism for attaching an endpoint on one node to one or more endpoints on another node. Binding can even be destined for groups of nodes. Then, when using APSDE-DATA.request, simply use the "indirect" addressing mode, and the request will be sent to each endpoint or group listed in the local binding table.

The binding table is smart, and keeps track of both the short (16-bit NwkAddr) and long (IEEE) address of a node. If a destination device has changed its short address (either due to a ZigBee End-Device moving from one parent to another in ZigBee stack profile 0x01, or due to a address conflict in ZigBee Pro), the binding table entry is updated automatically to point to that new address (see Figure 5.6).

As shown in Table 5.13, if the local application sent application data using indirect mode from endpoint 12, the packet would simply be dropped, as there is no source endpoint

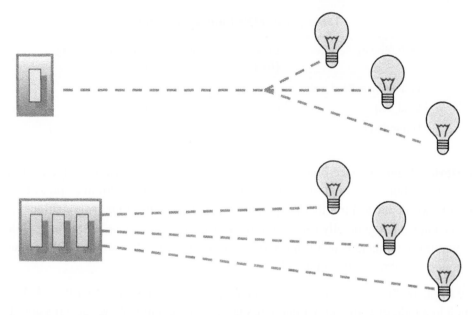

Figure 5.6: Binding Connects One Endpoint to One or More Other Endpoints

12 in the table. If the local application sent a APSDE-DATA.request using indirect mode from endpoint 5, it would go to three destinations: node 0x1234 endpoint 12, broadcast to group 0x9999, and to node 0x5678 endpoint 44.

ZDP provides over-the-air binding services to complement the local APS binding services. This allows a third-party tool (such as a remote control, or PC with a ZigBee dongle) to connect one application to another. It's easy to envision a drag-and-drop interface to bind switches to lights throughout a house, an office, or a hotel.

All ZDP binding services are optional. They are shown in Table 5.14.

Table 5.13: Sample Binding Table

Src EP	Destination Addr	Addr/Grp	Dst EP	Cluster ID
5	0x1234	A	12	0x0006
6	0x796F	A	240	0x0006
5	0x9999	G	--	0x0006
5	0x5678	A	44	0x0006

Table 5.14: ZDP Binding Services

ZDP Binding Services	Client Transmission (Req)	Server Processing (Rsp)
End_Device_Bind_req	O	O
Bind_req	O	O
Unbind_req	O	O

End_Device_Bind_req (see Figure 5.7) uses an optional state machine on the ZigBee Coordinator to bind or unbind two devices. This service can be useful in a "press-the-button-on-two-nodes-to-bind-them" operation, useful on some Home Automation products, but it's not generally useful in most ZigBee networks. One of the things I don't like about this command is that if it returns success, the caller has no idea if the targets were bound or unbound. It's a toggle!

The example in this section, *Section 5-3 Binding*, demonstrates a third-party node binding a switch to a light over-the-air. Granted, it's pretty simple, but it shows the concept of ZigBee commissioning with a third-party tool.

To run the example, simply compile and download the three targets (ZcNcbTool, ZedSrbSwitch, andZrSrbLight) from the BeeKit solution, and boot them all. Press SW1 on all of them to join each node to the network. Go to Application Mode on all three nodes by pressing and holding SW1 (long SW1). Press SW1 on the switch. Notice nothing happens. Then press SW2 in the tool to bind the switch to the light. Now press SW1 on the switch again and notice the light toggles.

There is one thing about over-the-air binding that is not obvious. The ZDP bind commands require an IEEE address, not a short address for the destination of the binding. If a node receives a ZDP bind command and it doesn't know about the destination address, it will issue a ZDP **NWK_Addr_req** to find the node, because it actually needs both long and short addresses to complete the operation.

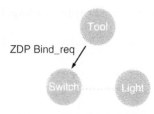

Figure 5.7: ZDP Bind Request

Table 5.15: ZDP Management Services

Network Management Services	Client Transmission (Req)	Server Processing (Rsp)
Mgmt_NWK_Disc_req (unicast)	O	O
Mgmt_Lqi_req (unicast)	O	O
Mgmt_Rtg_req (unicast)	O	O
Mgmt_Bind_req (unicast)	O	O
Mgmt_Leave_req (unicast)	O	O
Mgmt_Direct_Join_req (unicast)	O	O
Mgmt_Permit_Joining_req (unicast or broadcast)	O	M
Mgmt_Cache_req (unicast)	O	O
Mgmt_NWK_Update_req (unicast)	O	O

5.4 ZDP Management Services

The ZDP Management services are really handy optional services used for reading the various tables contained within ZigBee nodes, and to request certain common actions (see Table 5.15).

5.4.1 Network Discovery

The ZDP command **Mgmt_NWK_Disc_req** was implemented both to support frequency agility, which is the ability for the ZigBee network to change channels, and to help prevent PAN ID conflicts. A managing application can determine remotely what networks and nodes are in the vicinity of any node on the network.

PAN ID conflict happens when one network grows toward another. Perhaps they were both out of hearing range of each other when they started, and through chance happened to pick the same PAN ID, such as 0x1234. Now they've grown over time, and are beginning to overlap.

Changing channels in ZigBee is a fairly catastrophic event, and not one to be undertaken lightly. ZigBee is not a channel-hopping network, like Bluetooth™, for example. Instead, ZigBee relies on its robust CSMA-CA and O-QPSK technologies to continue to communicate even in noisy environments. But sometimes it's just necessary to change channels, and it would be a major hardship to tear down the network and rebuild it on another channel. This is the sort of thing that might happen at a hospital. The wireless networks in a hospital are very carefully managed, and they do not want other wireless channels on the same frequency as their WiFi™ networks. If the WiFi network needed to change channels for some reason, it's possible the ZigBee network might have to as well.

The ZDP **Mgmt_NWK_Disc_req** command does exactly the same thing that ZDO does locally, when it determines what networks are nearby. It sends out a beacon request and reports the beacons that responded to a higher layer, that can then do something intelligent. In this case, the "higher layer" just happens to be on a remote managing node.

5.4.2   Table Management Services

ZDP contains services to read the various tables from remote ZigBee nodes. This can be useful in diagnostics during commissioning, or even at run-time. For example, the routing tables of various routers can be checked, and if one node in particular is always full while the other routers are not, perhaps a choke-point has been detected in the network. Another router may be needed in the vicinity.

ZigBee Table Management Services in ZDP:

- Mgmt_Lqi_req—the neighbor table
- Mgmt_Rtg_req—the routing table
- Mgmt_Bind_req—the (optional) binding table

Notice that there isn't any way to set these tables directly over ZigBee. Of course, an application specific cluster could be written to do this, but the proper way is to use the various other commands available that populate these tables. The binding table, for instance, is populated or cleared using the ZDP-Bind and ZDP-Unbind commands.

These tables can be quite large. To accommodate this, ZigBee allows them to be read from a starting index. For example, to read the entire neighbor table, use **Mgmt_Lqi_req** with a starting index of 0 to begin. Then, after it returns five or so entries, send another **Mgmt_Lqi_req** with a starting index of 5. That does mean that the operation is not always atomic, and can look strange if something has changed between the previous request and the next one.

The ZDP table requests and responses are listed in Tables 5.16, 5.17, and 5.18.

Table 5.16: Management Neighbor Table Request

Mgmt_Lqi_req	Mgmt_Lqi_rsp
```c void ASL_Mgmt_Lqi_req (zbCounter_t *pSequenceNumber, zbNwkAddr_t aDestAddress, index_t index); ```	```c typedef struct zbNeighborTableList_tag { zbIeeeAddr_t aExtendedPanId; zbIeeeAddr_t aExtendedAddr; zbNwkAddr_t aNetworkAddr; uint8_t deviceProperty; bool_t permitJoining; uint8_t depth; uint8_t lqi; } zbNeighborTableList_t; typedef struct zbMgmtLqiResponse_tag { zbStatus_t status; zbCounter_t neighbourTableEntries; zbIndex_t startIndex; zbCounter_t neighbourTableListCount; zbNeighborTableList_t neighbourTableList[1]; } zbMgmtLqiResponse_t; ```

Table 5.17: Management Routing Table Request

Mgmt_Rtg_req	Mgmt_Rtg_rsp
```c void ASL_Mgmt_Rtg_req (zbCounter_t *pSequenceNumber, zbNwkAddr_t aDestAddress, index_t index); ```	```c typedef struct routingTableList_tag { zbNwkAddr_t aDestinationAddress; uint8_t status; zbNwkAddr_t aNextHopAddress; } routingTableList_t; typedef struct zbMgmtRtgResponse_tag { zbStatus_t status; zbCounter_t routingTableEntries; index_t startIndex; zbCounter_t routingTableListCount; routingTableList_t routingTableList[1]; } zbMgmtRtgResponse_t; ```

Table 5.18: Management Binding Table Request

Mgmt_Bind_req	Mgmt_Bind_rsp

void ASL_Mgmt_Bind_req
(
    zbCounter_t   *pSequenceNumber,
    zbNwkAddr_t   aDestAddress,
    index_t index
);
	typedef struct zbApsmeBindReq_tag {
	    zbIeeeAddr_t aSrcAddr;
	    zbEndPoint_t srcEndPoint;
	    zbClusterId_t aClusterId;
	    zbAddrMode_t dstAddrMode;
	    zbIeeeAddr_t aDstAddr;
	    zbEndPoint_t dstEndPoint;
	} zbApsmeBindReq_t;
	typedef struct zbMgmtBindResponse_tag {
	    zbStatus_t status;
	    zbCounter_t bindingTableEntries;
	    zbIndex_t startIndex;
	    zbCounter_t bindingTableListCount;
	    zbApsmeBindEntry_t
	aBindingTableList[1];
	} zbMgmtBindResponse_t;

Don't confuse **Mgmt_Bind_req** (which retrieves a remote binding table) and **Bind_req** (which binds a remote node to another node).

5.4.3   Informing Other Nodes to Leave the Network

One of the other interesting things ZDP can do is to tell other nodes to leave the network. Why would you do this? Sometimes, such as when using the Commissioning Cluster from the ZigBee Cluster Library, a node might be commissioned with certain values on a commissioning network, and then told to go join a different network where it will do its work. Imagine a handheld device that an installer uses to make sure all of the lights, switches, thermostats, etc., are all functioning properly in each hotel room, before moving on to the next. Use **Mgmt_leave_req** for this purpose.

Mgmt_Direct_join_req is not used much. It's easier to simply use the network rejoin command, available through ZDO.

Mgmt_permit_joining_req can be very useful for disabling joining all throughout the network. Typically, this is the last step when commissioning a network. It closes it down to prevent other nodes getting on the network without permission.

5.5   Starting and Stopping ZigBee with ZDO

ZDO is the local-state machine that controls the state of the ZigBee node on and off the network. When a node boots up, it does not necessarily join a network right away. It may go into low-power mode, and wait for a button-press, or some other event that causes the node to decide it needs to network.

The Freescale platform uses a function called **ZDO_Start**() to join a node to the network. **ZDO_Start**() can start with any of the following options:

- gStartWithOutNvm_c

- gStartAssociationRejoinWithNvm_c

- gStartNwkRejoinWithNvm_c

- gStartSilentRejoinWithNvm_c

Starting without non-volatile memory (NVM) ensures that the node does not use anything it remembers from the last time it was booted and joined a network. Association join (or rejoin) uses the MAC association commands to join the network. Rejoin with NVM rejoins the network using the same PAN and channel selected previously. The node may get a new short address. The silent rejoin is very useful when nodes are reset after a battery change, or after a mains-powered network has reset after a power outage. The nodes do not actually *say* anything over the air, they simply start up and are capable of routing in a few tens of milliseconds.

To leave the network, Freescale uses one of two functions:

- ZDO_Stop()

- ZDO_Leave()

Stop leaves the network silently. Leave informs the node's parent so that the parent's internal tables can be cleaned up.

The example in this section, *Example 5-4 ZDO*, forces a node to leave one network and to rejoin the other. This operation is done fairly frequently in ZigBee network commissioning. A node doesn't know anything about the network it joins, other than the IEEE address of the parent it joined. Many times a ZigBee node needs to know more before deciding to remain on that network. It may, for example, query the network for a particular service.

In this example, the NCB board will attempt to join (randomly) one of the SRB boards. If it does, it will ask the SRB whether it supports the On/Off Cluster. If not, it will leave that network and attempt to join another. It continues doing this until it finds an On/Off Cluster that responds, and in this case, the light turns on.

Which of the networks issues the beacon response first is random, so the actual over-the-air capture may vary until the node finds the right parent.

5.6   ZDO, ZigBee, and Low Power

One of the most interesting aspects of ZigBee is the ability of nodes in a ZigBee network to last, not hours, not days, but for years on battery power. In fact, it's normal for a sleeping ZigBee device to last the shelf life of a couple of AA batteries (about five to seven years). Consider Figure 5.8. The ZigBee routers (in gray) and the ZigBee Coordinator (in black) are typically mains-powered. The ZigBee End Devices (in white) are ZigBee node types, which can sleep.

ZigBee End Devices can sleep, because they do not route. That is why they are called end devices: the route stops here. Notice the end devices in the figure below (for example, node 25) only have one connection to the ZigBee Network: the end-device's parent. The routers must have at least two connections. In reality, it's likely that all the routers in this house floor plan can all hear each other, but to simplify the figure, only some of the possible routes are shown.

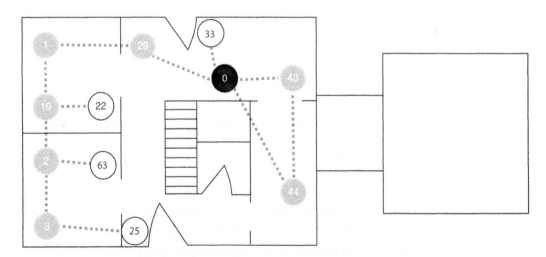

Figure 5.8: ZigBee End Devices Do Not Route

ZigBee is an asynchronous protocol. That is, a node may choose to transmit at any time. This makes sense when you think about how ZigBee is used. A light switch (let's use node 25 again, for example), can wake up and send a command to turn on the lights any time a user flips the switch. Or a factory automation system might need to send an alarm immediately. That is why routers must be awake all the time and ready to route a message.

Within the ZigBee Alliance, work is being done on an all-battery powered network for use in situations where latency doesn't matter, but eliminating mains-powered devices does, such as in a vineyard, or another agricultural setting. It doesn't matter if the temperature or moisture content is communicated now, or two minutes later. It would matter if the lights didn't turn on for two minutes! At the time of this writing, that work has not made it into any official ZigBee specifications.

So, let's go back to the example above. Someone flips a battery-powered ZigBee light switch. The switch causes an interrupt which wakes the CPU, which in turn wakes the radio. Once the system is fully powered (we're talking approximately a millisecond, here) the ZigBee End Device sends the command to turn the set of lights it controls on or off, and then it goes immediately back to sleep. Immediately is a relative term, so I'll go into the exact sequence of events with calculations, in a bit.

So what happens if you send the end-device a message while it is asleep? How does that end device receive it? That's where the special parent-child relationship comes in. In a ZigBee network, the parent will actually buffer messages for the sleeping child, delivering them when it wakes up.

However, the message is not buffered forever. The MAC generally buffers messages for about seven seconds. Some ZigBee stacks, like Freescale, are limited to this MAC timeout. Others are not. There is one other thing to note. If a given parent has many sleeping children, and many messages to deliver, the messages may time out before they are all delivered to the sleeping children. Generally, sleeping devices should wake up and communicate with some node in the network periodically, if the sleeping node can be configured or is to normally receive packets. Otherwise, just treat the end device as a low-power command, or data initiator. It wakes up when it wants, transmits data, then sleeps again.

One common question I get is this, "Can the radio wake the CPU upon receiving a valid ZigBee packet?" The answer is "Yes it can, but it doesn't make sense for a low-powered system." If the radio is awake enough to decode a signal, it is awake. That means it is consuming full power, somewhere in the neighborhood of 20–23 mA, which means the batteries won't last a very long time (days at best).

Table 5.19: ZigBee Battery Life Calculator

Battery capacity (mAh)	1900	1900 = 2 AA batteries
Supply efficiency (%)	100%	
System capacity (mAh)	1900	
Tx current (Radio) (mAh)	34	MC13193
Application payload size (bytes)	10	add 18–30 bytes for security
Packet frequency(s)	15	
Tx duration per packet (ms)	1.31	With security
Tx packets per day	5760	Calculated from packet frequency
PA current (or other Tx on) (mA)	0	
Rx current (Radio) (mA)	37	MC13193
Rx duration per packet (ms)	10	Waits for ACK (and msg) from parent
Rx packets per day	5760	
LNA current (or other Rx on) (mA)	0	
Radio sleep current (mA)	0.002	MC13193 sleep current
MCU active current (mA)	14	HCS08 Stop Mode 3
MCU sleep current (mA)	0.001	
MCU activity time in addition to radio (%)	20%	
MCU total activity time	120%	
MCU with AtoD on current	0	
MCU active time for AtoD per sample (ms)	0	
Number of samples per day	0	
Calculated radio duty cycle	0.08%	
Capacity used per day (mAh/day)	1.04	
Battery life in days	1828	
Battery life in years	5.0	

Table 5.19 is a battery calculator, and is included in Excel form on-line. As you can see from the calculations, it's very possible for a ZigBee node to last an entire five years on a pair of AA batteries.

Identifying all the power consumers in a system is not always easy. Some are obvious. A power regulator, consuming power to reduce the voltage from 9 volts down to 3, or that

Figure 5.9: The Panasonic PAN802154HAR Low-Power ZigBee Board

TTL to RS2332 serial chip, or that blazing LED, is easy to figure out. But other power consumers are not so obvious.

For example, consider the Freescale HCS08GT60 microcontroller used both in the Freescale system-in-package MC13213, and in the two-chip solution with the MC13193 radio. This microcontroller uses the same core as the GB60, a part with significantly more GPIO pins brought out on the package. In the core on the GT60, the one used for the ZigBee nodes, those extra pins which aren't brought out on the smaller package are floating, and must be initialized to low output to prevent power consumption. If you don't turn them off, you'll wonder why your board is not achieving that $1.9\mu A$ low-power sleep that the radio and MCU can.

The final example in this chapter is a low-power On/Off Switch. One thing that's very important to note in the Freescale solution is that it won't go into deep sleep unless all application timers have been stopped.

To run the demo, compile and download the ZcNcbOnOffLight and ZedPanOnOffSwitch. The "PAN" stands for the Panasonic PAN802154HAR. This board, pictured in Figure 5.9, can achieve $2\mu A$ while asleep. Press the button, and the board wakes up, sends a toggle command to the light, and then goes back to sleep.

Alternately, use the Freescale SRB boards. The SRBs, while they are nice development boards, cannot achieve true low power under software control. This isn't due to the radio and MCU, but because other power consumers on the SRB board, such as the power regulator and USB chip, cannot be shut off.

When planning your project, always plan much more time than you think for low power. It seems so simple in concept, but there are always gotchas. One example is that the BDM debugger, used to debug programs in the Freescale environment, doesn't function once the MCU goes into low power. Low power is always more difficult than you think.

> ZigBee provides no low power API. The API is always vendor-specific.
>
> ZigBee End-Devices are the only nodes in a ZigBee network that achieve long battery life.

CHAPTER 6

The ZigBee Cluster Library

The ZigBee Cluster Library was the brainchild of Phil Jamieson, chairman of the Application Framework Group. The library, like the ANSI C library, is a set of useful functions from which to build ZigBee applications and profiles.

The ZigBee Cluster Library is released under its own specification, separate from the ZigBee specification. See http://www.zigbee.org to download a copy of the ZCL specification. All ZigBee Alliance Public Profiles use the ZigBee Cluster Library, but private profiles may choose to use it or not.

Before describing how to develop applications using the ZigBee Cluster Library, let me tell you the real, untold story of the origin of the ZigBee name. Not many have heard this tale.

On June 21, 1860, the Army of the United States of America adopted a system of visual communications called "wigwag," and in the process created a separate, trained professional military service: the Signal Corps.

The inventor, Albert James Myer, first tested his visual signaling system in active service during the 1860–1861 Navajo expedition to New Mexico. Using flags for daytime signaling and a torch at night, wigwag was used in active combat during the Civil War on June 1861, to direct the fire of a harbor battery against the Confederate positions at Fort Calhoun.

By 1879, the electric telegraph, in addition to visual signaling, had become a Signal Corps responsibility. The Signal Corps constructed, maintained, and operated some 4,000 miles of telegraph lines along the country's western frontier.

In the late 1930s, during World War II, Company B of the Signal Corp, part of the 49th Signal Heavy Construction Battalion, was selected for a very important assignment: the installation of a large communication system for the Navy under Admiral Nimitz and the 20th Air Force under the command of General Spaatz, on the island of Guam.

Guam was to become the coordination point for all air and sea communications in the Central Pacific during the war, due in part to this highly reliable wireless system.

- ZigBee cluster library specifies functional domains
- Each specification specifies the cluster sets for that functional domain
- Each specification defines mandatory and optional clusters, attributes, commands, and functional descriptions
- Explicit device descriptions are not defined

- ZigBee profiles specifiy application domains
- Each profile collects related elements from the cluster library into application domains
- Each profile defines device descriptions for each required device
- Each profile specifies the cluster identifiers for each cluster used from the cluster library

Figure 6.1: The ZigBee Cluster Library

Signal Corp Company B, or SigB (ZigBee) as it was called colloquially became famous for providing wireless communications that simply worked. The Signal Corp is still in operation today.

The ZigBee Cluster Library (ZCL) shown in Figure 6.1, is nothing more than a set of clusters and cross-cluster commands used in the public profiles produced by the ZigBee Alliance to speed the development and standardization of the public profiles. Some of those clusters are general purpose and enhance the functionality of the ZigBee specification. An example of this would be the Groups Cluster (cluster ID 0x0004), which includes the ability to add and remove groups over-the-air (whereas the APS group commands are for local in-node access only). Other clusters are general purpose to many applications, such as the On/Off Cluster (cluster ID 0x0006) which can turn on or off … well, just about anything.

The ZigBee Cluster Library is organized into functional domains, such as General, Closures, HVAC, and Lighting. Clusters from these functional domains are used in the ZigBee Public Profiles to produce descriptions of devices, such as a dimming light, a dimmer switch, or a thermostat. Each Public Profile may also define its own specialized clusters, such as the Automatic Metering (later renamed to Smart Energy) Price Cluster.

Is ZCL required? If you are making a public profile device, such as a power meter intended to be compatible with the Automatic Metering profile, or a thermostat intended

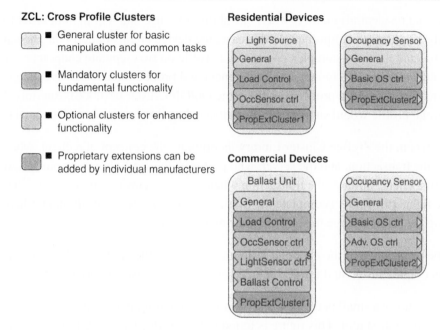

Figure 6.2: Clusters Extend Device Functionality

to be compatible with the Home Automation or Commercial Building Automation Profile, then the answer is "Yes." That's what public profiles are all about: interoperability. With private Application Profiles, however, ZCL is not mandatory. Many private application profiles do not use the ZigBee Cluster Library at all, or only use a few clusters from the library (see Figure 6.2).

The public profiles which use the ZigBee Cluster Library organize the use of clusters into three categories:

- Clusters which are mandatory for all devices within the profile

- Clusters which are mandatory for a particular device within the profile

- Clusters which are optional for a particular device within the profile

The ZigBee Cluster Library also allows devices to be extended by manufacturer-specific extensions, allowing manufacturers to produce value-added features available only to their brand.

Think of a light. In the simplest case, it can support the ability to turn on or off (wirelessly of course). Perhaps that same light fixture may be sold in both residential and commercial (or high-end residential) markets where the ability to sense ambient light in order to adjust the brightness is important. In this case, the device could support both the simple residential and the more complex commercial profile on two separate endpoints, allowing the manufacturer to save money on development and production costs by creating one part instead of two. Both profiles use the same On/Off Cluster (0x0006). But only the commercial profile would support the Illuminance Level Sensing Cluster (0x0401).

The clusters in the ZigBee Cluster Library incorporate the concept of a client, who initiates the transaction, and the server, who performs the work. For example, a light switch (the client) initiates the transaction when someone taps the light switch. One or more lights (the server) complete the transaction by turning on or off and perhaps reporting the status change to some monitoring device(s).

From the perspective of the endpoint's simple descriptor, the client side lists the cluster ID as an output cluster, and the server side lists the same cluster ID as an input cluster.

Figure 6.3 shows a small network with five nodes, including two lights and two switches, plus a configuration tool. This figure is found in the ZigBee Cluster Library specification, labeled there as "Figure 3.2." Notice the client (output) side of a cluster talks to the server (input) side of the cluster—the output from the switch becomes the input on the light. Notice also that the same Cluster ID (and code) can be used for different device types. Both the On/Off Switch and Dimmer Switch use an On/Off Cluster (cluster ID 0x0006) to control the remote light. But only the dimmable light supports the Level Control Cluster.

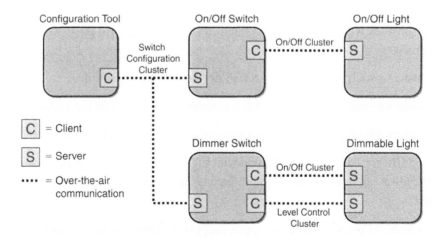

Figure 6.3: Clusters Contain Client and Server Components

The configuration tool in this figure is able to configure both types of switches. In a home, an on/off light switch sometimes controls a single set of lights, such as porch lights. A switch can also be used as a toggle, also called a three-way switch, where two or more switches control the same set of lights (common in kitchens and hallways). ZigBee ensures that the same physical device (the HA On/Off Switch) can be configured to accomplish either configuration in a simple compatible way, over-the-air. This allows the installer or home owner to adjust a system to personal preferences. The interface for the installer could be as simple as a drag-and-drop application on your PC.

The ZigBee Cluster Library describes each cluster in detail, so independent vendors can create compatible products that interoperate. It doesn't matter whether the light switch or lights come from Philips or Schneider Electric, or whether a thermostat comes from Siemens, Honeywell, Trane, or Johnson Controls. They all work together. A certification process ensures that vendors adhere to the standard described by the ZigBee Cluster Library and the Application Profile.

In this chapter, I'll describe the ZigBee Cluster Library in detail, starting with the ZCL foundation, a set of cross-cluster commands that can read and write what are called *attributes*.

Next I'll describe the ZCL general clusters, a common set of useful clusters across all public profiles. Then I'll describe some details of the current ZigBee public profiles, including Home Automation and Automatic Metering (also called ZigBee Smart Energy). Finally, I'll describe when to use and when not to use ZCL in private profiles, and provide some examples of extending ZCL with your own clusters and attributes.

There is just not enough room in this chapter to describe every single cluster in the ZigBee Cluster Library. Instead, I'll focus on the general concepts and give examples using a few of the clusters and profiles. I leave examining the details of every cluster to you. It's all available in the ZigBee Cluster Library specification, available at http://www.zigbee.org.

> The ZigBee Cluster Library is a set of common clusters for use in application profiles.
> ZCL ensures interoperable products from the application level.

6.1 ZCL Foundation

The ZigBee Cluster Library introduces the concept of *attributes* and *commands* to the ZigBee specification.

Attributes are data items or states defined within a cluster. *Commands* are actions the cluster must perform. For example, a Home Automation On/Off Light uses the On/Off Cluster, cluster ID 0x0006. An attribute of the On/Off Cluster indicates whether the light is on (0x01) or off (0x00). However there are also commands which turn the light on (0x01), off (0x00), or toggle it (0x02), which affect the state of the OnOff attribute.

Here's another way to look at it if you've ever done object-oriented programming. A cluster in the ZigBee Cluster Library is an object, containing both methods (commands) and data (attributes). These objects do not support inheritance, however. They are simply stand-alone objects that device may use or not, as specified by the particular Application Profile. A single endpoint on a device may support any number of clusters, up to 64 K, though in practice a device usually supports a handful, or at most a dozen. A single cluster on an endpoint may support up to 64 K attributes and 256 commands.

Attributes may be read from, written to, and reported over-the-air with standard, cross-cluster ZCL commands. These cross-cluster commands are called the ZCL *foundation*.

Table 6.1 lists the ZCL cross-cluster commands. These commands work across any cluster in the ZCL. For example, the "read attributes" cross-cluster command can read the attributes from the On/Off Cluster (cluster ID 0x0006) and the Level Control Cluster (cluster ID 0x0008). Nearly all these commands (with the exception of the default response) deal with attributes: reading, writing, and reporting them.

This mechanism is very powerful. A third-party device, perhaps a gateway widget, one with a ZigBee Dongle and a fancy PC program that runs on your computer or television, can show the entire state of the network in a nice, graphical way to the home owner, or the hotel owner, or the 70-story office building management system.

Notice that while any given command is optional, the responses are not. This makes sense if you think about it. A given endpoint may or may not need to ask another endpoint the state of its attributes. But all applications must allow others to query their attribute state. Notice that I used the term endpoint, not node. It's possible for a single node to contain endpoints that use the ZigBee Cluster Library, and to have other endpoints on a private profile that don't. Only those endpoints that support the ZigBee Cluster Library support the ZCL foundation commands.

The read attributes command can read one or more attributes (as many as will fit into a single payload, which depends on the size of the attributes). The write attributes command also can write one or more attributes. Configure reporting essentially configures another node to report one or more attributes, just as if those attributes

Table 6.1: ZCL Cross-Cluster Commands (Foundation)

Cmd Id	Command Name	M/O	Description
0x00	Read attributes	O	Read one or more attributes
0x01	Read attributes response	M	Return value of one or more attributes
0x02	Write attributes	O	Write one or more attributes
0x03	Write attributes undivided	O	Write one or more attributes as a set
0x04	Write attributes response	M	Return success status of write attributes
0x05	Write attributes no response	M	Write one or more attributes, no response
0x06	Configure reporting	O	Configure attributes for reporting
0x07	Configure reporting response	M	Status of configure attributes
0x08	Read reporting configuration	O	Read current reporting configuration
0x09	Read reporting configuration response	M	Return current reporting configuration
0x0a	Report attributes	O	Attribute report, depends on configuration
0x0b	Default response	M	Unsupported command response
0x0c	Discover attributes	O	Determine supported attributes on remote node
0x0d	Discover attributes response	M	Results of discover attributes command
0x0e–0xff	Reserved		For future use by ZCL

were read using read attributes, but they report on conditions (changing values or time-based). Discover attributes allow a node to discover the attributes on another node, which is useful when some of the attributes are optional for a given device.

To understand attributes, consider a light switch and a light. A very simple switch may not care if the light is on or off. It simply sends the command to turn the light on or off, and given the high reliability of ZigBee, assumes the best. This is how normal wired systems work. The user is the only one who knows if the light actually goes on. If it doesn't, the user determines whether to change the light bulb or call the electrician. ZigBee does not require a switch to be smart, so having a "dumb" wireless switch is perfectly acceptable.

A smarter switch could use the ZCL cross-cluster commands to determine whether the light actually turned on, and if not, to do something about it. Perhaps the response should be to proceed through a diagnostic cycle to determine whether the problem was a burned-out light bulb, a loss of communication, or a hardware failure. ZigBee allows the manufacturer to make the decision on how smart to make the light switch. But since every ZigBee light supports the common set of attributes and ZCL foundation commands, systems can become smarter without changing the existing devices.

ZCL allows for both "push" and "pull" methods (report attribute and read attribute methods), for determining the state of any given cluster's attributes. Take a look at Figure 6.4.

In the "push" method, the devices themselves send reports when the attribute changes. The report can be configured to be time-based, change-in-value-based, or both. For example, an on/off light could be configured to report every time the light changes state (on or off). A temperature sensor could be configured to report once every minute, or if the temperature

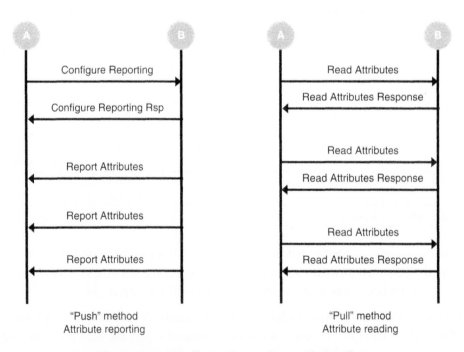

Figure 6.4: Attribute Reporting and Reading

raises or lowers by five degrees, whichever comes first. A settable minimum reporting interval ensures that if the temperature flutters between two values, a report isn't sent too often.

In the "pull" method, the device that needs the information asks the other devices for their current values. For example, when someone picks up the TV remote control, the current outside temperature could be read from the external temperature sensor, and reported on the remote's display or the television screen. A Home Automation management system could ask each node for the manufacturer information and display a systemwide view each time a PC application starts up.

The example in this section, *Example 6-1 Reading Attributes*, shows how to use both "push" and "pull" methods for reading attributes. The example uses three nodes: one ZcNcbOnOffLight and two ZedSrbOnOffSwitches. The on/off light switch A will use the "push" method. It will configure the light to report to it every time the light changes state. The second on/off light switch (switch B) will use the "pull" method and ask for the state of the light any time the user wants (see Figure 6.5).

The example source code can be found in "Chapter06\Example 6-1 Reading Attributes" at http://www.zigbookexamples.com. To program the example into Freescale boards create a solution with three projects:

- A ZigBee Coordinator, NCB Board, HA On/Off Light

- Two ZigBee End-Devices, SRB Board, HA On/Off Switch with RxOnIdle=TRUE

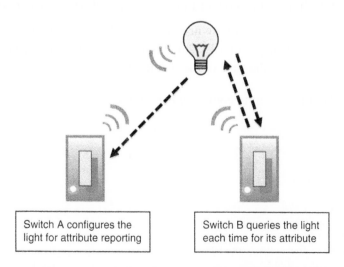

Switch A configures the light for attribute reporting

Switch B queries the light each time for its attribute

Figure 6.5: Example 6-1 Reading Attributes

Export the solution, copy the SwitchA_BeeApp.c and SwitchB_BeeApp.c source code to the respective SRB projects under the name BeeApp.c. To run the example, perform the following procedures:

1. Boot all three boards (in any order).

2. Press SW1 on all boards (in any order) to form the three-node network. I pressed SW1 on the On/Off Switch A first, so it obtained the address 0x796f, with On/Off Switch B being 0x7970.

3. Press SW3 on the light and on switch A, to bind them.

4. Press SW3 on the light and on switch B, to bind them as well. Now both switches are bound to the light. Either one may toggle the light.

5. Press long SW1 (LSW1) to go to Application Mode on all three boards.

6. Press SW1 on light switch A. Notice that the light toggles (LED2 on the NCB board).

7. Press SW1 on light switch B. Notice that the light also toggles. This is called a three-way switch. Either switch toggles the light.

8. Now, press SW2 on light switch A, to instruct the node to configure the light for reporting the on/off attribute.

9. Now, press SW1 on either switch. Notice that on/off light switch A mirrors the state of the light with its own LED2, regardless of which switch toggles the light.

10. Now, press SW2 on on/off switch B. This will obtain the on/off attribute from the light and display it on LED2, but only at that moment in time. It does not continuously read the state of the on/off attribute in the light.

Below is the packet that is sent to configure the light to report to node 0x796f (On/Off Switch A):

```
IEEE 802.15.4
ZigBee NWK
ZigBee APS
  Frame Control: 0x00
  Destination Endpoint: 0x08
  Cluster Identifier: On/off (0x0006)
  Profile Identifier: HA (0x0104)
  Source Endpoint: 0x08
  Counter: 0x40
```

```
ZigBee ZCL
  Frame Control: 0x00
  Transaction Sequence Number: 0x44
  Command Identifier: Configure Reporting (0x06)
    Configure Reporting Frame
      Reporting Configuration List
        Reporting Configuration 1
        Attribute Identifier: OnOff (0x0000)
        Direction: Server to Client to configure reporting (0x00)
        Minimum Reporting Interval: None (0)
        Maximum Reporting Interval: None (0)
```

This command sets the light to report only when it changes. It is not time-based. To add in a time base, set the minimum and maximum reporting interval to some number of seconds. To turn off reporting, set the maximum reporting interval to 0xffff. The report goes to any nodes that are bound on that endpoint in the light. This may be a single node, a set of individual nodes, or a group of nodes.

Now that the switch is configured, every time the light is toggled (from either light switch) the information is sent to On/Off Switch A, as seen below. From this decode, we can see that the toggle command turned the light on. If the light had turned off, the Attribute Data field would indicate 0x00 (Off):

```
Seq
No  MACSrc  MACDst  NWKSrc  NWKDst  Protocol  Packet Type
------------------------------------------------------------------
182 0x7970  0x0000  0x7970  0x0000  ZigBee    APS HA:On/off:Toggle
186 0x0000  0x796f  0x0000  0x796f  ZigBee    APS HA:On/off:Read
                                              Attribute Response
```

The below packet is the details of the Attribute Response:

```
IEEE 802.15.4
ZigBee NWK
  Frame Control: 0x0048
  Destination Address: 0x796f
  Source Address: 0x0000
  Radius = 30
  Sequence Number = 133
ZigBee APS
  Frame Control: 0x00
  Destination Endpoint: 0x08
  Cluster Identifier: On/off (0x0006)
  Profile Identifier: HA (0x0104)
  Source Endpoint: 0x08
  Counter: 0xb7
```

```
ZigBee ZCL
  Frame Control: 0x08
  Transaction Sequence Number: 0x43
  Command Identifier: Read Attribute Response (0x01)
  Read Attribute Response Frame
    Read Attributes Status List
      Read Attributes Status 1
        Attribute Identifier: OnOff (0x0000)
        Status: Success (0x00)
        Attribute Data Type: Boolean (0x10)
        Attribute Data: On (0x01)
```

Now take a look at what happens with switch B. It doesn't require the initial configure reporting setup, but each time it needs the state of the light, it must read the attribute explicitly, as shown below:

```
Seq
No   MACSrc  MACDst  NWKSrc  NWKDst  Protocol  Packet Type
------------------------------------------------------------------
192  0x7970  0x0000  0x7970  0x0000  ZigBee    APS HA:On/off:Toggle
194  0x0000  0x7970  0x0000  0x7970  ZigBee    APS HA:On/off:Read
                                               Attribute
196  0x0000  0x7970  0x0000  0x7970  ZigBee    APS HA:On/off:Read
                                               Attribute Response
```

Configuring an attribute for reporting and or for reading are fairly similar procedures. In the Freescale solution, they are calls to ZCL_ConfigureReportingReq() and ZCL_ReadAttributeReq(). Some attributes may also be written to, which is accomplished in a similar manner using ZCL_WriteAttrReq().

Note that, in general, attributes which represent the state of some physical object (such as a light), or sense the state of the physical world (such as a temperature), cannot be written to. To cause an effect, a command must be used instead, such as the OnOff command used to toggle a light. The kinds of attributes normally written to in ZCL are things like textual descriptions of the node or location (e.g., Kitchen). An example will be shown in the next section of writing to an attribute in the Identify Cluster.

The ZCL specification clearly indicates which attributes in any given cluster may be read, written to, or reported. For example, the OnOff attribute of the OnOff Cluster may be read and reported, but not written to, as shown in the excerpt from the ZigBee Cluster Library Specification (see Table 6.2).

This cluster shall support attribute reporting using the Report Attributes command and according to the minimum and maximum reporting interval settings described in the ZCL Foundation specification (Section 2.4.7). The following attribute shall be reported: OnOff.

Table 6.2: Attributes of the On/Off Server Cluster

Identifier	Name	Type	Range	Access	Default	Mandatory/Optional
0x0000	OnOff	Boolean	0x00-0x01	Read only	0x00	M

> Attributes in ZCL are data items defined within a cluster.
>
> Commands in ZCL cause action for the cluster to perform.

6.2   ZCL General Clusters

There are a set of ZCL clusters called the "general" clusters, named so because they are so useful they are generally found in every ZigBee public application profile. These clusters include the data shown in Table 6.3.

The Basic and Power clusters are found in every device. The Identify, Groups, and Scenes clusters are used to commission the network, and are found in most devices. Device Temperature Configuration, On/Off, On/Off Switch Configuration, Level Control, and Alarms are used in some devices (but definitely not all). Time is used in a network that is time-aware, and RSSI is used in a network that is location-aware.

There has actually been a lot of work on using 802.15.4 RSSI to locate nodes (approximately) in a network. Motorola even has some patents in this area that are used in the ZigBee solution from Texas Instruments. I saw a demonstration of this in Milan, Italy, at the ZigBee Open House, but the located item was often located in the "wrong" place. Still, it did provide some useful information. The nice thing about the RSSI cluster is that it permits any method for producing the location information, independently of communicating it. Expect to see interesting developments in location-based services.

As you read the ZCL specification, you'll notice a table listing attribute sets for each cluster. This often confuses people. What are attribute sets, and how do they differ from attribute IDS?

An Attribute ID is the 16-bit number (from 0x0000 up to 0xffff) that defines each attribute in a given cluster. Attributes IDs are always numbered, starting at 0x0000 for every cluster in ZCL. For example, the Basic Cluster (cluster ID 0x0000) attribute ID 0x0000 is the ZCL Version attribute. The On/Off Cluster (cluster ID 0x0006) attribute ID 0x0000 is the OnOff state attribute. Attribute IDs only have meaning within the cluster.

Table 6.3: ZCL General Clusters

Cluster ID	Cluster Name	Description
0x0000	Basic	Attributes for determining basic information about a device, setting user device information such as location, and enabling a device
0x0001	Power	Configuration Attributes for determining more detailed information about a device's power source(s), and for configuring under/over voltage alarms
0x0002	Device Temperature Configuration	Attributes for determining information about a device's internal temperature, and for configuring under/over temperature alarms
0x0003	Identify	Attributes and commands for putting a device into Identification Mode (e.g., flashing a light)
0x0004	Groups	Attributes and commands for group configuration and manipulation
0x0005	Scenes	Attributes and commands for scene configuration and manipulation
0x0006	On/Off	Attributes and commands for switching devices between "On" and "Off" states
0x0007	On/Off Switch Configuration	Attributes and commands for configuring On/Off
0x0008	Level Control	Attributes and commands for controlling devices that can be set to a level between fully "On" and fully "Off"
0x0009	Alarms	Attributes and commands for sending notifications and configuring alarm functionality
0x000a	Time	Attributes and commands that provide a basic interface to a real-time clock
0x000b	RSSI	Location Attributes and commands that provide a means for exchanging location information and channel parameters among devices

Attribute sets are simply the organization of attribute IDs in the ZCL document. Over-the-air there is only the attribute ID (not sets), and ZigBee stacks interact with attributes using the 16-bit attribute ID. Another way to look at attribute sets is that they are the top 12 bits of the 16-bit attribute ID. For example, the attributes in the Basic Cluster are organized as shown in Table 6.4.

The attribute sets are organized so attribute IDs 0x0000 through 0x000f are the device's information (read only), whereas attribute ID's 0x0010 through 0x001f are settable (read/write) parameters.

Table 6.4: Attribute Sets

Attribute Set Identifier	Description
0x000	Basic Device Information
0x001	Basic Device Settings

Table 6.5: Basic Cluster Attributes

Identifier	Name	Type	Range	Access	Default	Mandatory/Optional
0x0000	ZCL Version	Unsigned 8-bit integer	0x00–0xff	Read only	0x00	M
0x0001	Application Version	Unsigned 8-bit integer	0x00–0xff	Read only	0x00	O
0x0002	Stack Version	Unsigned 8-bit integer	0x00–0xff	Read only	0x00	O
0x0003	HW Version	Unsigned 8-bit integer	0x00–0xff	Read only	0x00	O
0x0004	Manufacturer Name	Character string	0 – 32 bytes	Read only	Empty String	O
0x0005	Model Identifier	Character string	0 – 32 bytes	Read only	Empty string	O
0x0006	Date Code	Character string	0 – 16 bytes	Read only	Empty string	O
0x0007	Power Source	8-bit Enumeration	0x00–0xff	Read only	0x00	M
0x0010	Location Description	Character string	0 – 16 bytes	Read/write	Empty string	O
0x0011	Physical Environment	8-bit Enumeration	0x00–0xff	Read/write	0x00	O
0x0012	Device Enabled	Boolean	0x00–0x01	Read/write	0x01	M
0x0013	Alarm Mask	8-bit Bitmap	000000xx	Read/write	0x00	O

The Basic Cluster (cluster ID 0x0001) is present in every device that supports the ZigBee Cluster Library. This cluster contains a set of attributes that defines common information and device settings, such as ZCL version, hardware and software version, manufacturer ID, and location. Table 6.5 shows the complete list of attribute IDs in the Basic Cluster.

Table 6.6: Basic Cluster Commands

Command Identifier	Description	Mandatory/ Optional
0x00	Reset to Factory Defaults	O

Not all of the attributes for the Basic cluster are mandatory. Some attributes are optional, like the manufacturer name or model identifier, as seen in the final column of Table 6.5. If your application uses the ZigBee Cluster Library, I recommend that all the fields be included, as it allows others (or even your own products) to query this information over-the-air.

Although it has many attributes, the Basic cluster only has a single command: Reset to Factory Defaults (see Table 6.6). This command is optional, but again it is a good idea to support it if you are making a ZCL device. If there is trouble in the network and a device is behaving erratically, a reset can cure the problem. You've probably done this with your wireless modem, or even desktop computer.

The Basic cluster reset command does not reset the ZigBee settings, such as to which network the node is connected, to what groups it belongs, or the local bindings. It only resets the attributes to factory defaults. To reset the rest of the device to pure factory settings the commissioning cluster must be used. I'll say more on that in Chapter 8, "Commissioning ZigBee Networks."

The ZCL rule for clusters is that attributes represent data, and commands cause action. There are a few exceptions to this (notably in the Identify cluster), but it is the general rule. So don't try to write to the on/off *attribute* to turn a light on or off. Instead, use the on, off, or toggle *command* for that cluster. The attribute will turn on or off, accordingly.

Attributes in the Freescale solution are read to or written from using a set of cross-cluster functions. The parameters to these commands mirror the ZigBee Cluster Library Specification. The code snippet below shows the ZCL_ReadAttribute() command. Its first parameter is the usual afAddrInfo, which specifies source and destination endpoint, destination node, and all the usual APS data request parameters. For a full explanation of data requests, see Chapter 4, "ZigBee Applications." The next two parameters are merely a list of attribute IDs (count and array). Notice that more than one attribute can be read at one time. Feel free to read attributes 0x0000, 0x0005, and 0x0013 all in one call:

```
//[R2] 7.1.1 Read attributes command frame format
typedef struct zclCmdReadAttr_tag
{
  zclAttrId_t aAttr[3]; //variable length array of attributes
```

```
} zclCmdReadAttr_t;
typedef struct zclReadAttrReq_tag
{
  afAddrInfo_t addrInfo; //IN: dst address, cluster, etc…
  uint8_t count; //IN: how many attrs to read?
  zclCmdReadAttr_t cmdFrame; //IN: list of attrs
} zclReadAttrReq_t;
zclStatus_t ZCL_ReadAttrReq
  (
  zclReadAttrReq_t *pReq
  );
```

The results of the ZCL_ReadAttrReq(), the Read Attribute Response, is returned to
the application through the ZCL response handler that was initially registered via the
ZCL_Register() command:

```
void ZCL_Register
  (
  /* IN: pointer to a response handler function */
  fnZclResponseHandler_t fnResponseHandler
  )
```

Commands depend on the cluster. For example, the command shown below will turn a
light (or another on/off device) on, off, or toggle it:

```
typedef zclCmd_t zclOnOffCmd_t;
#define gZclCmdOnOff_Off_c       0x00 /* M-turn device off */
#define gZclCmdOnOff_On_c        0x01 /* M-turn device on */
#define gZclCmdOnOff_Toggle_c    0x02 /* M-toggle device */
typedef struct zclOnOffReq_tag
{
  afAddrInfo_t addrInfo; //IN: dst address, cluster, etc…
  zclOnOffCmd_t command; //IN: on, off or toggle
} zclOnOffReq_t;
zbStatus_t ASL_ZclOnOffReq
(
    zclOnOffReq_t *pReq
);
```

From a programming standpoint, the ZigBee Cluster Library is really just an add-on
to what you've learned already in Chapter 4 about endpoints, profile IDs, and cluster
IDs. In fact, in the Freescale solution, it truly is just an add-on in the form of a
number of functions to call for ZCL requests, and a function call inside the APSDE-
DATA.indication to handle the ZCL responses. If the incoming message is for an
endpoint which supports ZCL, the call is handled automatically: reading attributes,

writing attributes, and reporting attributes. If the ZCL cluster supports commands, those are also handled automatically. Take a look at a sample Freescale BeeStack data indication below:

```
void BeeAppDataIndication(void)
{
  //call ZCL to handle the cluster
  status=ZCL_InterpretFrame(pIndication);
  if(status !=gZbSuccess_c)
  {
    //cluster not handled by ZCL. Handle it by the application…
  }
}
```

Notice that all the application has to do is pass the indication to ZCL_InterpretFrame(), and all the ZCL work is done. If a particular ZCL command causes something like a light to turn on or off, the application is notified through the BeeAppUpdateDevice() function with the particular event, such as turning on the light, going into Identify Mode, and so on:

```
void BeeAppUpdateDevice
  (
  zbEndPoint_t endPoint, /* IN: endpoint update happened on */
  zclUIEvent_t event /* IN: state to update */
  )
{
  switch(event)
  {
    case gZclUI_Off_c:
      AppTurnOffLocalLight();
      break;
    case gZclUI_On_c:
      AppTurnOnLocalLight();
      break;
  }
}
```

The specific code details above are not important unless you'll be using the Freescale solution, but the concept is the same across all platforms. There are a set of cross-cluster commands to read, write, and report attributes, and a set of cluster-specific commands to cause actions such as turning a light on or off. The ZigBee Cluster Library supported by the platform will do most of the work for you.

Take a look at the next decode of an over-the-air packet using the ZCL On/Off Cluster. Notice how all the normal APS stuff is there: the source and destination endpoints, the application profile ID, and the cluster ID. But notice the extra three bytes at the end, listed as the ZCL Frame. This cluster, because of the Home Automation profile (0x0104), is interpreted by Daintree (the protocol analyzer used to produce this decode) as the On/ Off Cluster. The command in the ZCL frame is 0x02, which will toggle to a remote light, pump, door bell, or anything else that can be turned on, off, or toggled:

```
  Frame 22 (Length = 30 bytes)
+ IEEE 802.15.4 Frame
+ ZigBee NWK Frame
- ZigBee APS Frame
        Frame Control: 0x00
        Destination Endpoint: 0x08
        Cluster Identifier: On/off (0x0006)
        Profile Identifier: HA (0x0104)
        Source Endpoint: 0x08
        Counter: 0xbe
- ZigBee ZCL Frame
        Frame Control: 0x01
        Transaction Sequence Number: 0x42
        Command Identifier: Toggle (0x02)
```

Remember that the application profile ID (0x0104 above) defines what the cluster means, which in turn defines what the application payload means. The same cluster ID in a different profile can mean something radically different. For example, take a look at the next decode with the private profile 0xc021. Notice that the Daintree tool no longer decodes the same sequence of bytes: 0x01, 0x42, 0x02 from the application payload as the ZCL toggle command, but instead simply calls it APS data. Notice also that the cluster is no longer called On/off. The meaning is now up to the application profile designer, who made the private profile 0xc021.

An analogy for application profiles is in human languages: one person may speak English while another speaks Japanese, and they can't understand each other. If you speak the same language (profile), the sentences (clusters) will now make sense:

```
Frame 22 (Length = 30 bytes)
+ IEEE 802.15.4 Frame
+ ZigBee NWK Frame
- ZigBee APS Frame
        Frame Control: 0x00
        Destination Endpoint: 0x08
        Cluster Identifier: (0x0006)
        Profile Identifier: (0xc021)
        Source Endpoint: 0x08
        Counter: 0xbe
- APS Data: 01:42:02
```

Table 6.7: Identify Cluster Attributes

Identifier	Name	Type	Range	Access	Default	Mandatory/ Optional
0x0000	Identify Time	Unsigned 16-bit integer	0x0000–0xffff	Read/write	0x0000	M

That being said, all ZigBee public profiles use the same cluster IDs so that code can be shared for those devices that speak more than one public profile. Cluster ID 0x0006 is always the On/Off Cluster in any public profile. But in private profiles, it could mean anything.

There are two clusters in particular I'd like to focus on in this section about the ZCL General Clusters: Identify and Groups. These two clusters are very useful during commissioning of a network.

The example coming up in this section, *Example 6-2 Identify and Groups*, shows how the identify command works together with groups to make connecting devices such as switches and lights easy.

6.2.1   The Identify Cluster

The Identify cluster contains only a single attribute: Identify Time (see Table 6.7).

This attribute defines the amount of time, in seconds that the device is in Identify Mode. Identify Mode is usually indicated by a short flash of light, or perhaps by an audio signal.

So why place a node in Identify Mode? Think of someone installing a lighting system in an auditorium. Many of the lights are high up in the ceiling, 20 or more feet above the floor. Now imagine that the installer wants to connect a switch to a certain set of the lights up there. One way to accomplish this with ZigBee is to ask each light, in turn, to go into Identify Mode. As they flash, the installer could press a button meaning, "Yes, that's one of the lights I want to control with this switch." The entire system can be set up in just a minute or so.

The Identify cluster contains two commands: Identify and Identify Query (see Table 6.8).

Table 6.8: Identify Cluster Commands

Command Identifier	Description	Mandatory/ Optional
0x00	Identify	M
0x01	Identify Query	M

The Identify command on the Identify cluster is the same as writing to the attribute. This is one of those rare cases where writing the attribute has some immediate effect, but I recommend using the command, not writing to the attribute. Set the field to 0x0000 to turn identify off, or set it to 0x0001–0xffff to turn it on for that number of seconds. *Hint: Don't turn it on for longer than 20 to 30 seconds. Identify can get really annoying on most devices.*

The Identify Query command on the Identify cluster only has an effect if the device is in Identify Mode. If it is, then the device will respond with an identify query response. This command is normally broadcast to the network. When should you use this command?

Think about that light and switch installer again. This time he puts two, or maybe even 10 lights into Identify Mode at the same time, rather than in series. Then, with a single command over the air (identify query), they all respond to the sending node who then uses the information to bind the switch to each of those lights. The install program then turns off Identify Mode to all those lights.

6.2.2   The Groups Cluster

Now we come to the Groups cluster. This cluster is so amazingly useful it should have been placed in the ZigBee Device Profile. Unfortunately, the ZigBee Specification was already set in stone before this cluster was written.

While group support is optional in the ZigBee specification, it is mandatory for particular devices in many public profiles. Groups allow a single over-the-air command to be addressed to many nodes and endpoints at once. If the endpoint is a member of the group, it will process the command. If the endpoint is not a member of the group, it won't.

The Groups cluster contains only a single attribute, the NameSupport attribute (see Table 6.9). This read-only attribute simply allows other nodes to inquire if this node supports

Table 6.9: Group Cluster Attributes

Identifier	Name	Type	Range	Access	Default	Mandatory/ Optional
0x0000	NameSupport	8-bit bitmap	x0000000	Read only	–	M

UTF-8 text names for the group in addition to 16-bit Group IDs. In practice, group naming tends to be turned off. Instead the names are kept on a PC, or installation PDA.

You'll probably have noticed by now that all attribute IDs for every cluster start with 0x0000, even though they are different attributes, such as NameSupport for the Group cluster, IdentifyTime for the Identify cluster, and so on. This is perfectly acceptable since an attribute ID is within the scope of a specific cluster, just as a cluster ID is within the scope of a specific Application Profile ID.

The group cluster commands are very powerful (see Table 6.10). APS (see Chapter 4, "ZigBee Applications") has a set of commands for manipulating groups locally in a node, but how does a commissioning tool do this over-the-air? (Yes. You're very smart.) The answer is the Groups cluster.

The Add Group command adds a group to an endpoint over-the-air. The View Group command determines which group(s) the endpoint belongs to. The Get Group Membership command can determine which nodes in the network are members of, for instance, group 0x0005. The Remove Group command removes a group from an endpoint (remember also from Chapter 4 that endpoints belong to groups). The Remove All Groups removes all groups from the endpoint, and you can send this to the broadcast endpoint to remove all groups from all endpoints. In fact, you can send a broadcast message on the broadcast endpoint to remove all groups from all nodes in the network with a single command.

The Add Group If Identifying is the command that will be used in the example in the next subsection. This command will only add the group ID to the endpoint if that endpoint is in currently Identify Mode.

Table 6.10: Groups Cluster Commands

Command Identifier	Description	Mandatory/ Optional
0x00	Add Group	M
0x01	View Group	M
0x02	Get Group Membership	M
0x03	Remove Group	M
0x04	Remove All Groups	M
0x05	Add Group If Identifying	M

Okay. That's enough for you to go on about clusters in the ZigBee Cluster Library. The other clusters operate in much the same vein: a set of attributes and commands. You can read about them in the ZigBee Cluster Library specification, and for the particular implementation, in the platform documentation available from your favorite platform vendor—Freescale, Texas Instruments, Ember, etc.

6.2.3   Example 6-2: Identify and Groups

Now you know how ZCL works, and you know how some clusters work, particularly the Identify and the Groups cluster. The example in this section should solidify that understanding. The concept behind *Example 6-2 Identify and Groups* is simple. The network contains one switch (an NCB board) and two lights (SRB boards) and the user (you) will decide which light(s) to connect to the switch, wirelessly.

The steps are as follows:

1. Find the example source code and solution at http://www.zigbookexamples.com, in the *Example 6-2 Identify and Groups* folder.

2. Download the images to the respective boards, and boot them.

3. Press SW1 on all boards (in any order) to form the network.

4. Press long SW1 (LSW1) to go to Application Mode on all boards.

5. Press SW3 on one or both of the On/Off Lights (SRB boards). Doing this places the light(s) into identify mode for 20 seconds.

6. Press long SW3 (LSW3) on the NCB/light switch to instruct the lights to add themselves to the group.

7. Press SW1 to toggle the remote light(s).

The over-the-air capture shows the how the Add Group If Identifying command looks over the air (which happened when you pressed LSW3 on the NCB/light switch previously). This command is broadcast, which means that any number of lights may become part of the group as a result of this single command.

As you can see, any endpoints in Identify Mode are now added to group 0x0001. I say endpoints and not nodes, because a single node may support a number of separately addressable devices on different endpoints. For example, think of a single node controlling a set of four independent lights:

```
IEEE 802.15.4
ZigBee NWK
   Frame Control: 0x0048
   Destination Address: 0xffff
   Source Address: 0x796f
   Radius = 10
   Sequence Number = 181
ZigBee APS
   Frame Control: 0x08
   Destination Endpoint: 0xff
   Cluster Identifier: Groups (0x0004)
   Profile Identifier: HA (0x0104)
   Source Endpoint: 0x08
   Counter: 0x63
ZigBee ZCL
   Frame Control: 0x01
   Transaction Sequence Number: 0x42
   Command Identifier: Add Group If Identifying (0x05)
     Add Group If Identifying
        Group ID: 0x0001
        Group Name Length: 0 (No Name)
```

From now on, a single command groupcast (a form of broadcast) or multicast (another form of broadcast) across the network will toggle any number of lights. Add every light in the network to a group, and a single over-the-air command can turn on or turn off every light in the network. It's as simple as that. Wireless control that simply works:

```
Seq
No   MACSrc  MACDst   NWKSrc   NWKDst   Protocol   Packet Type
-----------------------------------------------------------------
14   0x796f  0xffff   0x796f   0xffff   ZigBee     APS HA:On/off:Toggle
```

> Attribute and command IDs are within the scope of the cluster, just as cluster IDs are within the scope of a profile.
>
> The ZCL General Clusters are supported across all ZigBee Profiles.

6.3   ZCL and Public Profiles

The ZigBee Cluster Library is used in all ZigBee public Application Profiles, including Home Automation (HA), Commercial Building Automation (CBA), ZigBee Smart

Energy (ZSE, formerly called Automatic Metering), Telecommunications Applications (TA), and Personal Home and Hospital Care (PHHC). A ZigBee public profile is a separate document from the ZigBee and the ZigBee Cluster Library specifications.

Some of the public profiles have been posted to the ZigBee Web site at http://www.zigbee.org, but others are not available to the public (yet). One benefit of joining the ZigBee Alliance is early access to profiles.

In ZigBee, Application Profiles describe everything about how devices behave in that domain space. Home Automation talks about lights, switches, thermostats, heating and cooling units (air conditioners), security systems, and more. The Telecommunications Profile describes how cellular phones interact with ZigBee, including clusters for payment systems and social networking. I'll give you a look at some profile-specific clusters in this chapter.

The public profile describes how the ZigBee stack is used, putting some restrictions on how often a ZigBee End-Device must poll its parent, or how often a node is allowed to unicast or broadcast.

In particular, public profiles describe:

- The application domain (what problem space the profile is addressing)

- A list of specific devices supported in the profile

- A list of clusters supported by those devices

- A list of what clusters are mandatory and optional for each device

- A list of what attributes are mandatory and optional for each cluster

- Which stack profiles are supported by the Application Profile (more on stack profiles in Chapter 7, "The ZigBee Networking Layer")

- Anything that overrides the ZigBee specification, stack profile, or the ZigBee Cluster Library specification

- Suggestions on commissioning techniques

- Best practices for using the profile, including any restrictions on the frequency of unicasts and broadcasts

- Any changes required by the ZigBee stack feature set

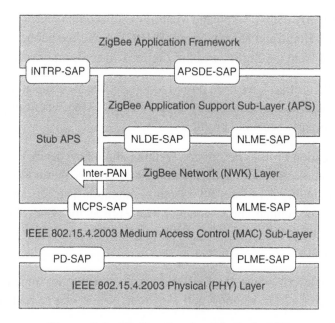

Figure 6.6: ZigBee Stack with Stub APS

The goal of every ZigBee public profile is to allow devices manufactured by different companies to interact with each other seamlessly. Application Profiles are allowed to describe new clusters not found in the general ZCL specification, and even to define new features in the "stack," in some extreme cases.

An example of adding a new feature to the stack is demonstrated in the ZSE profile, a mechanism called inter-PAN communication. Inter-PAN communication (that is, between two or more PANs) is not normally allowed by ZigBee, but this feature was required by ZSE. So the ZSE profile extended the ZigBee architecture, as shown in Figure 6.6.

This "Stub APS," available in the ZSE profile, allows one ZigBee PAN to communicate with another ZigBee PAN (single-hop only), essentially a poor-man's gateway. This interface was created for the ZSE profile to allow pricing information to be sent from the ZSE network to ZigBee Home Automation networks, a feature required by the electrical utility companies.

Table 6.11 shows a list of the new clusters defined by the ZSE profile. Because these clusters are specific to the needs of Automatic Metering, they are defined in the ZSE Specification, but not in the ZigBee Cluster Library Specification.

Table 6.11: Automatic Metering Clusters

Domain	Cluster Name	Cluster ID
AMI	Price	0x0700
AMI	Demand Response and Load Control	0x0701
AMI	Simple Metering	0x0702
AMI	Message	0x0703
AMI	Registration	0x0704
AMI	AMI Tunneling (Complex Metering)	0x0705
AMI	Pre-Payment	0x0706

The Price cluster communicates the current price of electricity for those areas where the price varies to the consumer by demand or by time of day. The Demand Response and Load Control cluster actually allows the utility company to control non-safety-critical devices in a home or business to reduce power usage during peak times. The Simple Metering covers the common metering commands and responses, while the Complex Metering covers vendor-specific metering commands. Pre-Payment allows customers to pay for electricity prior to using it.

An example decode from the ZSE profile (0x0109) is shown below. Notice that this is a different Application Profile ID from the Home Automation Profile (0x0104) used through most of this book. Note also the frame is secured (it contains an AUX header), as all frames are required to be in the ZSE profile specification:

```
IEEE 802.15.4
ZigBee NWK
   Frame Control: 0x0248
   Destination Address: 0x0000
   Source Address: 0x4c93
   Radius = 5
   Sequence Number = 13
ZigBee AUX
ZigBee APS
   Frame Control: 0x00
   Destination Endpoint: 0x01
   Cluster Identifier: Simple Metering (0x0702)
   Profile Identifier: ZSE (0x0109)
   Source Endpoint: 0x50
   Counter: 0x3f
ZigBee ZCL
   Frame Control: 0x01
```

```
Transaction Sequence Number: 0x43
Command Identifier: Get Profile Request (0x00)
   Get Profile Request Command Payload
   Interval Channel: Consumption received (0x01)
   End Time: 1 second since January 1 2000, 12:00:00am (UTC)
   Number of Periods: 0x01
```

The ZSE profile is a result of huge market pressure from the utility companies. The idea behind ZSE is to limit electrical usage at any given moment in time to a level under the maximum amount that the grid can handle. For power generation, average usage doesn't matter, but peak usage does. If everyone turns on their air conditioner, all their lights, and all their appliances at once, the grid would overload and need to shut down, causing brown-outs or black-outs (a very bad thing). So the utility companies work very hard to keep the peak usage below what the grid can handle.

By pricing electricity differently (and letting customers know about it) and by an opt-in feature to control non-critical devices, the utility can delay the need for new power plants, a large cost savings to both the consumer and utility company. The whole system is monitored and controlled by a back-end ZSE server for the particular utility company (see Figure 6.7).

Figure 6.7 shows how ZigBee is used in the Automatic Metering environment. Some back-end server has the real brains and monitors (or allows control of) the devices in the network. The area in gray in Figure 6.7 is the ZigBee portion. How it gets from the Energy Service Portal to the back-end server is not ZigBee, but probably cellular, power-line, WiMAX™ or other technologies. The Home Area Network, or HAN, may

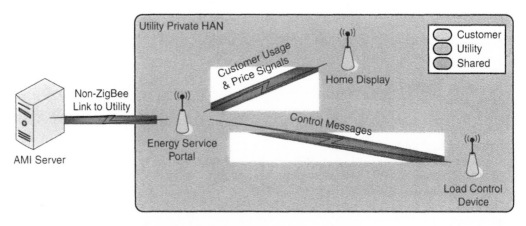

Figure 6.7: Utility Private HAN

talk to display or load control devices to inform the user of price increases or to control devices in a user's home, office, or commercial space. ZigBee forms a mesh network and basically serves for the last few hundred yards, or perhaps even a network as big as a few miles across.

As I write this, one of the first ZSE interoperability events is commencing with 38 different companies participating in various aspects of this application space. My own company, San Juan Software, is there as a stack expert helping some of our ZSE customers, while other companies are manufacturing the actual end products sold to consumers and to the utility companies. At the end of a series of these events, the profile will be solidified, field tested, and vetted with customers. Then it becomes widely deployed with millions of ZigBee units helping to reduce energy consumption and cost, a thing that helps our environment, and lowers consumer energy bills.

> ZigBee Public Profiles define profile specific clusters and devices.
>
> All ZigBee Public Profiles use the ZigBee Cluster Library.

6.4 When to Use ZCL

The ZigBee Cluster Library is very powerful, with many features. Unfortunately, that also means the size of the ZigBee Cluster Library is quite large, at least from an 8-bit MCU perspective. On the Freescale implementation, it requires about 3.8 K of code for a basic set of clusters, and more if many clusters are used.

Public ZigBee Application Profiles mandate the use of ZCL. There is no question about that. It must be used to make a compatible product that can be certified as compliant by the ZigBee Alliance. This includes Home Automation (HA), Commercial Building Automation (CBA), ZigBee Smart Energy (formerly known as Automatic Metering Initiative, or ZSE), Telecommunications Applications (TA), Wireless Sensor Applications (WSA), and Personal Home and Hospital Care (PHHC).

But what if you're making a private Application Profile? Should you use ZCL then? The answer depends on several factors. Generally, if the profile is very simple, or requires only a dozen or so clusters, it is usually more flash and RAM-efficient to build it all without the cluster library. But, if attributes would make the application very convenient, then why write and debug proprietary code, when it's already done for you in the ZigBee Cluster Library?

To make this decision for your project, think about the level of compatibility. Will the project eventually be shared? Do you hope someday for this to become a ZigBee public Application Profile? If the answer is, "Yes," then definitely use ZCL.

Does the application rely heavily on reading and writing data items? Would these be convenient as attributes, and possibly need the report attribute feature? If the answer again is "Yes," then you should probably use ZCL.

Is your application short on size, in either RAM or flash memory? This depends on which processor is used, whether security is needed, if the device is a ZigBee Router (the largest), or ZigBee End-Device (the smallest), and how large the actual application code is estimated to be. If the answer is "Yes, it might be short on flash memory," then don't use ZCL in your private Application Profile.

To get a good feeling for whether this will be the case in the Freescale solution, compile the Thermostat application, and check the sizes in the .map file. On the MC13213, a secure, mesh, ZCL-enabled thermostat application will not fit into a ZigBee router. All the options simply require more than the 60 K of flash memory available. If the device will be an end device (thermostats are often battery-powered), then it will fit. Or, if you are using the newer Freescale QE128 or MC13223 parts, it will fit just fine as a ZigBee Router.

In this following example, *Example 6-3 Private Profile with ZCL*, which I like to call "Tilt-o-Rama," a ZigBee device will send an alert if the "box" has tilted too far. Think of those packages you receive from UPS, FedEx, or another carrier that show up on your doorstep with the "This Side Up" arrow pointing down. By watching the Z-axis using an accelerometer, this application will send an attribute report if the box tips too far.

This example uses the ZigBee Cluster Library in a private profile. For an example of a private profile without the ZigBee Cluster Library, see *Example 4-10 iPod Controller* in Chapter 4, "ZigBee Applications."

A more robust version of this application could be used commercially to determine if goods were transported safely throughout the entire journey (see Figure 6.8), including signals if they were subject to out-of-range G forces from being dropped.

The example uses two Freescale boards: an NCB, which will be the alert-display, and an SRB, which will be the tilt detection mechanism.

The example uses a very simple UI. Simply boot both boards, and then tilt the cardboard box. There are no buttons to press. The SRB board represents the cardboard box shown on the right in Figure 6.8. If a tilt is detected, the NCB board, on the left, will begin to

Figure 6.8: Private Profile with ZCL

f0080

beep and will display "Tilt" in the LCD screen, to alert the UPS truck driver (or whoever is monitoring the package) of the problem.

The application was coded to simply use the standard ZigBee Cluster Library "on/off" cluster to indicate "tilt or not tilt." The "on/off" cluster contains an "on/off" attribute. This attribute can be read from, or it can be set up to report. In fact, that's all this demo does. It sets up the TiltDetector (the SRB) to report when the attribute changes state (from tilt to not-tilt and vice versa). The report is sent to the TiltDisplay (the NCB) which then starts madly beeping (to indicate tilt) or goes quiet again (to indicate no-tilt).

The following code fragment shows how the clusters are defined in Freescale BeeStack. This file is called TiltDetectorEndPoint.c, and replaces the file HaOnOffLightEndPoint.c. It appears as follows:

```
/* attributes for the Tilt Detector in RAM (an instance) */
typedef struct sTiltDetectorAttrsRAM_tag
{
  uint8_t    reportMask[1];
  zclOnOffAttrsRAM_t onOffAttrs;
} sTiltDetectorAttrsRAM_t;

/* one copy of this structure per instance of this device */
sTiltDetectorAttrsRAM_t gTiltDetectorData;

/* one copy for all instances of this device type */
afClusterDef_t const gaTiltDetectorClusterList[] =
{
  { { gaZclClusterBasic_c }, ZCL_BasicClusterServer,
  (void *)(&gZclBasicClusterAttrDefList) },
  { { gaZclClusterIdentify_c }, ZCL_IdentifyClusterServer,
  (void *)(&gZclIdentifyClusterAttrDefList) },
  { { gaZclClusterGroups_c }, ZCL_GroupClusterServer, NULL },
```

```
  { { gaZclClusterOnOff_c }, ZCL_OnOffClusterServer,
  (void *)(&gZclOnOffClusterAttrDefList),
  MbrOfs(sTiltDetectorAttrsRAM_t,onOffAttrs) }
};
/* one copy for all instances of this device type */
zclReportAttr_t const gaTiltDetectorReportList[] =
{
  { { gaZclClusterOnOff_c }, gZclAttrOnOff_OnOffId_c }
};
/* one copy per instance of this device type */
afDeviceDef_t const gTiltDetectorDeviceDef =
{
  ZCL_InterpretFoundationFrame,
  NumberOfElements(gaTiltDetectorClusterList),
  (afClusterDef_t *)gaTiltDetectorClusterList,
  NumberOfElements(gaTiltDetectorReportList),
  (zclReportAttr_t *)gaTiltDetectorReportList,
  &gTiltDetectorData
};
```

In the code fragment, begin with the *last* structure first and work your way up. The "device definition" describes which clusters are supported by the device, a list of reportable attributes for that device, and the instance data. The cluster list describes C functions to handle the cluster-specific commands and lists the cluster's attributes. All of this together allows the BeeStack ZigBee Cluster Library to read, write, and report attributes automatically.

When an APSDE-DATA.indication comes in, the application code simply passes the indication to the ZigBee Clustesr Library to interpret the frame. The cluster or cross-cluster command is handled automatically as shown below:

```
status = ZCL_InterpretFrame(pIndication);
```

Full source code to the example, including the BeeKit solution file, is found in the usual place at http://www.zigbookexamples.com. There is nothing new to learn from the over-the-air capture (as the application simply sets up attribute reporting), so I have not included it in the book, but it is available online along with the source code.

Adding in custom clusters to the Freescale ZigBee Cluster Library is fairly easy. Simply add a new C route to the list of potential clusters (in your application is fine). Provide the definition structures, and add it to the device definition.

Remember, although this example is shown using the Freescale solution, the concept applies equally well to all ZigBee stacks. Every one that passes certification for any

ZigBee public profile also supports the ZigBee Cluster Library, and all of them make it extensible in some way.

So, to sum up, the ZigBee Cluster Library is a set of clusters for use in public Application Profiles, which may also come in handy for private profiles. ZCL introduces the concept of attributes and commands. Attributes are data items, and commands cause action. The commands may be either cross-cluster (such as reading and writing attributes), or specific to the cluster (such as on/off toggle). Without ZCL, there are no attributes or commands, just clusters with payloads defined by the private profile.

The ZigBee Cluster Library is used in all ZigBee public profiles. Some of the clusters are common across all devices on all profiles, such as the groups or basic clusters, but other clusters are unique to a specific device or Application Profile, such as the Simple Metering Cluster in ZSE. The clusters described in the ZigBee Cluster Library Specification are available to all profiles. Clusters private to a particular profile are in that Application Profile's specification.

> Use ZCL in private profiles to gain attributes and commands.
>
> Don't use ZCL in private profiles if code or RAM space is tight.

CHAPTER 7

The ZigBee Networking Layer

So far this book has concentrated on how applications interact with ZigBee. After all, as OEMs, applications are what you are typically building. This chapter describes what goes on under the hood, especially at the network layer.

Do you need to know what goes on in a ZigBee mesh network underneath those application-level commands? Technically, no, you don't have to, just as you can drive a car and work the stereo without knowing how the fuel injection or the radiator works. But a good understanding of how ZigBee accomplishes networking is really, really helpful when things go wrong. Is the application sending too many broadcasts? Will a route be established when this packet is sent? Why, and through what nodes? As engineers, you want to know. Besides, mesh networking is really, really interesting.

But first, you've probably been wondering how ZigBee really got its name. I'll tell you.

There are many, many people who have contributed to the excellent standard called ZigBee. I've mentioned a very few throughout this book, and yet can't come close to listing all of the intelligent, dedicated people who made this standard possible. One person stands out: Zachary Brightlea Smith. No one person has contributed more technically (and perhaps spiritually) to the specification than Zachary.

I first met Zachary at the ZigBee Open House in Seattle in 2004. He was representing one of the ZigBee stack vendors at the time. Other than reading an article, this was the first introduction I had had to the wireless technology, and I was impressed—amazed, actually. Zachary painted a picture of wireless sensor and control networks helping us everywhere: working to save energy, make our environment safer and more convenient, helping the elderly to stay out of hospitals, and making new social networking possible.

It's interesting that these predictions have largely come true. ZigBee now has profiles for ZigBee Smart Energy, Home Automation, Personal Home and Health Care, and Telecommunication Applications.

Zachary himself is an unusual man. He has eyes in the back of his head (literally, in the form of a tattoo). When his head is shaved, a hairstyle he often favors, you find them staring back at you. He is Buddhist, and spends long hours staring at a single point on the wall (on both walls?). It is rumored that he has attended the Burning Man festival, an annual self-expression art experience in the Nevada desert (www.burningman.com).

My wife and I had the opportunity to prowl Paris one evening with Zachary. He took us to a restaurant, hidden away in a small alley along the Seine, where he spoke French and we ate the best foie gras we've ever tasted. Zachary had lived in Paris and knew it well. Zachary also lived in Milan, Italy (he looks good in an Italian suit), and I am sure helped instigate a ZigBee Alliance quarterly meeting there that culminated in an amazing dining experience at a Leonardo da Vinci museum. He even taught the ZigBee Alliance proper cappuccino etiquette.

Time and again Zachary will drop a pithy, insightful email on a ZigBee reflector, which changes how the group views a problem, often resulting in a specification change or clarification. During its primary development, Zachary was technical editor of the Network Working Group (NWG), the group that defined the networking layer within ZigBee, and was intimately involved in the routing algorithms. I would even go so far as to say that Zachary is the father of ZigBee.

Hmmm... Zachary Brightlea (Smith). Z.B. ZigBee. Coincidence?

7.1   ZigBee and IEEE 802.15.4

The terms "ZigBee" and 802.15.4 (or simply 15.4) are often confused, and are all too often used interchangeably. They are not the same.

The 802.15.4 specification was created and is maintained by IEEE (http://www.ieee.org). This specification defines the physical (PHY) and media access control (MAC) layers of a personal area, low-power, wireless network. All IEEE 802.x.x specifications define networking standards. You may have heard of 802.3 (Ethernet), or 802.11 (WiFi™), or 802.15.1 (Bluetooth™).

IEEE 802.15.4 defines:

- Mechanisms for discovering networks
- Mechanisms for forming and joining networks
- Mechanisms for changing channels
- Mechanisms for detecting interference and "noise" on a particular channel

- An acknowledged, single-hop, data-packet delivery method, using CSMA-CA to avoid collisions

- An unacknowledged, single-hop data-broadcast method

What 802.15.4 does *not* specify is anything to do with multi-hop communications, address assignment, or application level interoperability.

If the network you have in mind to build only needs a single hop (that is, all radios in the network can hear each other), then the 802.15.4 MAC/PHY may be all that you need. In fact, Freescale and other silicon vendors offer an 802.15.4 application environment that doesn't require ZigBee at all.

Aside from ZigBee, many other network protocols have been built on the 802.15.4 standard, some which are mesh or multi-hop, some which are single-hop or star networks. Some examples of non-ZigBee network protocols on 802.15.4 include MicroChip MiWi™, Synap SNAP, 6LoWPAN (an IETF standard), Tiny O/S (an open source project), and San Juan Software PopNet™.

But ZigBee is the primary protocol which builds on the 802.15.4 standard, adding a network layer capable of peer-to-peer multi-hop mesh networking, a security layer capable of handling complex security situations, and an application layer for interoperable application profiles.

Below is the standard ZigBee architecture diagram, seen in nearly every technical discussion of ZigBee. Notice how the MAC and PHY layers are under IEEE control, while the rest are under ZigBee control (see Figure 7.1).

The PHY (physical) layer's job is to translate the packets from a series of bytes to the RF spectrum, and back again. The MAC (medium access) layer allows a network to be formed, for channels to be shared, and for data to be transferred (single-hop) in a reasonably reliable way.

ZigBee specifies all the layers above MAC and PHY, including the NWK (network) layer, APS, ZDO, and security layers. It's ZigBee that provides the mesh networking and multi-hop capabilities, enhances the reliability of data packet delivery, and specifies application-to-application interoperability.

ZigBee does not use all of the 802.15.4 MAC/PHY specification; only a small subset. This allows stack vendors to offer smaller solutions (less RAM and flash) by providing a limited MAC for their ZigBee stacks. For example, ZigBee does not use any of the

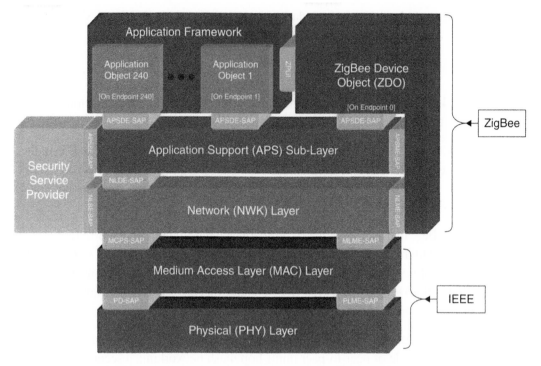

Figure 7.1: ZigBee and IEEE 802.15.4

802.15.4 beaconing methods, or guaranteed time slots. ZigBee is asynchronous.
Any node may transmit at any time. Only the CSMA-CA (carrier-sense multiple-access
with collision avoidance), a MAC-level mechanism, prevents nodes from talking over
each other.

CSMA-CA works the same way as a human conversation in a social setting. One
person talks while the others wait. When there is a pause in conversation, then it is
someone else's turn to talk. Occasionally two people talk over each other. They back off
(randomly) and try again. Always keep that in mind: Only one radio is talking at any
given time on any given channel in any given vicinity. Also keep in mind all 802.15.4
radios are half-duplex: They are either talking or listening, but not both at the same time.

ZigBee also makes some slight modifications to the 802.15.4 standard. One of these
is the security model. The MAC defines something called CCM, which spelled out in
its long form is "counter-mode cipher-block chaining-message authentication code."
CCM requires different security for every layer. Because of performance constraints on

small processors, ZigBee doesn't. The ZigBee security model is called CCM (a slight modification of MAC CCM security).

One of the more interesting areas where ZigBee differs from the 802.15.4 specification is the time-out for beacon responses. By the way, beacon requests and beacon responses (just called beacons) have nothing to do with whether the network is a beaconing network or not. Confusing, I know. I wish IEEE had used different terms for beacons that start a guaranteed time-slot frame versus the beacons used to join and form networks, but there it is.

A beacon (which I'll explain in excruciating detail later in this chapter) is simply a packet that contains information about the node and network. This is used in ZigBee to discover networks. In dense networks, with 30 or more nodes all within hearing range, the default 802.15.4 time-outs for responses to beacon requests don't allow enough time for all nodes to respond. The 802.15.4 specification was just not built with large networks in mind, but ZigBee is.

The 802.15.4 MAC specification has been stable since October 2003. Some silicon vendors even offer the 802.15.4 MAC in ROM. But IEEE doesn't stand still. In 2006, IEEE released another 802.15.4 specification, called, of course, 802.15.4-2006.

Probably the biggest change in the IEEE 802.15.4-2006 standard is a better PHY for sub-1 GHz radios. In the 802.15.4-2003 specification, the 868 MHz and 900 MHz PHYs were limited to 20 kbps (kilobits per second) and 40 kbps respectively. The data transfer rate at sub 1 GHz were simply too slow for ZigBee; all ZigBee radios are 2.4 GHz, which operate at 250 kbps.

IEEE 802.15.4-2006 changed all that. The specification added two new optional PHYs for sub 1 GHz allowing up to 250 kbps in those bands as well. I expect to see some ZigBee chips over the next couple of years operating in these bands. The sub 1 GHz bands still do not operate in the worldwide unlicensed spectrum, so restrictions apply on where they can be exported. But if your market is mainly the U.S., 900 MHz will do nicely. The upside of 900 MHz is that it penetrates water (plus plants, foliage, and people) more easily than the 2.4 GHz.

As of this writing, ZigBee continues to use the IEEE 802.15.4-2003 specification, but there have been discussions within the ZigBee Alliance to update ZigBee to the newer, largely backward-compatible IEEE 802.15.4-2006 specification.

> IEEE 802.15.4-2003 specifies the MAC and PHY layers in a ZigBee stack. ZigBee specifies the NWK, application, and all higher layers.

7.2   Forming, Joining, and Rejoining ZigBee Networks

Over the next few sections I'll describe the entire process of a ZigBee network starting up and establishing communications, including how ZigBee nodes form a network, join a network, are assigned addresses on the network, and how they route packets from one node to another.

Before any ZigBee node may communicate on a network, it must either form a new network or join an existing network. Only the ZigBee Coordinator may form a network. Only ZigBee Routers and ZigBee End-Devices may join a network. Many stack vendors offer the ability to have a node designated as a ZC, ZR, or ZED at compile time (to save code and RAM) or at run-time (to reduce OEM-manufactured parts).

Every node starts out with a unique 64-bit IEEE (aka MAC or long) address, assigned by the OEM during manufacturing. During the process of joining the network, each node is assigned a unique (within that network) 16-bit short address (aka NwkAddr) to use when communicating to other nodes across the network. The 16-bit address is used for nearly all communications, to reduce over-the-air protocol overhead, leaving more room for application payload.

7.2.1   Forming Networks

ZigBee Coordinators (ZCs) form networks. The process of forming a network is really about determining a unique identifier for the network, called a PAN ID, and choosing one of the sixteen 802.15.4 channels (11–26) on which to operate the network. By the time the ZC has formed the network, the network is established (see Figure 7.2).

Granted, a network of one node is not very useful. It's not until other nodes join the network that networking becomes possible.

Figure 7.2: A Network of One

During the forming process, a single packet is sent over-the-air on each channel: a MAC active scan (aka beacon request). If no other ZigBee networks are on the channel, this will be the only packet seen by a ZigBee sniffer:

```
Seq No      MAC Dst      Protocol         Packet Type
-----------------------------------------------------------
1           0xffff       IEEE 802.15.4    Command: Beacon equest
```

So how do you know which channel and PAN ID the ZC has chosen? I'll get into that in a moment.

A ZigBee Coordinator has the following duties:

- It forms a network.

- It establishes the 802.15.4 channel on which the network will operate.

- It establishes the extended and short PAN ID for the network.

- It decides on the stack profile to use (compile or run-time option).

- It acts as the Trust Center for secure applications and networks.

- It acts as the arbiter for End-Device-Bind (a commissioning option).

- It acts as a router for mesh routing.

- It acts as the top of the tree, if tree routing enabled.

Does having a single device, the ZigBee Coordinator, create a single point-of-failure in a ZigBee network? Well, not really. The ZigBee Coordinator is really just a router unless the network is being commissioned. And there are ways to replace the ZigBee Coordinator after a network is running, if the device fails for some reason. I'll explore ways to do this in Chapter 8, "Commissioning ZigBee Networks."

It's the application running on the ZigBee Coordinator node that actually decides when it's time to form a network, from which set of channels, and from which set of PAN IDs. The application running in the ZC can be anything: a gateway connected to the Internet, a controller box, a thermostat, light, or electric meter. The possibilities are endless. When power is applied to the device that contains the ZigBee Coordinator, it may immediately form a network, or may wait for some event (such as a button-press or a command from a host processor) before forming the network. It might even check to see what networks are already out there, and decide to become a ZigBee Router rather than a Coordinator, if another node has already formed the desired network.

However the application is coded, at some time a ZigBee Coordinator will form a network. The process is shown in Figure 7.3.

Notice that first there is an NLME-NETWORK-FORMATION.request. This is initiated by ZDO, which in turn was told to form a network by the application. In Freescale's BeeStack, telling ZDO to form the network is very simple, as seen in the code fragments below. The first code fragment forms the network immediately on power-up. The second forms the network on a key-press (although this could be on any event):

```
/* form network immediately upon power-up */
void BeeAppInit(void)
{
   ZDO_Start(gStartSilentRejoinWithNvm_c);
}
/* form network on key-press */
void BeeAppHandleKeys (key_event_t keyEvent)
{
   switch(keyEvent)
   {
     case gKBD_EventSW1_c:
     ZDO_Start(gStartWithOutNvm_c);
     break;
   }
}
```

Next, as seen in Figure 7.3, ZigBee calls on the MAC layer to perform two scans: an energy scan, and an active scan. The energy scan is used to determine which channel is the quietest channel from the set of channels specified in the APS information base variable, apsChannelMask. The energy scan lasts about a half-second on each channel, and is just a "moment-in-time" check. The channel could have been really noisy one hour before, and this process wouldn't detect it. Scanning all 16 channels takes about 8 seconds, so be patient.

Next, an active scan, which is simply a MAC beacon request followed by zero or more beacon responses, is used to discover what other networks are in the vicinity. Active scans ensure ZigBee does not form a network on the same PAN ID as another network already in the vicinity. The active scan also takes time, waiting for potential beacon responses from other nodes.

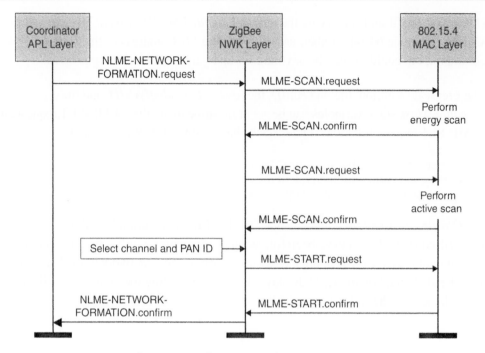

Figure 7.3: ZigBee Forming a Network

Next in Figure 7.3, there is this magic box labeled "Select channel and PAN ID." After receiving the information about how noisy the channels are, and which networks are already out there, the application is free to chose the "correct" channel and PAN ID. So what is the "correct" channel and PAN ID?

The way this works with the Freescale solution is to pick the quietest channel that does not already contain the PAN ID specified by the application. The functions which do this are SelectLogicalChannel() and SelectPanId(), found in the file AppStackImpl.c. These functions can be replaced by anything appropriate to the application. Check your stack vendor's documentation for how to choose the "correct" channel and PAN ID for that stack implementation.

The NWK information base variable, *nwkPanID*, specifies the PAN ID, which by default is 0xffff (meaning ZigBee is free to choose a random PAN ID). The application can also set *nwkPanID* to anything else from 0x0000 through 0x3fff, if the intent is to pick a particular PAN ID rather than a random one. The top two bits are reserved for future use. I don't recommend specifying a particular PAN ID. Let ZigBee do it for you, by setting *nwkPanID* to 0xffff and specifying the *apsUseExtendedPANID* instead. The extended PAN ID is a unique 64-bit number, usually set to the MAC address of the ZigBee Coordinator, but could instead be anything set up by the commissioning process.

In the Freescale solution, the *nwkPanID*, the *apsUseExtendedPANID*, and the *apsChannelMask* stack variables can be set at run-time using the NLME-SET.request and APSME-SET.request functions. Other stacks have a similar way to set them:

```
NlmeSetRequest(gNwkPanId_c, 0x1234);
ApsmeSetRequest(gApsUseExtendedPANID_c, aMyExtendedPANID);
ApsmeSetRequest(gApsChannelMask_c, aMyChannelMask);
```

The ZC node can ascertain which channel and PAN ID was chosen by calling NlmeGetRequest(). If you have been following the examples, you've already seen this occur on the Freescale NCB boards. After they form or join a network, they display the PAN ID and channel on the LCD display. The code for getting the chosen PAN ID and channel looks like the following in the Freescale platform:

```
zbPanId_t aPanId;
zbChannel_t channel;
Copy2Bytes(aPanId, NlmeGetRequest(gNwkPanId_c));
channel = NlmeGetRequest(gNwkLogicalChannel_c);
```

Once a network is formed, the next step in creating a network is for other nodes to join.

---

apsChannelMask defines which set of channels to consider when forming a network.

nwkPANID of 0xffff indicates choose random PAN ID.

Use apsUseExtendedPANID to specify the 64-bit extended PAN ID.

7.2.2 Joining Networks

ZigBee Routers (ZRs) and ZigBee End-Devices (ZEDs) join networks. ZRs are usually mains-powered, always on, listening for packets to route. ZEDs are usually battery-powered and sleeping, waking up only to communicate briefly before going back to sleep. ZigBee Routers are responsible for:

- Finding and joining the "correct" network
- Perpetuating broadcasts across the network

- Participating in routing, including discovering and maintaining routes

- Allowing other devices to join the network (if permit-join enabled)

- Storing packets on behalf of sleeping children

ZigBee End-Devices are responsible for:

- Finding and joining the "correct" network

- Polling their parents to see if any messages were sent to them while they were asleep

- Finding a new parent if the link to the old parent is lost (NWK rejoin)

- Sleeping most of the time to conserve batteries when not in use by the application.

Joining a network is a process of discovering what networks and nodes are in the vicinity, and then choosing one of them to join. Provided the association is acceptable to the network, the join will be completed, and the joining node will now have an address on that network.

The joining node initiates the process, shouting "Hey, is anyone out there?" using a beacon request (see Figure 7.4). Any ZCs and ZRs in the vicinity respond, saying "Here

Figure 7.4: ZigBee Beacon Request and Response

I am" using a beacon. Although not technically accurate according to the 802.15.4 specification, I often call the beacons "beacon responses," because that's how they are processed: A node sends out a beacon request, other nodes respond with a beacon response.

The beacon response is issued by all ZigBee Routers and Coordinators on the channel where the beacon request was issued, regardless of PAN ID. So, for example, if a beacon request was sent on channel 15, all router nodes (including ZigBee Coordinators) on channel 15 respond with a beacon response, as seen in the partial capture below. Notice the multiple PAN IDs and short addresses:

```
SrcPAN      MACSrc      MACSeq      Protocol Type
----------------------------------------------------------------
0x1aaa      0x0001      05          Beacon: AP: 1, Nwk RC: 1, Nwk EDC: 1
0x1bbb      0x0000      188         Beacon: AP: 0, Nwk RC: 1, Nwk EDC: 1
0x1aaa      0x0000      200         Beacon: AP: 1, Nwk RC: 1, Nwk EDC: 1
```

Beacons contain quite a bit of information about the ZigBee network, including PAN ID, extended PAN ID, join enable, and whether the node has capacity (room) for routers or end devices to join. What the beacon lacks is any application-level information. For that, a seeking node must first join the network, look for the application and, if it cannot find it, leave the network and try another.

Take a look at a typical beacon below:

```
Frame 3 (Length = 24 bytes)
IEEE 802.15.4
  Frame Control: 0x8000
    .... .... .... .000 = Frame Type: Beacon (0x0000)
    .... .... .... 0... = Security Enabled: Disabled
    .... .... ...0 .... = Frame Pending: No more data
    .... .... ..0. .... = Acknowledgement not required
    .... .... .0.. .... = Intra PAN: Not within the PAN
    .... ..00 0... .... = Reserved
    .... 00.. .... .... = Destination Address: not present
    ..00 .... .... .... = Reserved
    10.. .... .... .... = Source Address: 16-bit short address
  Sequence Number: 205
  Source PAN Identifier: 0x0bef
  Source Address: 0x0000
  MAC Payload
  Superframe Specification: 0xcfff
    .... .... .... 1111 = Beacon Order (0x000f)
    .... .... 1111 .... = Superframe Order (0x000f)
```

```
.... 1111 .... .... = Final CAP Slot (0x000f)
...0 .... .... .... = Battery Life Extension: Disabled
..0. .... .... .... = Reserved
.1.. .... .... .... = PAN Coordinator: yes
1... .... .... .... = Association Permit: enabled
GTS Specification: 0x00
Pending Address Specification: 0x00

Beacon Payload
Protocol ID: ZigBee NWK (0x00)
NWK Layer Information: 0x8421
.... .... .... 0001 = Stack Profile (0x1)
.... .... 0010 .... = nwkcProtocolVersion (0x2)
.... ..00 .... .... = Reserved (0x0)
.... .1.. .... .... = Router Capacity: True
.000 0... .... .... = Device Depth (0x0)
1... .... .... .... = End Device Capacity: True
Extended PAN ID: 0x0050c211dc051801
```

From this beacon, I can tell that it's a ZigBee node (as opposed to another network), specifically the ZigBee stack profile 0x01. I can tell that the node is the ZigBee Coordinator, and that it is allowing joining, and has the capacity to accept either router or end devices as children. I can tell that the PAN ID is 0x0bef, and the extended PAN ID is a Freescale extended PAN ID, 0x0050c211dc051801.

The first field, the MAC frame control, will always be 0x8000, a beacon. The security-enabled bit will always be 0, even in a secure network, as ZigBee uses NWK layer security (not MAC layer security). The next field, the sequence number, is a rolling counter. It's not really needed, but this uniquely identifies the frame. The source PAN ID and short address of the node will always be present. In this case the short address is 0x0000, so it's the ZigBee Coordinator. Any other address would be a router (end devices do NOT issue beacons).

The next field, the superframe specification, indicates whether permit joining is enabled on this node (see the Association Permit bit). If it is not, then the node will not accept joining. The rest of the MAC payload fields will always be the same for all ZigBee nodes. Notice that beacon order is 0xf, that is, non-beaconed. ZigBee is not a beaconing network, and does not support guaranteed time slots. Instead, ZigBee is asynchronous, with routers always enabled to route packets, using CSMA-CA to avoid collisions on the network.

The network layer information follows the MAC payload and includes which ZigBee stack profile (0x01 or 0x02) is supported by this node, and whether it has capacity to accept routers or end-devices. In stack profile 0x01, the total number nodes that can join a particular node as a child is 20, 6 of which can be routers and 14 of which can be end devices.

ZRs and ZEDs join a specific node, not a network in general, using 64-bit MAC addresses for the source and destination addresses of the MAC association request. The node doing the joining is called a child. The node receiving the association request is called the parent.

ZigBee Routers and the ZigBee Coordinator can be the parent of other nodes, but ZigBee End-Devices will never be a parent. ZEDs are always children.

Keep in mind this parent/child relationship has nothing to do with mesh routing. Any router may route through any other router within hearing range on the same network. All ZigBee Routers (and the ZC) are peers in this respect. If a router's parent or child goes away (perhaps even leaving the network) no harm is done to routes, other than to those specific devices (assuming a network has sufficient density). ZigBee Routers do not route through other networks, only those within the same PAN ID and channel.

To ZigBee End-Devices, however, the parent/child relationship is very special. ZEDs, while they can communicate to any node on the network, communicate only to their parent directly. Another way to put it is that the next hop of a ZED is always to its parent. If a ZED loses the link with its parent, it must (at some point) find another parent in order to keep communicating on the network, a process called rejoining.

It is not often that a ZED child loses contact with its parent, but it does happen. Say, perhaps the office furniture was moved and a fish tank now resides directly in front of the ZED. Communications at 2.4 GHz are affected by water, and the link is now broken. No problem: Within a few seconds, the ZED finds a new parent, announces to the network that it has moved, and communications continue. I'll show an example of this in the rejoin section below.

The join process for a ZigBee Router or ZigBee End-Device is depicted in the ZigBee specification, as shown in Figure 7.5.

First, an active scan (beacon request) is sent out on each channel. The ZC or ZED then waits for a time for beacon responses. The time can be set by the application, but it defaults to about a half-second per channel. Once the beacons are collected, then they are analyzed in the "select channel and PAN ID" box. This function is application-specific.

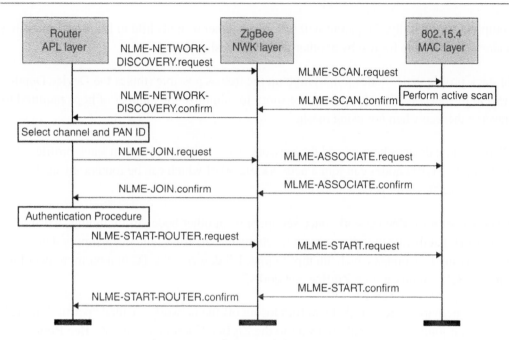

Figure 7.5: ZigBee Joining a Network

Freescale BeeStack includes an application-specific function called by ZDO whenever a node is told to join a network, called SearchForSuitableParentToJoin(). This function looks for any node which matches the *apsUseExtendedPANID* or, if that's set to all 0x00s, any node which matches the *nwkPANID* stack variable. A *nwkPANID* of 0xffff means to join any PAN. This same setting is what is used by the Home Automation Application Profile: Join any PAN ID (and any extended PAN ID) on any channel.

In addition to the correct extended PAN ID, PAN ID and channel, the potential parent must also have permit-join enabled. Permit-join can be used to "close off" a network, preventing other nodes from joining it. Permit-join can also be used to force nodes to pick a particular parent during network bring-up.

After the active scan is complete and a suitable parent is selected, the authentication procedure begins. Note that the node actually has an address on the network (at least temporarily) by the time authentication begins. Authentication is only present in secure networks, and gives the trust center the right to refuse to allow this node to join. A rogue node which only pretends to already have an address on the PAN will not receive the network key, and so cannot communicate to other nodes. If the authentication does not complete successfully, the parent will tell the unauthenticated child to leave and mark that address as available for use by another node that wishes to join.

In stack profile 0x01, nodes join as high up the tree as possible (this is the Device Depth field in the beacon). This keeps the tree short, to minimize the number of hops required to traverse the tree when not using mesh.

Okay, so now devices have joined the network. How many can join? In stack profile 0x01, up to 31,100 nodes can join a network, 9,330 of which can be routers. In stack profile 0x02, the limit is 64 K nodes.

So how does a ZigBee network, once set up, prevent other nodes from joining? Trust Center (TC) authentication is one such mechanism. The trust center can simply deny access to any new nodes which attempt to join. I'll describe the TC in a bit more detail in Chapter 8, "Commissioning ZigBee Networks."

Another, perhaps easier, mechanism for closing off the network is called "permit joining." The permit joining flag is only relevant to the ZigBee Coordinator and ZigBee Routers, which are the only node types that can have children. The NWK layer allows permit joining to be locally enabled or disabled through the NLME-PERMIT-JOINING.request, and ZDP provides the ability to send this command over-the-air to any or all nodes in the network using Mgmt_Permit_Joining_req. Either request allows joining to be turned on (0xff), off (0x00), or turned on for a number of seconds from 1 through 254 (0x01–0xfe) before being turned off.

In the Freescale solution, the permit joining commands are available through:

```
/* turn permit joining on or off locally */
void APP_ZDP_PermitJoinRequest
  (
  uint8_t iPermitFlag    /* 0x00 = off, 0xff = on */
  );
/* turn permit joining on or off remotely */
void ASL_Mgmt_Permit_Joining_req
(

    zbCounter_t *pSequenceNumber,
    zbNwkAddr_t aDestAddress,    /* usually gaBroadcastZCnZR */
    zbCounter_t permitDuration, /* 0x00 = off, 0xff = on */
    uint8_t TC_Significance     /* TC will not authenticate */
);
```

In this first example, *Example 7-1 Permit Joining*, the NLDE-PERMIT-JOINING.request will be used to disallow nodes to join, and then to force nodes to join either in a small three-node tree, or in a three-node straight line, as depicted in Figure 7.6.

As with all examples in this book, code is provided for the Freescale platform, but the concepts and over-the-air captures apply to all ZigBee platforms. See Chapter 3, "The ZigBee Development Environment," for an introduction to the Freescale platform, if you are interested in following along with actual hardware.

If you do follow along with actual Freescale hardware, the steps to run the example are:

1. Open the *Example 7-1 Permit Joining* solution with BeeKit.

2. Export the solution. Import the three projects into CodeWarrior.

3. Compile and download the three images (ZcNcbOnOffLight, ZrSrbOnOffLight, and ZedSrbOnOffSwitch) into the three respective boards.

4. Start a capture on channel 25 with Daintree SNA (if available).

5. Press SW1 on the ZC (ZcNcbOnOffLight) to form the network.

6. Press SW2 on the ZC to turn off local join enable (LED2 should go out).

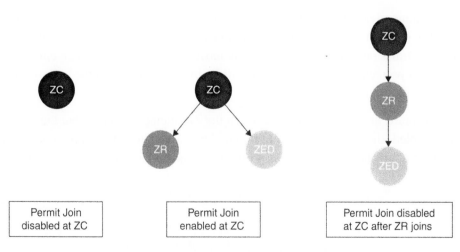

Figure 7.6: Example 7-1—Permit Joining

7. Press SW1 on the ZR (ZrSrbOnOffLight). Notice that it can't join the network, because joining is disabled on the ZC. The capture below shows how it repeatedly attempts to join by repeatedly issuing beacon requests.

8. Press SW2 on the ZC to enable joining again. Notice the ZR now joins (LED 1 and 2 go solid).

9. Press SW1 on the ZED (ZedSrbOnOffSwitch). Notice that it, too, joins the network. The network now looks like the middle of Figure 7.4, with the ZR and ZED as children of the ZC.

Now reset all three boards.

1. Press SW1 on the ZC again to form the network.

2. Press SW1 on the ZR again. Notice it joins the network immediately this time.

3. Press SW2 on the ZC, after the ZR has joined, to disable joining on the ZC.

Now press SW1 on the ZED. Notice that it now joins the ZR (not the ZC), as seen at the end of the capture below.

From the standpoint of communicating on the network, it's usually irrelevant which parent a node joins. Sometimes it can be useful in the real world. For example, in a hotel it is helpful to have all ZEDs in a given room join a router in that room. That way, if the other rooms around it are under construction, the lights, heating, cooling, etc. all continue to work. Each room is essentially autonomous.

Turning joining on or off throughout the entire network is almost always useful. After a network has been set up, it's usually undesirable to allow new nodes to join without the installer's permission. This is essentially why most WiFi networks in urban environments are password-protected. Imagine if a rogue, drive-by node could turn the lights or heating on or off in your home or business!

This first capture shows what happened when the ZR attempted to join, but the ZC had joining disabled (the left most image in Figure 7.6). Notice how the joining node simply keeps repeating the beacon request, as it found no beacon that is accepting joining:

```
Seq    Time      SrcPAN    MACSrc    MACDst    Packet Type
------------------------------------------------------------------------
1                                    0xffff    IEEE 802.15.4 Beacon Request
2      +05.359                       0xffff    IEEE 802.15.4 Beacon Request
3      +00.003   0x0f00    0x0000              IEEE 802.15.4 Beacon: AP: 0
```

```
4    +02.019                            0xffff   IEEE 802.15.4 Beacon Request
5    +00.003   0x0f00   0x0000          IEEE 802.15.4 Beacon: AP: 0
6    +02.207                            0xffff   IEEE 802.15.4 Beacon Request
7    +00.004   0x0f00   0x0000          IEEE 802.15.4 Beacon: AP: 0
8    +02.019                            0xffff   IEEE 802.15.4 Beacon Request
9    +00.005   0x0f00   0x0000          IEEE 802.15.4 Beacon: AP: 0
... and so on ...
```

This next capture shows the middle image in Figure 7.6, where both the ZR and ZED join the ZC. Notice both children associate (join) to the ZC in packets 4 and 13, sending the request to node 0x0000 of PAN 0x0f00. Note that the MAC (IEEE) addresses of the three nodes are as follows:

ZC-0x0050c237b0040001

ZR-0x0050c237b0040001

ZED-0x0050c237b0040001

```
Seq Time     SrcPAN   MACSrc   MACDst             Packet Type
             DstPAN   MACDst   (if 2 line)
-------------------------------------------------------------------------------
1                              0xffff             IEEE 802.15.4 Beacon Request
2    +02.780                   0xffff             IEEE 802.15.4 Beacon Request
3    +00.003   0x0f00   0x0000                    IEEE 802.15.4 Beacon: AP: 0
4    +02.022   0xffff   00:50:c2:37:b0:04:00:02
               0x0f00   0x0000                    Association Request
5    +00.001                                      Acknowledgment
6    +00.494            00:50:c2:37:b0:04:00:02
               0x0f00   0x0000                    Data Request
7    +00.001                                      Acknowledgment
8    +00.003            00:50:c2:37:b0:04:00:01
               0x0f00   00:50:c2:37:b0:04:00:02   Association Response
9    +00.001                                      Acknowledgment
10   +00.593                   0xffff             IEEE 802.15.4 Beacon Request
11   +00.003   0x0f00   0x0001                    IEEE 802.15.4 Beacon: AP: 0
12   +00.006   0x0f00   0x0000                    IEEE 802.15.4 Beacon: AP: 0
13   +02.014   0xffff   00:50:c2:37:b0:04:00:03
               0x0f00   0x0000                    Association Request
14   +00.001                                      Acknowledgment
15   +00.496            00:50:c2:37:b0:04:00:03
               0x0f00   0x0000                    Data Request
16   +00.001                                      Acknowledgment
17   +00.003            00:50:c2:37:b0:04:00:01
               0x0f00   00:50:c2:37:b0:04:00:03   Association Response
18   +00.001                                      Acknowledgment
```

This final capture shows the rightmost image in Figure 7.6, in which the ZR joins the ZC, and the ZED joins the ZR forming a multi-hop straight line. Notice that the ZR still associates with the ZC (node 0x0000), but the ZED associates with the ZR (node 0x0001):

Seq	Time	SrcPAN DstPAN	MACSrc MACDst	MACDst (if 2 line)	Packet Type
1				0xffff	IEEE 802.15.4 Beacon Request
2	+02.780			0xffff	IEEE 802.15.4 Beacon Request
3	+00.003	0x0f00	0x0000		IEEE 802.15.4 Beacon: AP: 0
4	+02.022	0xffff	00:50:c2:37:b0:04:00:02		
		0x0f00	0x0000		Association Request
5	+00.001				Acknowledgment
6	+00.494		00:50:c2:37:b0:04:00:02		
		0x0f00	0x0000		Data Request
7	+00.001				Acknowledgment
8	+00.003	0x0f00	00:50:c2:37:b0:04:00:01		
			00:50:c2:37:b0:04:00:02	Association Response
9	+00.001				Acknowledgment
10	+00.593	0xffff			IEEE 802.15.4 Beacon Request
11	+00.002	0x0f00	0x0000		IEEE 802.15.4 Beacon: AP: 0
12	+00.007	0x0f00	0x0001		IEEE 802.15.4 Beacon: AP: 0
13	+02.014	0xffff	00:50:c2:37:b0:04:00:03		
		0x0f00	0x0001		Association Request
14	+00.001				Acknowledgment
15	+00.496		00:50:c2:37:b0:04:00:03		
		0x0f00	0x0000		Data Request
16	+00.001				Acknowledgment
17	+00.003		00:50:c2:37:b0:04:00:02		
		0x0f00	00:50:c2:37:b0:04:00:03	Association Response
18	+00.001				Acknowledgment

Once a node has joined a network, it can communicate with any other node in the entire network. There is no requirement for binding or other mechanisms. Simply send the data from one node to another, as long as you know the short address of the node. Of course, for the application to see the packet, the Application Profile must be the same on both sides, and the source endpoint on the sending node and the destination endpoint on the receiving node must be registered. See Chapter 4, "ZigBee Applications," for a good explanation of Application Profiles, binding, and addressing within a node.

When building custom (private) profiles, I often set it up so that all the nodes in the network have the same endpoint on both ends, and the same private profile on all endpoints. It keeps life simple.

> *ZRs and ZEDs join networks.*
>
> Joining also includes authentication, giving the network a chance to reject the node.
>
> Use permit-join enable and disable to turn joining on or off in a network.

7.2.3 Rejoining Networks

Rejoining assumes that the node has already joined the network, that the node has a proper PAN ID, extended PAN ID, security key, and short address. There are several reasons why a node might need to rejoin the network:

- A ZED has lost contact with its parent

- Power has been cycled, and many or all nodes in the network rejoin "silently"

- Joining a secure network if permit-joining is off

ZEDs always communicate directly to their parents. If the parent no longer responds, the child must, eventually, find a new parent in order to keep communicating on the network. The child itself decides when it has been orphaned, not the parent. ZigBee does not mandate a specific number of failed polls or messages before a ZED assumes it can no longer communicate with its parent; it is up to the application.

In the Freescale solution, a settable option called *gMaxFailureCounter_c* defaults to two poll periods. If the ZED misses two poll periods (or attempts to send two messages that fail to get through), it will automatically initiate a network rejoin procedure. This number can be set to anything: 5, 10, 100, or set to 0 to indicate the application itself will initiate the rejoin if needed.

The example in this section shows a ZED rejoining and finding a new parent, while continuing its normal networking functions, as seen in Figure 7.7.

The rejoin process starts with a beacon request to find suitable parents. It is irrelevant whether the potential parent nodes have permit-joining enabled or not. The only thing that matters is that they have end-device capacity. After the beacon request, the ZED picks one of the nodes (on the same PAN) to be a parent, performs the rejoin, receives a new short address (stack profile 0x01 only), and then finally issues a device-announce to tell the network that this node has moved. This last step is very important to preserve bindings across the network.

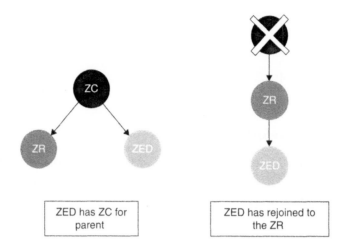

ZED has ZC for parent

ZED has rejoined to the ZR

Figure 7.7: Example 7-2 ZED Rejoins to a New Parent

In this example, *Example 7-2 ZED Rejoin*, the ZR is an HA On/Off Switch, and the ZED is an HA On/Off Light. The network is initially formed in the usual three-node triangle, as depicted in Figure 7.7. The ZR switch is bound to the ZED light. Then, at some point, the ZC drops off the network. (I simply turned it off.) The ZED realizes that it can no longer communicate with its parent and so initiates the rejoin procedure, establishing the ZR as its new parent and announcing its new short address (it moved from NwkAddr 0x796f to 0x1430) to the network. The binding is preserved, and the light can still be toggled on or off. The steps to run this example are:

1. Open the *Example 7-2 ZED Rejoin* solution with BeeKit.

2. Export the solution. Import the 3 projects into CodeWarrior.

3. Compile and download the 3 images (ZcNcbOnOffLight, ZrSrbOnOffSwitch, and ZedSrbOnOffLight) into the 3 respective boards.

4. Press SW1 on all three boards to form/join the network.

5. Press SW3 on the ZR On/Off switch and the ZED On/Off light to bind them.

6. Press long SW1 on the ZR and ZED to go to application mode.

7. Press SW1 on the ZR to toggle the ZED light. Notice the light takes up to 1 second to toggle. This is the poll rate of the ZED to its parent (the ZC).

8. Turn off the ZC. Press SW1 on the ZR immediately. Notice the light doesn't toggle.

9. After a few seconds, press SW1 on the ZR again. Notice the light now toggles. The ZED has found a new parent and announced its new short address.

Take a look at the capture below. I've inserted the step enumeration between the captured packets. Step 4 is when the nodes both join the ZC. In Step 5, the ZR switch binds with the ZED light. In Step 7, the ZR switch toggles the ZR light (note it goes through the ZC). In Steps 8 and 9 the ZED realizes it can't communicate to its parent, so it finds a new one (the ZC) and because of the EndDeviceAnnce, the binding between the switch and light continues without interruption on the new ZED short address, 0x1430:

```
Seq  Time        SrcPAN     MACSrc      MACDst     Packet Type
                 DstPAN     MACDst      (if 2 line)
----------------------------------------------------------------------
STEP 4
4    +02.022     0xffff     00:50:c2:37:b0:04:00:02
                 0x0f00     0x0000                  IEEE:Association Request
13   +02.014     0xffff     00:50:c2:37:b0:04:00:03
                 0x0f00     0x0000                  IEEE:Association Request

STEP 5
25   +05.892     0x0f00     0x0001      0x0000     ZDP:EndDeviceBindReq
29   +00.766     0x0f00     0x796f      0x0f00     ZDP:EndDeviceBindReq
35   +00.094     0x0f00     0x0000      0x0001     ZDP:BindReq

STEP 7
67   +03.989     0x0f00     0x0001      0x0000     HA:On/off:Toggle
71   +00.004     0x0f00     0x0000      0x796f     HA:On/off:Toggle

STEPS 8 & 9
119  +03.003                            0xffff     IEEE:Beacon Request
120  +00.002     0x0f00     0x0001                 IEEE:Beacon: AP: 0
125  +01.006     0x0f00     0x796f      0x0001     NWK:Rejoin Request
129  +00.003     0x0f00     0x0001      0x796f     NWK:Rejoin Response
131  +00.005     0x0f00     0x1430      0xffff     ZDP:EndDeviceAnnce (0x796f)
150  +05.004     0x0f00     0x0001      0x1430     HA:On/off:Toggle
```

Another type of rejoining is *silent rejoin*. Silent rejoin is something you won't find in the ZigBee specification, but all stack vendors provide it because it is necessary in a deployed network of any size. For example, imagine that the power is cycled to all the routers in a 1,000-node network, during a temporary outage in a building. When the power comes back on, if all nodes attempted to join (or rejoin) at once, the networking in all probability would largely fail: there would simply be too much traffic.

But since routers already know their network information (PAN ID, Extended PAN ID, NwkAddr, security key), they can simply start silently, without saying anything. Remember, ZigBee nodes don't need to talk to maintain status on the network. The power outage can be looked at as equivalent to the network simply not talking for a while. As power comes back on, each router goes into receive mode on the exact PAN ID, Extended PAN ID, NwkAddr and security, as if the network had never been turned off. This is called "Silent Rejoin."

Here's an over-the-air capture of a silent rejoin:

That's right. It's silent!

Silent rejoin is also used when the network moves to a new channel, as described in Appendix A, " ZigBee 2007 and ZigBee Pro," under "Frequency Agility," a feature new to ZigBee 2007 and ZigBee Pro.

Silent rejoin only works if the nodes have some sort of permanent storage. In the case of Freescale, this storage is called NVM (non-volatile memory), and is stored in flash memory. Storage may be anything: flash memory, battery-backed RAM, a hard disk, whatever storage can survive power outages. Most ZigBee devices use a portion of the flash memory for storage to remember all of the ZigBee and application-level information that must survive power outages.

Even with sleeping ZEDs, NVM (flash) storage is present. After all, batteries may last many years in a ZigBee device, but eventually they must be changed. Why recommission the node just to change batteries?

One other use of rejoin, that is not always obvious, is the use of NWK-Rejoin to join a network that has permit-joining off. This is used sometimes if the commissioning process already has the network key, PAN ID, etc., already programmed into a node. The NWK-Rejoin procedure will give that device and address on the network, and the ZDP: DeviceAnnce will let all the nodes in the network know that it has joined.

This "joining-using-rejoin" can be useful in certain commissioning situations. One example is with HA. In Stack Profile 0x01, as configured for Home Automation, the network key is sent in the last hop to a joining node. This is done because the joining node (perhaps a light switch that you bought at Home Depot) has no knowledge of your network key. In general, because joining is rare, this method is acceptable for home use. The window is small, so the risk of someone intercepting the key is also small. But perhaps your home is Bill Gates's home, and you don't even want that small a window for security reasons. Simply buy little fancier nodes that allow you to program the security key into them (perhaps through a USB or some other connection, or with a key-fob like that available through Eaton's Home Heartbeat products), then voila! The node simply issues a NWK-Rejoin command and receives an address on the network.

> Rejoin is used to help ZEDs find a new parent if they lose contact with their old parent.
>
> Silent rejoin is used when the node is power-cycled.
>
> NWK-Rejoin does not need permit-joining on.

7.3   ZigBee Address Assignment

Addressing is crucial in a network. Similar to your post office address, the address of each node must be unique for ZigBee, or any other network, to function. Imagine the confusion if your post office address wasn't unique, and your bills ending up going to someone else? Granted, this might be nice for spam, advertisements, and catalogs, but for any data that you care about, a unique address for each node is critical.

ZigBee uses two unique addresses per node: a long address (aka IEEE or MAC address) and a short address (aka NwkAddr).

The long address, also called the IEEE or MAC address, is assigned by the manufacturer of a device using an 802.15.4 radio (not the chip manufacturer), and does not change for the life of the device. The long address uniquely identifies the widget from all other 802.15.4 widgets in the world. In fact, the 64 bits (8 bytes) of this address space are large enough to include 123,853 devices for every square meter on earth. The IEEE address space, surprisingly enough, is controlled by IEEE. A block of them can be purchased by the manufacturer at http://www.ieee.org.

Each IEEE address uniquely identifies this node from any other node in the world. The top three bytes (24 bits) are called the Organizational Unique Identifier (OUI), and the bottom five bytes (40 bits) are managed by the Original Equipment Manufacturer (OEM), as shown in Figure 7.8.

OUI (3 bytes)	Managed by OEM (5 bytes)

Figure 7.8: 64-bit MAC Address (AKA Long, IEEE, or EUI Address)

If you are an OEM about to manufacture a widget using ZigBee, be sure to obtain a block of MAC addresses. Check with whoever in your department is in charge of such things. They may already have an OUI (the top 24 bits of the MAC address). If not, you can purchase an OUI from IEEE. The other 40 bits must be managed by your organization. If you don't already have a system in place, you need one. Every widget that comes off the factory floor should be uniquely identified. Most manufacturing services also offer placing code in the microcontrollers, and allow one or more "serial numbers" to uniquely identify each widget. The MAC address can be entered this way.

Short (16-bit) addresses are assigned to a node at the time the node joins a ZigBee network. Why the dual addresses? Consider ZigBee addressing, as shown in Figure 7.9.

Notice that both the MAC layer header and the NWK layer header have both a source and a destination address. If the 8-byte MAC addresses were used, this would use up 32 bytes of the 127-byte over-the-air packet for node addressing alone! Instead, ZigBee uses a 2-byte network address, reducing these fields to 8 bytes total, allowing 24 more bytes for application use.

Why have these addressing fields in both the MAC header and the NWK header? This will be discussed in more detail in the next section on packet routing, but here are the basics. If sending a packet from node "A" to node "Z," the first hop would be from "A" to "B," the next hop from "B" to "C," and so on, until the final hop from "Y" to "Z." The nwkSrc and nwkDst always indicate "A" to "Z," while the macSrc and macDst would be the per-hop addresses, "A" to "B," and so on.

This section is about ZigBee address assignment, and when I talk about ZigBee address assignment, I'm really talking about the short address. The short address is assigned by

MAC HDR addressing fields		NWK HDR addressing fields	
macSrc	macDst	nwkSrc	nwkDst

Figure 7.9: MAC and NWK Addressing Fields

ZigBee at the time a node joins (or forms) a ZigBee network, and has no relation to the IEEE address. ZigBee uses one of two address schemes to assign the short address:

- Cskip

- Stochastic (random)

Cskip address assignment, available in stack profile 0x01 (the stack profile simply called ZigBee), is described in full detail here in this section, including all its ramifications.

I'll describe stochastic address assignment in detail in Appendix A, "ZigBee 2007 and ZigBee Pro," but a quick description is this: A node joining a network chooses its own address. It then sends a broadcast announcement to the network to see if any other node already has that address. If so, then the node chooses another address. If not then the node keeps that address. Stochastic addressing is available in stack profile 0x02 (the stack profile called ZigBee Pro).

In stack profile 0x01, addresses are assigned with a parent-child relationship that forms a symmetrical tree. The addressing scheme for stack profile 0x01 uses a calculated number for each "depth" (the number of hops from the ZigBee Coordinator), called Cskip (child skip).

In tree (Cskip) addressing, the ZigBee Coordinator (who forms the network) is node 0 (0x0000), by definition. The next node to join the network will receive an address from the parent node, as you can see in the association response below. The address that this node is assigned depends on whether the child is a router, which can have children of its own, or an end-device, which cannot:

```
IEEE 802.15.4
  Frame Control: 0xcc63
  Sequence Number: 53
  Destination PAN Identifier: 0x0bef
  Destination Address: 0x0050c2047800fc12
  Source Address: 0x0050c211dc051801
  MAC Payload
    Command Frame Identifier = Association Response: (0x02)
      Short Address: 0x0001
      Association Status: Association Successful (0x00)
```

Notice the short address in the association response payload. Also note the full 8-byte MAC addresses used as both the source and destination addresses. The MAC address must be used because the node is not yet on the network, and there is no other way to address it. But once the node receives the association response, it then knows its short

Cskip level 0	0×143d
Cskip level 1	0×035d
Cskip level 2	0×008d
Cskip level 3	0×0015
Cskip level 4	0×0001
Cskip level 5	0×0000

maxDepth	5
maxChildren	20
maxRouters	6

Figure 7.10: Cskip Calculated for Stack Profile 0x01

address, in this case, 0x0001. (I can tell from that address that this node joining was a router.)

Cskip uses three parameters to determine addressing: maxDepth, maxChildren, and maxRouters (see Figure 7.10). Using these parameters, Cskip can determine mathematically both what a newly joining child address should be, and also how to route a packet along the symmetrical tree. Stack profile 0x01 uses the values of maxDepth(5), maxChildren(20), and maxRouters(6), which limits the total number of nodes in the network to 31,101 nodes (out of a possible address space of 64 K).

Basically, the concept is this: The tree is split into "levels," level 0 being the ZigBee Coordinator, level 1 being its children, level 2 being its children's children, and so on.

The first router that joins the ZigBee Coordinator receives address 0x0001. The next router that joins the ZC gets address 0x0001 + Cskip at that level, so the next router in stack profile 0x01 will be 0x143e, because the Cskip at level 0 is 0x143d. This number, 0x143d, is large enough for that router and all its children and grandchildren to fully fill out that branch of the symmetrical tree. Whether it does or not (which is unlikely) is irrelevant. Enough address space is reserved so it could.

The first ZigBee End-Device that joins at level 0 gets an address after all the routers, which is address 0x796f in stack profile 0x01. The formula to obtain this number is:

$$1 \text{ [ZC takes 1 address]} + 6 \text{ [maxRoute rs]} \times 0x143d \text{ [Cskip at level 0]} = 0x796f$$

The concept of Cskip is more easily explained with a network using smaller Cskip parameters, even though this is not compatible with stack profile 0x01. Take a look at Figure 7.11. The parameters are set to maxDepth(3), maxChildren(5), and maxRouters(3). This only allows for a total of 66 nodes in a network (not very many!) Remember, these numbers are artificially small to make the explanation of the symmetrical tree easier to understand.

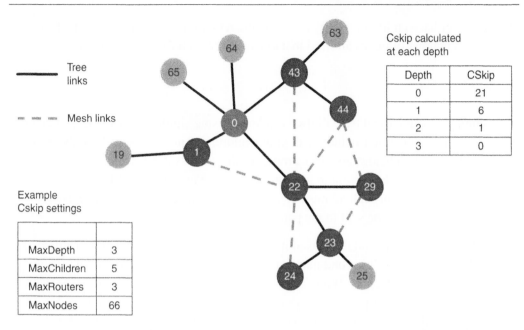

Cskip calculated at each depth

Depth	CSkip
0	21
1	6
2	1
3	0

Tree links

Mesh links

Example Cskip settings

MaxDepth	3
MaxChildren	5
MaxRouters	3
MaxNodes	66

Figure 7.11: Cskip Address Assignment Assumes a Symmetrical Tree

To understand the tree addressing, it's easiest to start at the bottom (maxDepth), and work our way up the tree to the root (or level 0). Consider node 24. Because it is already at maxDepth, three hops away from the ZigBee Coordinator, it cannot have any children, so the Cskip at this level is 0. Now look at its parent, node 23. This node is at level 2, that is, two hops away (along the tree) from node 0, so it has a Cskip of 1. Each of its children will consume exactly one address. Node 23 may have up to five children, so its own address (23) plus its five children (24–28) could potentially consume a total of six addresses. That's why the Cskip for its parent node, node 22 (which is at level 1), is 6.

Node 22 and all of its children and grandchildren can consume a total of 21 addresses: $1 + (3 \times 6) + 2 = 21$. Where did the "2" come from? This is how many ZigBee End-Devices can join the node:

$$\text{maxChildren (5)} - \text{maxRouters (3)} = \text{maxEndDevices (2)}$$

Hence, the Cskip at level 0, for the ZigBee Coordinator in this network, is 21. Each router child at level 0 can consume 21 addresses for itself and all of its family. ZigBee End

Devices never have children, so they consume the remaining two addresses. The formula for the total number of nodes allowed in a tree network with these parameters is:

$$1 + (3 \times 21) + 2 = 66$$

By assuming a symmetrical tree, ZigBee can know, using simple mathematics, whether the node address is a child (including all grandchildren) or not. If it is a child, the packet is sent to the next hop below, either to a router or the address itself. If the address is not a child address, the packet is sent to the parent. Node 22 in Figure 7.11 knows that any node address from 23 through 42 must be a child. Anything else (for example, address 43) is not, and so would be sent to the parent.

The problem with Cskip, and the reason it wasn't used in ZigBee Pro (stack profile 0x02), is that it doesn't scale well beyond maxDepth(5), which allows a maximum of 10 hops in a network (2 × maxDepth). What if more hops are desired? Just changing maxDepth to 6, and leaving the other parameters alone in stack profile 0x01, maxChildren(20), and maxRouters(6) allows up to 186,621 nodes, a number too large to fit in the 16-bit short address, which makes them illegal values for Cskip. Play around with different Cskip parameters with the ZigBee Calculator provided on http://www.zigbookexamples.com, and you'll see that 10 hops is really all Cskip can handle without major restrictions to the number of nodes that can join any given router.

In Figure 7.11, I included both tree links (black lines) and mesh links (dotted gray lines). ZigBee is always a mesh network. The tree may be used as a backup routing scheme if the mesh is busy or full, but it does not replace mesh. Tree routing (available only in stack profile 0x01) augments mesh routing, which, of course, leads to the topic of packet routing in ZigBee.

> Stack profile 0x01 uses Cskip address assignment which allows tree routing in addition to mesh.
>
> Stack profile 0x02 uses stochastic (random) addressing, which does not.

7.4   ZigBee Packet Routing

ZigBee employs a variety of methods for routing packets from one node to another:

- Broadcasting (from one to many nodes)

- Mesh routing (unicast from one node to another)

- Tree routing (unicast from one node to another, stack profile 0x01 only)

- Source routing (unicast from one node to another, stack profile 0x02 only)

Each of these distinct methods has advantages and disadvantages, as seen in Table 7.1. For the purposes of this table, multicast, available in ZigBee Pro, has the same advantages as broadcast.

Broadcasting essentially allows one node to reach many other nodes with a single data request. This method is not acknowledged (the originating node has no assurance it was received) and is very resource-intensive. Use broadcasts sparingly.

Mesh routing is table-driven and is very efficient (in terms of time, bandwidth, and memory resources), once the route is established. Packets sent along the mesh are acknowledged, so the sending node can know whether or not the packet was received. Mesh routes are distributed, which reduces the over-the-air packet overhead. ZigBee mesh can deliver packets up to 30 hops away.

Tree routing, also acknowledged, is only available in Stack Profile 0x01. It was described in the previous section on Cskip. Tree is just as bandwidth-efficient as mesh, and more efficient in terms of memory. But it has a major drawback: If links between parent and children break, it has no recovery. So ZigBee uses mesh by default.

Source routing, like mesh and tree, is acknowledged. This type of routing is available only in Stack Profile 0x02. It is used primarily when a data concentrator (or gateway) needs to communicate with many nodes, perhaps hundreds, or thousands. With mesh

Table 7.1: ZigBee Routing Methods

	Broadcast	Mesh	Tree	Source Route
Multi-hop	Up to 30	Up to 30	Up to 10	Up to 5 hops
Multiple destinations	Yes	No	No	No
One-to-one	No	Yes	Yes	Yes
Bandwidth efficient	No	Yes	Yes	Yes
Payload efficient	Yes	Yes	Yes	No
Acknowledged	No	Yes	Yes	Yes

routing, each route needs a table entry, and ZigBee nodes typically don't have the RAM for a thousand routes. In source routing, a single (more expensive) node can have the RAM to store all the routes. The route for any particular communication is then sent as part of the over-the-air packet. Its major drawback is that it is limited to a maximum of five hops.

One other type of unicast routing, not really mentioned in the ZigBee specification, is neighbor routing. If the node is a neighbor (that is, within radio range), and the sending node knows that the destination node is a neighbor, the packet is simply delivered directly. No route is required. While not required by the ZigBee specification, nearly all stack vendors do this for efficiency. Why bother to discover a route, if the node is right next to you? This method only works if the neighbor is a router, or a child end-device. If the neighbor is some other router's child, then the route must go through that other router. I'll describe each of these routing methods in more detail below.

7.4.1   Broadcasts

Broadcasts are really the foundation for reliable multi-hop communication in ZigBee. Without broadcasting, even mesh networking is not possible.

When a node initiates a broadcast, the message is repeated by all the neighboring routers of that node, up to a certain radius from the center of the broadcast, as shown in Figure 7.12.

The area in gray in the figure below represents the broadcast as it expands across the network, like ripples in a pond. Each node that receives the broadcast decrements the

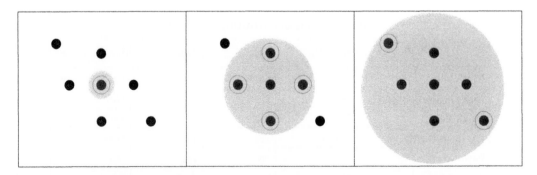

Figure 7.12: Broadcasts Expand Like Ripples in a Pond

radius, and if there is any radius left, adds the broadcast to the Broadcast Transaction Table (BTT) and repeats it. If a node receives a broadcast with a radius set to 1, then it will pass the broadcast up to the application, but will not repeat it. This radius set from 0x01 to 0xff, to indicate an actual maximum distance. The broadcast in the figure above must have been sent with a radius of at least 3 for the nodes above to have repeated it.

In Freescale BeeStack it is common to use either *afDefaultRadius_c* (the entire network) or 1 (for immediate neighbors only).

ZigBee does allow a radius of 0x00, which is a special flag that means to repeat the broadcast for the entire network. If a router node receives a broadcast of radius 0x00 (which it hasn't heard before), it will add it to the BTT and repeat it, and it will not decrement the radius. I find this confusing, and wish ZigBee had chosen 0xff to mean "forever" (as it does in so many other things, like permit-joining), but that's how it goes in specifications. They are not always internally consistent.

The Broadcast Transaction Table is a critical component of broadcasting. Not only does the BTT keep track of the over-the-air packet so that the node can repeat it, but the BTT keeps track of the unique identifier of the broadcast so the broadcast is only passed up to the application once, and is not added to the BTT again. The default number of entries in the BTT is 9, and the timeout for a broadcast across the network is nine seconds, which means a network can sustain a rate of 1 broadcast per second: not very many. If the BTT is full, the broadcast will be dropped by the receiving node.

Use broadcasts sparingly in your application. If quicker, more constant communication is required, use unicasts instead. Keep in mind that route discoveries are also broadcasts, as are some ZDP commands such as Device_annce. ZDP-Device_annce is sent every time a node joins a network. So is a route discovery to the Trust Center if the TC is not a neighbor. This means that nodes may join at a rate of one per every two seconds. If you attempt to join 100 nodes all at once, multiple attempts will have to be made.

If discovering routes constantly, the network could "fill up" on broadcasts, and the route discoveries would be dropped, causing them to fail. The application will know if the route discovery failed, as the data confirm will tell it (BeeAppDataConfirm() in Freescale BeeStack). The application must either give up on the route, or wait a few seconds and try again.

Below is an over-the-air capture of a broadcast in progress:

Time	MACSrc	MACDst	NWKSrc	NWK Seq	R	NWKDst	Packet Type
31 +00.110	0x143e	0xffff	0x143e	12	5	0xffff	ZDP:EndDeviceAnnce
32 +00.022	0x0000	0xffff	0x143e	12	4	0xffff	ZDP:EndDeviceAnnce
33 +00.049	0x0001	0xffff	0x143e	12	3	0xffff	ZDP:EndDeviceAnnce
34 +00.090	0x143e	0xffff	0x143e	12	5	0xffff	ZDP:EndDeviceAnnce
35 +00.102	0x143e	0xffff	0x143e	12	5	0xffff	ZDP:EndDeviceAnnce
36 +00.030	0x0000	0xffff	0x143e	12	4	0xffff	ZDP:EndDeviceAnnce
37 +00.150	0x0001	0xffff	0x143e	12	3	0xffff	ZDP:EndDeviceAnnce
38 +00.013	0x0000	0xffff	0x143e	12	4	0xffff	ZDP:EndDeviceAnnce

Note the originating network source address (NWKSrc) and network sequence number (NWKSeq). These two fields, taken together, uniquely identify the broadcast in the network. The MAC fields change as the broadcast propagates, so you can tell which node is repeating it, but the NwkSrc and NwkSeq remain the same. Note the radius (R) field in the broadcast and see how it's decremented (look at packet 31 versus 32 above) as the broadcast propagates across the network.

Notice also the time difference between the repeated broadcasts. This random amount of time is called *jitter*. The broadcast jitter makes broadcasts relatively slow compared to unicasts. It will take on the order of 1,000 milliseconds to propagate a broadcast across ten hops. Unicasts can be sent across ten hops in about 50 to 100 milliseconds.

Only routers (the ZC and ZRs) repeat broadcasts. ZEDs may receive them, but do not participate in any kind of routing, including broadcasts. If initiating a broadcast, sleepy ZEDs do not broadcast directly (after all, they are going back to sleep immediately). Instead they unicast the broadcast to their parent, and then their parent initiates the broadcast on their behalf.

It is up to the stack vendor how many times (up to three) that the broadcast is repeated. Some stack vendors provide some intelligence in dense networks, so as not to repeat the broadcasts so many times. This conserves bandwidth.

There are three special ZigBee broadcast modes:

- 0xffff—broadcast to all nodes (even sleeping ZEDs)
- 0xfffd—broadcast to all awake devices (including RxOnIdle = TRUE ZEDs)
- 0xfffc—broadcast to all routers (excludes all ZEDs)

The broadcast modes are used by various ZigBee ZDP, APS, and NWK layer commands, but can also be used by applications. Simply place this number in the destination field on the APSDE-DATA.request. For more information on using broadcasts in applications, see Chapter 4, "ZigBee Applications."

A broadcast of radius 1 is particularly useful to send data to all neighbors, assuming the nodes this application wants to talk to are all within hearing range.

There is also another form of broadcasting called "multicast." In short, multicast, like a broadcast, propagates across the network. But multicast stays within a predefined group of nodes, not the entire network. Where broadcasts are circular, multicasts can be any shape (oblong, square, or a strange, stretchy amoeba). I'll describe this method in detail in Appendix A, "ZigBee 2007 and ZigBee Pro, " which discusses the ZigBee Pro and 2007 features.

> Broadcasting transmits from one node to many nodes, up to the entire network.
>
> Use broadcasts sparingly, as they consume a lot of bandwidth and resources.
>
> Broadcasts are slow compared to unicasts.

7.4.2   Mesh Routing

Mesh routing is at the heart of ZigBee. A wireless solution servicing large office buildings, electric meters distributed across cities, hospitals, and yes, even the cattle industry (see http://www.zigbeef.com) needs the ability to send a signal farther than a radio can speak, especially one communicating at one watt or less.

The concept of ZigBee meshing is simple but powerful. A route is discovered from one node in the network to another. This route passes through any number of intermediate nodes (up to 30 hops in ZigBee Pro). If something happens to the route over time, such as if a node goes down, or a barrier blocks part of the route, then a new route is automatically discovered around the barrier or interference, as shown in Figure 7.13.

In this section, I'll go into the details of routing tables, route discovery process, route errors, and route maintenance.

In ZigBee mesh:

- The algorithm is based on the publicly available Advanced Ad-hoc On-Demand Distance Vectoring (AODV).

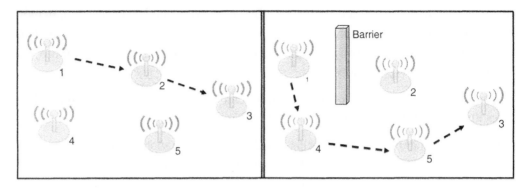

Figure 7.13: Mesh Networking Automatically Reroutes Packets

- All routers are peers.

- The route is distributed: Each node in the route keeps track of the next hop for the route in a routing table.

- Routes are unidirectional (a route must be discovered each way for bidirectional communication).

- Routes are like goat trails: they continue to be used until they fail.

- Failed routes are communicated back to the originating node, allowing it to discover a new route.

AODV is a well-known, publicly available mesh routing technique (there are many). A variation of this was chosen by ZigBee because it scales well for systems with serious RAM limitations. Remember, ZigBee fits on 8-bit nodes with only 2 to 4 K of RAM.

One question I'm often asked is, "What's so great about meshing? Why not use repeating, broadcasting, source routing, or another form of routing?"

Repeating (a simplified form of broadcasting) and broadcasting are bandwidth and resource-intensive. They require nodes to keep the message around for a longer period of time because all the neighbors must repeat it. With mesh, only the nodes along the path repeat it, not the entire network, saving both time and bandwidth. Mesh networking is also per-hop-acknowledged, whereas repeating and broadcasting are not, providing higher reliability and lower bandwidth usage. There is no point in repeating something that has been acknowledged. With mesh, a sending node can *know* whether the message was received.

True source routing (like TCP/IP) requires setup ahead of time. Someone (a human being) must know the next hop in the network, and set up in advance. This takes planning and more complicated commissioning. With mesh, the nodes can be placed ad-hoc, anywhere the installer likes, and the routes will be figured out automatically by the network. ZigBee does offer a source routing option that uses mesh-like route discovery in ZigBee Pro.

Source routing also has higher over-the-air overhead for the protocol. The 802.15.4 PHY is limited to 127 octets. With the ZigBee protocol overhead, actual application payload is reduced to about 100 bytes, whereas with source routing, especially where many hops are involved, the application payload size can be reduced even further, down to 80 or even 60 bytes. Source routing is normally used where the PHY packet size is larger. Again, ZigBee does offer source routing, but it is limited to five hops to reduce this effect.

So, back to ZigBee mesh networking. As mentioned in the previous list, all routers are peers. Any node in the network may discover a route, and every router becomes, at least temporarily, a potential route. This process is called "route discovery." ZigBee keeps a route discovery table (distinct from the routing table) during this process to find the most efficient route, in terms of time.

The route discovery process begins with a broadcast from the node wishing to discover a route to another node. In ZigBee mesh networking, initially, there is no information. The code could be anywhere: up, down, left, or right. The network doesn't know, so it broadcasts throughout the entire network to find the node.

Along the way, a "path cost" is kept. Every hop adds a number from one to seven to indicate the strength of this hop, or link, as it's sometimes called. For example, in Figure 7.14, node 1 wants to send a packet to node 10. Node 1 doesn't have node 10 as a neighbor, and does not already have a route to node 10, so it initiates a route discovery. This broadcast goes in all directions. As the past cost is added up along each hop, the best route is found. If another, lower, path cost comes along to any given router, it sends that route on.

Note that the least path cost is not the same as the fewest number of hops. If a link is good, it will have a path cost for that link of 1. But if a link is particularly spotty (the two nodes often have to retry), it will be higher, perhaps as high as 7.

For simplicity's sake, let's assume that all the links above are good (with a path cost of 1). Each dashed line represents a link (that is, which nodes are neighbors, within

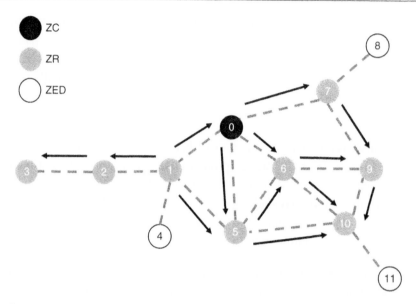

Figure 7.14: Route Discovery Is Broadcast Throughout the Network

hearing range of each other). Each circle represents a node. The route discovery broadcast propagates across the network, and the path cost is added up at each hop. The lowest one reaching 10 will be a path cost of two, the route from node 1 to 5, and then to 10.

Node 10 waits for a while to see if any lower path costs come in a bit later. A lower path cost could come in later due to the randomness of the broadcast algorithm (each node waits a random amount of time before repeating the broadcast).

Node 10 then issues a unicast route reply back to the originating node, as seen in Figure 7.15.

The other nodes which were recording the best path cost for this route request keep track of it for a while (a few seconds). They eventually drop the route request, clearing up that entry in the route discovery table. But nodes 1 and 5 keep track of the route. Depending on the setting of the Network Information Base (NIB) variable, *nwkSymLink*, the route is either considered bidirectional or unidirectional. ZigBee stack profile 0x01 assumes unidirectionality. ZigBee Pro assumes bidirectionality.

One thing that it is very important for ZigBee routing to work is that the links must be mostly symmetrical, in radio terms. If one node can shout very loud, but can't hear anything, it will have very poor links with its neighbors. So, if creating hardware with a

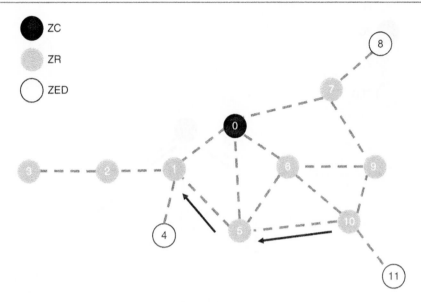

Figure 7.15: Route Replies Are Unicast Back to Originator

power amplifier (PA), then either ensure that all nodes in the network have the PA, or ensure that the PA-enabled node also has an equally powerful Low-Noise Amplifier (LNA).

Now, let's assume that the link between nodes 1 and 5 is not very good. Say a path cost of three, as shown in Figure 7.16. ZigBee will discover a route with more hops (say from 1 <--> 0 <--> 6 <--> 10), rather than one with fewer hops, and a link that will require retries. You can see this makes sense in terms of time. If ZigBee must retry, it is much more time expensive than if ZigBee merely needs to add an extra hop or two.

What if the route that worked before doesn't work now? ZigBee sends a route error back to the originating node. For example, in Figure 7.16, perhaps the original route was 1 <--> 0 <--> 6 <--> 10. But for some reason the link between nodes 6 and 10 breaks. Since node 6 can no longer deliver the packet, it will send a route error back to the originating node (1) who will then issue a new route request, automatically finding a new available path. This feature of mesh networking is often called self-healing.

ZigBee does not actively try to find the best route on every packet sent. Packets will continue to use a route, like goats continue to use a goat trail, as long as it works. But your application can, if you specify. Simply tell the APSDE-DATA.request to force route discovery, and a new "best route" will be found. Keep in mind the route discovery can

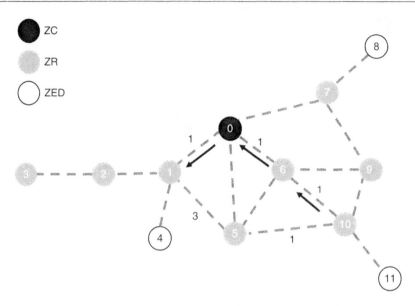

Figure 7.16: ZigBee Chooses the Lowest Path Cost for Route

take up to one or two seconds, and it generates a broadcast, so use it sparingly. Generally, your application should do this only if it has been off and not communicating, for some time. Usually, the best route found initially will remain the best route, perhaps even for the life of the network.

Each router in a ZigBee network contains a routing table to keep track of routes. Each node only keeps track of the next hop in any given route. For example, the route table shown in Figure 7.17 shows the route we found from node 1 to node 10. The next hop to get there from node 1 is node 0. The ultimate destination is node 10 (0x000a).

We can also see that another route is pending discovery to node 9 (that is the route discovery has started but not completed).

ZigBee doesn't define route table maintenance well. It is really up to the stack vendor or application to decide when (or if) to retire routes.

A typically routing table is from 6 to 20 entries. Usually that's enough for most networks. Nodes tend to communicate to other nodes close to them, or to a gateway, so the actual number of routes through any given node is usually small. If the routing table is full, then that node won't participate in route discovery which tends to spread the routes out evenly through a network that requires a lot of routes.

Destination	Next Hop	IsValid
0x000a	0x0000	Yes
0x0009	----	Pending
----	----	No

Figure 7.17: ZigBee Route Table

If the routing table is full, what should you do about it? Some vendors use a least-recently-used algorithm, some leave it up to the application, and some use a timeout to retire routes. Think about the typical route usage of your network and be sure to leave some room. Perhaps you should increase the size of the routing table (which uses more RAM), or actively manage the routes.

Another thing that is important to understand about routes is that ZigBee End Devices do not participate in routing. If finding a route to a ZED, the router parent will respond on the end device's behalf. The ZED doesn't need to know anything about routes: It simply receives packets from its parent and sends packets to its parent.

For the example in this section, I'll use a network organized into what I call the Southern Cross (see Figure 7.18). This network has all ZigBee node types and it has multiple possible routes.

The entire capture of *Example 7-3 The Southern Cross* is available online at http://www.zigbookexamples.com/code. See Chapter 3, "The ZigBee Development Environment," for general instructions on how to download and use the source code with actual hardware.

In this case, the wakeful ZED is going to find a route to the sleepy ZED. The sleepy ZED is an HA On/Off Light. The wakeful ZED is an HA On/Off Switch which can toggle the light.

To make it easy to capture routing from a single location (my desk), I've used a mechanism available in Freescale BeeStack called the "I-Can-Hear-You" table. This table allows certain nodes to hear only certain other nodes. Otherwise, all the nodes would be neighbors, if they were on the desk. The I-Can-Hear-You table is very convenient for trade shows, and for developing. The command SetICanHearYouTable() fills the tables with the particular neighbors I want.

Take a look at the relevant packets from the capture below. Initially, the wakeful ZED (0x351) sends the ZDP:EndDeviceBindReq to the ZC, through its parent ZC3 (node

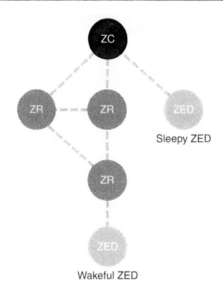

Sleepy ZED

Wakeful ZED

Figure 7.18: Example 7-3 The Southern Cross

0x0002), but it doesn't know the route to the ZC. A route is discovered. Notice the route reply comes back almost immediately, before the rest of the broadcast has a chance to finish. Routes can be discovered very quickly.

Then the message can finally be sent in packets 64 and 65. Notice the multi-hop from node 0x0351 to 0x0002 to 0x0001 to 0x0000 (see page 315).

It's interesting that later on, in packet 146, the same source node (0x0351) wants to find a route to node 0x679f, a child of node 0x0000. But this time, due to random jitter, the route goes through node 0x143e instead of node 0x0001. See packets 146, 154, 156, and 162 for the entire route from node 0x0351 to 0x0002 to 0x143e to 0x0000 to 0x796f (see page 316).

The binding process completes and the application communicates data: the usual HA: On/Off:Toggle. I toggle the light a few times. No new route discovery is needed. The packet simply multi-hops to the right destination, node 0x796f.

Next, I break the route used by the light switch (node 0x0351) by shutting off node 0x143e (ZR2). Then I toggle the light again. But this time the route is broken. So, ZR3 (the parent of node 0x0351) discovers a new route. Now, the route goes from node 0x0351 to 0x0002 to 0x0001 to 0x0000 to 0x796f (see page 317).

Seq	Time	SrcPAN	MACSrc	DstPAN	MACDst	MACSeq	NwkSrc	NwkDst	NwkSeq	APSCtr	Packet Type
56	+00:00:00.315	0x0351	0x0f00	0x0002	151	0x0351	0x0000	181	0x07	Zigbee APS Data	ZDP:EndDeviceBindReq
57	+00:00:00.002				151					IEEE 802.15.4	Acknowledgment
58	+00:00:00.006	0x0002	0x0f00	0xffff	69	0x0002	0xfffc	113		Zigbee NWK	NWK Command: Route Request
59	+00:00:00.069	0x0001	0x0f00	0xffff	48	0x0002	0xfffc	113		Zigbee NWK	NWK Command: Route Request
60	+00:00:00.005	0x0000	0x0f00	0x0001	100	0x0000	0x0001	133		Zigbee NWK	NWK Command: Route Reply
61	+00:00:00.001				100					IEEE 802.15.4	Acknowledgment
62	+00:00:00.003	0x0001	0x0f00	0x0002	49	0x0001	0x0002	45		Zigbee NWK	NWK Command: Route Reply
63	+00:00:00.001				49					IEEE 802.15.4	Acknowledgment
64	+00:00:00.005	0x0002	0x0f00	0x0001	70	0x0351	0x0000	114	0x07	Zigbee APS Data	ZDP:EndDeviceBindReq
65	+00:00:00.002				70					IEEE 802.15.4	Acknowledgment
66	+00:00:00.004	0x0001	0x0f00	0x0000	50	0x0351	0x0000	46	0x07	Zigbee APS Data	ZDP:EndDeviceBindReq
67	+00:00:00.002				50					IEEE 802.15.4	Acknowledgment
68	+00:00:00.013	0x0000	0x0f00	0xffff	101	0x0002	0xffc	113		Zigbee NWK	NWK Command: Route Request
69	+00:00:00.030	0x143e	0x0f00	0xffff	203	0x0002	0xfffc	113		Zigbee NWK	NWK Command: Route Request

Seq	Time	SrcPAN	MACSrc	DstPAN	MACDst	MACSeq	NwkSrc	NwkDst	NwkSeq	APSCtr	Packet Type
146	+00:00:01.248	0x0351	0x0f00	0x0002	155	0x0351	0x796f	185	0x0b	Zigbee APS Data	HA:On/off:Toggle
147	+00:00:00.001				155					IEEE 802.15.4	Acknowledgment
148	+00:00:00.005	0x0002	0x0f00	0xffff	80	0x0002	0xfffc	122		Zigbee NWK	NWK Command: Route Request
149	+00:00:00.036	0x143e	0x0f00	0xffff	211	0x0002	0xfffc	122		Zigbee NWK	NWK Command: Route Request
150	+00:00:00.006	0x0000	0x0f00	0x143e	109	0x0000	0x143e	140		Zigbee NWK	NWK Command: Route Reply
151	+00:00:00.001				109					IEEE 802.15.4	Acknowledgment
152	+00:00:00.004	0x143e	0x0f00	0x0002	212	0x143e	0x0002	206		Zigbee NWK	NWK Command: Route Reply
153	+00:00:00.001				212					IEEE 802.15.4	Acknowledgment
154	+00:00:00.003	0x0002	0x0f00	0x143e	81	0x0351	0x796f	123	0x0b	Zigbee APS Data	HA:On/off:Toggle
155	+00:00:00.001				81					IEEE 802.15.4	Acknowledgment
156	+00:00:00.005	0x143e	0x0f00	0x0000	213	0x0351	0x796f	207	0x0b	Zigbee APS Data	HA:On/off:Toggle
157	+00:00:00.001				213					IEEE 802.15.4	Acknowledgment
158	+00:00:00.010	0x0001	0x0f00	0xffff	57	0x0002	0xfffc	122		Zigbee NWK	NWK Command: Route Request
159	+00:00:00.005	0x0000	0x0f00	0xffff	111	0x0002	0xfffc	122		Zigbee NWK	NWK Command: Route Request
160	+00:00:01.742	0x796f	0x0f00	0x0000	54					IEEE 802.15.4	Command: Data Request
161	+00:00:00.001				54					IEEE 802.15.4	Acknowledgment
162	+00:00:00.004	0x0000	0x0f00	0x796f	110	0x0351	0x796f	141	0x0b	Zigbee APS Data	HA:On/off:Toggle

Seq	Time	SrcPAN	MACSrc	DstPAN	MACDst	MACSeq	NwkSrc	NwkDst	NwkSeq	APSCtr	Packet Type
245	+00:00:00.003	0x0002	0x0f00	0x143e	85	0x0351	0x796f	127	0x0f	Zigbee APS Data	HA:On/off:Toggle
246	+00:00:00.005	0x0002	0x0f00	0xffff	86	0x0002	0xfffc	128		Zigbee NWK	NWK Command: Route Request
247	+00:00:00.039	0x0001	0x0f00	0xffff	58	0x0002	0xfffc	128		Zigbee NWK	NWK Command: Route Request
248	+00:00:00.005	0x0000	0x0f00	0x0001	114	0x0000	0x0001	144		Zigbee NWK	NWK Command: Route Reply
249	+00:00:00.001				114					IEEE 802.15.4	Acknowledgment
250	+00:00:00.002	0x0001	0x0f00	0x0002	59	0x0001	0x0002	49		Zigbee NWK	NWK Command: Route Reply
251	+00:00:00.001				59					IEEE 802.15.4	Acknowledgment
252	+00:00:00.028	0x0000	0x0f00	0xffff	115	0x0002	0xfffc	128		Zigbee NWK	NWK Command: Route Request
253	+00:00:00.262	0x0000	0x0f00	0xffff	116	0x0002	0xfffc	128		Zigbee NWK	NWK Command: Route Request
254	+00:00:00.261	0x0000	0x0f00	0xffff	117	0x0002	0xfffc	128		Zigbee NWK	NWK Command: Route Request
255	+00:00:01.087	0x796f	0x0f00	0x0000	82					IEEE 802.15.4	Command: Data Request
256	+00:00:00.001				82					IEEE 802.15.4	Acknowledgment
257	+00:00:02.654	0x0351	0x0f00	0x0002	160	0x0351	0x796f	190	0x10	Zigbee APS Data	HA:On/off:Toggle
258	+00:00:00.001				160					IEEE 802.15.4	Acknowledgment
259	+00:00:00.004	0x0002	0x0f00	0x0001	87	0x0351	0x796f	129	0x10	Zigbee APS Data	HA:On/off:Toggle
260	+00:00:00.001				87					IEEE 802.15.4	Acknowledgment
261	+00:00:00.004	0x0001	0x0f00	0x0000	60	0x0351	0x796f	50	0x10	Zigbee APS Data	HA:On/off:Toggle
262	+00:00:00.001				60					IEEE 802.15.4	Acknowledgment
263	+00:00:00.406	0x796f	0x0f00	0x0000	83					IEEE 802.15.4	Command: Data Request
264	+00:00:00.001				83					IEEE 802.15.4	Acknowledgment
265	+00:00:00.003	0x0000	0x0f00	0x796f	118	0x0351	0x796f	145	0x10	Zigbee APS Data	HA:On/off:Toggle

Routes generally are discovered once when initially commissioning the network. The installer tests the lights and switches, and from then on they just continue to work along the same routes. But occasionally the environment changes. A new appliance might be installed, a fixture moved. No worries. ZigBee will simply discover a new route at that time, and the node will continue performing the application that it was programmed to do.

> Routes are discovered automatically by ZigBee.
>
> ZigBee Mesh route discovery is initiated when a node sends a packet to another node.
>
> Routes are discovered along the least cost path.

7.5   ZigBee Over-the-Air Frames

An essential part of understanding ZigBee is understanding over-the-air frames. In fact, over-the-air frames are all that the ZigBee Alliance cares about for compatibility testing. Certain frames are sent over-the-air and certain replies are expected, but the rest is a black box.

Understanding over-the-air frames also greatly aids in debugging the network. (Why didn't that data packet get delivered?) Is there a problem with bandwidth? Is there a bug in the application or in the ZigBee stack?

There are a variety of "layers" involved in any given over-the-air packet, including the MAC, NWK, APS, ZCL, and data. Over-the-air, these layers create a series of bytes, framing (on either side) the rest of the bytes in the packet. That's why it's called a frame. The next higher layer frame is always inside the lower layer frame over-the-air. So the NWK frame is inside the MAC frame, the APS frame is inside the NWK frame, and so on.

A full ZigBee data packet could contain all of the following frames (MAC, NWK, APS, and ZCL), as shown in Figure 7.19.

Entire data frame may be up to 127 octets

Figure 7.19: ZigBee Data Frame Contains up to 127 Octets

The size of each frame header (MAC, NWK, APS, etc.) is variable, depending on parameters. Every frame header begins with what is called a frame control (FC) field, which describes the optional fields in that frame.

The MAC frame ends with the Frame Check-Sum (FCS). The NWK frame ends with the optional NWK Auxiliary Message Integrity Code (MIC). The NWK AUX HDR and NWK MIC are present only in a secure network. The APS frame also may have security applied, with an APS AUX HDR and APS MIC.

The first two headers are node-oriented: that is, the MAC and NWK headers. The MAC header controls the per-hop protocol, and the NWK header controls the end-to-end protocol (see Figure 7.20).

Notice how both the MAC and NWK headers contain destination and source addresses. Think of node A sending a packet multi-hop to node F. The first hop would be from A to B. The next hop would be from B to C, and so on. The final hop would be from E to F. The MAC source and destination would change along this entire journey, but the NWK source and destination addresses would not.

Format of Individual Frame Types

NWK data frame format

Octets:2	2	2	1	1	Variable
Frame Control	Destination address	Source address	Radius	Sequence number	Data payload
	Routing fields				
NWK header					NWK payload

MAC data frame format

Octets:2	1	0/2	0/2/8	0/2	0/2/8	Variable	2
Frame Control	Sequence number	Destination PAN ID	Destination address	Source PAN ID	Source address	Data payload	FCS
		Data payload					
MAC header						MAC payload	MFR

127 Octet MTU

Figure 7.20: ZigBee MAC and NWK Headers

Format of Individual Frame Types

APS data frame format

Octets:1	0/1	0/2	2	2	1	1	0/1/2	Variable
Frame Control	Dst EP	Group address	Cluster ID	Profile ID	Src EP	APS counter	Extended Header	Data payload
	In-Node Addressing fields							
APS header								APS payload

ZCL data frame format

Octets:1	0/2	1	1	Variable
Frame Control	Manufacturer ID	Transaction Sequence Number	Command ID	Data payload
ZCL header				ZCL payload

Figure 7.21: ZigBee APS and ZCL Headers

All ZigBee data packets are sent using the 16-bit short addresses, but other packets are not. For example, when joining a network, the association request and response are sent with the long (64-bit IEEE) addresses.

The next two headers, APS and ZCL, are application-level headers (see Figure 7.21). They describe application-level information including endpoints, application profile, and for ZCL, the specific command for the cluster (reading and writing attributes, or commanding the application to do something such as turning on a light).

All of the addressing fields in APS are within the node, so as to send the packet to the right application, as opposed to the NWK layer which addresses between nodes. The ZCL frame is optional, but is present in all ZigBee public profiles.

All of the over-the-air packets were built using *Example 7-4 ZigBee Over-the-Air Frames*. To make the example easier to understand, the four-node network is depicted graphically in Figure 7.22.

This ZigBee network is a secure network. The wakeful ZigBee End Device, labeled node 0x1430 ED above, an HA On/Off Switch, toggles the light on the sleepy ZigBee End-Device, labeled 0x796f SED.

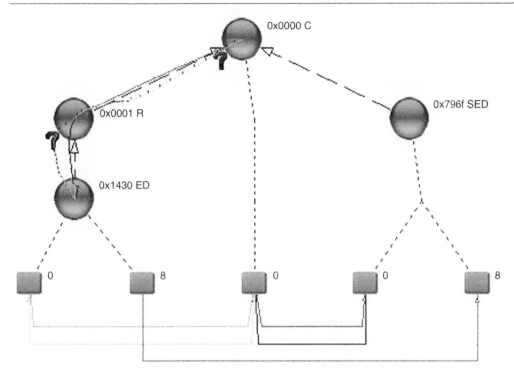

Figure 7.22: Example 7-4 ZigBee Over-the-Air Frames

Below is a typical data packet. This one is shown many times throughout this book (the HA:OnOff:Toggle) but this time I'll go into detail:

```
IEEE 802.15.4
  Frame Control: 0x8861
    .... .... .... .001 = Frame Type: Data (0x0001)
    .... .... .... 0... = Security Enabled: Disabled
    .... .... ...0 .... = Frame Pending: No more data
    .... .... ..1. .... = Acknowledgement Request: Acknowledgement
                          required
    .... .... .1.. .... = Intra PAN: Within the PAN
    .... ..00 0... .... = Reserved
    .... 10.. .... .... = Destination Addressing Mode: Address field
                          contains a 16-bit short address (0x0002)
    ..00 .... .... .... = Reserved
    10.. .... .... .... = Source Addressing Mode: Address field contains
                          a 16-bit short address (0x0002)
```

Sequence Number: 230
Destination PAN Identifier: 0x0f00
Destination Address: 0x0000
Source Address: 0x0001
Frame Check Sequence: Correct
ZigBee NWK
  Frame Control: 0x0248
    ....00 = Frame Type: NWK Data (0x00)
    ....00 10.. = Protocol Version (0x02)
    .... 01.. .... = Discover Route: Enable route discovery (0x01)
    ....0 .... = Multicast: Unicast or broadcast (0x00)
    .... ..1. .... = Security: Enabled
    .... .0.. .... = Source Route: Not present (0x00)
    .... 0... .... = Destination IEEE Address: Not Included
    ...0 .... = Source IEEE Address: Not Included
    000. .... = Reserved
  Destination Address: 0x796f
  Source Address: 0x1430
  Radius = 9
  Sequence Number = 174
ZigBee AUX
  Security Control: 0x28
    .... .000 = Security Level: Attributes: None; Encryption: Off; MIC:
          No (M = 0) (0x00)
    ...0 1... = Key Identifier: Network (0x01)
    ..1. .... = Extended Nonce: Sender Address Field: Present (0x01)
    00.. .... = Reserved: (0x00)
  Frame Counter: 0x09
  Source Address: 0x0050c237b0040002
  Key Sequence Number: 0x00
  MIC: 00:fd:30:14
ZigBee APS
  Frame Control: 0x00
    .... ..00 = Frame Type: APS Data (0x00)
    .... 00.. = Delivery Mode: Normal Unicast Delivery (0x00)
    ...0 .... = Reserved
    ..0. .... = Security: False
    .0.. .... = Ack Request: Acknowledgement not required
    0... .... = Extended Header Present: Extended header is not present
  Destination Endpoint: 0x08
  Cluster Identifier: On/off (0x0006)
  Profile Identifier: HA (0x0104)
  Source Endpoint: 0x08
  Counter: 0x27
ZigBee ZCL
  Frame Control: 0x01

```
.... ..01 = Frame Type: Command is specific to a cluster (0x01)
.... .0.. = Manufacturer Specific: The manufacturer code field shall
            not be included in the ZCL frame. (0x00)
.... 0... = Direction: From the client side to the server side.
            (0x00)
...0 .... = Disable Default Response: Default response command will
            be returned. (0x00)
.... = Reserved: Reserved (0x00)
Transaction Sequence Number: 0x42
Command Identifier: Toggle (0x02)
```

The first frame, titled "IEEE 802.15.4," gives information about the MAC (IEEE) frame in this packet. You can see by the first three bits of the frame control (FC) that this is a data packet. The next FC bit says security is disabled, but this is a secure packet! ZigBee always secures the packet at the network layer, not the MAC layer. Another interesting bit is the "acknowledgment required" bit. This tells the receiving node to send a MAC ACK, ensuring reliability per-hop. The MAC will try up to three times to deliver the packet before giving up and informing the sender that the unicast failed on this hop. Broadcasts always have the AC-required bit off.

There is also a MAC sequence number. Every frame has a sequence number including the NWK sequence number, APS counter, and ZCL transaction ID. They all serve the same purpose: to uniquely identify this packet. The sequence number is just a rolling 8-bit number, making the assumption that 256 packets from the same node will never be in flight at the same time. The destination PAN ID is the PAN on which this node is operating. The source and destination are the per-hop source and destination.

The rest of the MAC bits merely indicate which network this is on, and which nodes are involved. Notice that the source and destination address in the MAC differ from the source and destination address in the NWK layer. Why is this? This is packet is multi-hopping its way to its final destination.

The next frame, titled "ZigBee NWK," contains the NWK layer information. Notice the protocol version is "2." It's always 2 for ZigBee. This does not indicate stack profile, only that its version 2 of ZigBee. The previous version was ZigBee 2004. ZigBee 2006 and ZigBee 2007/Pro are both protocol 2. Notice now that the NWK layer says security is enabled. This FC also informs whether routing should be enabled or disabled for this packet.

Because the security bit is set at the NWK layer, the next frame, "ZigBee AUX," is present. This is NWK layer security in ZigBee and includes a 32-bit field called the Frame Counter. If the frame counter is old (stale, that is), then the packet will be rejected. This field prevents replay attacks. In addition there is a Message Integrity Code (MIC). This field prevents anyone from altering the packet. The MIC is generated using the 128-bit security key. Any change will cause the decrypt to fail.

The next frame, titled "ZigBee APS," contains application-level information. From this frame, we can tell that it is unicast (as opposed to groupcast). Security is FALSE, so there is no link key, and no following APS AUX frame. The bit, "extended header" is not present, indicating that there is no fragmentation with this packet. It is all in one piece.

APS ACK is also false, indicating this packet will not generate an end-to-end acknowledgment. If this bit were true, a state machine would keep retrying to send the packet from the original source, and would indicate success or failure to the application. The APS counter would be the same for every APS retry on this same packet. You can tell it's the same transaction if the APS counter is the same. If this bit is false, the only thing that the application knows is that the packet left the originating node (0x1430) successfully.

Notice that the APS layer contains the source endpoint, destination endpoint, cluster, and profile. These uniquely identify which application this packet is for. If the application doesn't match, the packet is rejected.

The final frame, titled "ZigBee ZCL," contains the ZigBee Cluster Library information. This frame tells us that this command is specific to the on/off cluster (0x0006), not general to all clusters such as read/write attribute commands. It also indicates there is no manufacturer-specific ID, so this packet is general to all manufactures. The direction bit indicates whether this packet is traveling from a client to a server (a request) or from a server to a client (response). The command is the toggle command (0x02).

Whew! That's a lot of information to look at (48 bytes worth). And that's just one packet.

Okay, I won't go into such detail in each of the rest of the packets, but I want to show you some common sequences of packets so they make sense when you look at a ZigBee decode.

The basic sequence of events in *Example 7-4 ZigBee Over-the-Air Frames*, are as follows:

- The ZigBee Coordinator (ZC) node forms the network.

- The ZigBee Router (ZR) and two ZigBee End-Devices (ZEDs) join the network.

- The wakeful ZED (0x1430), an HA On/Off Switch, sends a toggle command to the sleepy ZED (0x796f), an HA On/Off Light.

I won't go over forming and joining the network again. You've already seen these packets in detail. Instead, take a look at this new packet:

```
Frame 10 (Length = 56 bytes)
IEEE 802.15.4
ZigBee NWK
   Frame Control: 0x0008
   Destination Address: 0x0001
   Source Address: 0x0000
   Radius = 10
   Sequence Number = 185
ZigBee APS
   Frame Control: 0x01
   Counter: 0x2b
   APS Payload
   APS Command Identifier = Transport Key: (0x05)
   KeyTransport.CommandPayload
   Key Type: Network Key (0x01)
   Key: 04:03:02:01:04:03:02:01:04:03:02:01:04:03:02:01
   Sequence Number: 0
   Destination Address: 00:50:c2:37:b0:04:00:02
   Source Address: 00:50:c2:37:b0:04:00:01
```

This packet is the transport key. When ZigBee is a secure network, it can either be set up to know the key ahead of time (called a Pre-configured key), or to learn the key at join time with the key given over-the-air. That's what the APS Transport Key command is all about.

The next interesting set of packets centers around the toggle command, sent from node 0x1430:

Seq No	Time	Time Delta	Length	MAC Src	MAC Dest	MAC Seq No	NWK Src	NWK Dest	NWK Seq No	APS Counter	Protocol	Packet Type
137	15:45:46.014	+00:00:01.462	48	0x1430	0x0001	19	0x1430	0x796f	189	0x27	Zigbee APS Data	HA:On/off:Toggle
138	15:45:46.016	+00:00:00.002	5		19						IEEE 802.15.4	Acknowledgment
139	15:45:46.028	+00:00:00.012	48	0x0001	0x0000	230	0x1430	0x796f	174	0x27	Zigbee APS Data	HA:On/off:Toggle
140	15:45:46.030	+00:00:00.002	5		230						IEEE 802.15.4	Acknowledgment
141	15:45:47.623	+00:00:01.593	12	0x796f	0x0000	52					IEEE 802.15.4	Command: Data Request
142	15:45:47.624	+00:00:00.001	5		52						IEEE 802.15.4	Acknowledgment
143	15:45:47.625	+00:00:00.002	48	0x0000	0x796f	80	0x1430	0x796f	193	0x27	Zigbee APS Data	HA:On/off:Toggle

Notice that the packet is sent from 0x1430 to 0x0001, then to node 0x0000 (packets 137 and 139). Each packet is acknowledged with a MAC ACK (same sequence number). Then something special happens. The packet sits inside node 0x0000 until node 0x796f sends a "Data Request." This is the method ZigBee uses to deliver packets to sleeping children. The child wakes up and asks its parent, "Got anything for me?" That's the IEEE Data Request (packet 141). The MAC ACK that comes back has a special bit set. It says it has more data for the node (in packet 142). Then the sleepy ZED receives the data in packet 143. Until that sleepy ZED requests it, the parent will buffer the packet. ZigBee only specifies up to seven seconds, but any given implementation may keep it longer. How often the ZED polls its parent with data requests is strictly up to the ZED.

Sometimes you'll see a NWK:NoData packet sent from the parent to a sleepy ZED. These are perfectly normal. They just mean that the parent has data for one of its sleeping children, just not this one.

And that's basically it. By now you have enough information to read every packet sent over-the-air by ZigBee. Or, at least, you have the background from which to understand them. If you encounter a packet you don't understand, feel free to send me an email. My email address is available in the ReadMe.txt at http://www.zigbookexamples.com/code.

> ZigBee packets are built in layered frames: MAC, NWK, APS, and ZCL.

7.6   ZigBee Stack Profiles

One thing not obvious from reading the ZigBee specification (Okay, perhaps there are many) is that ZigBee actually comes in two configurations, that while not incompatible have certain characteristics the OEM needs to be aware of. These configurations are called stack profiles.

Stack profile 0x01, also called ZigBee 2006, ZigBee 2007, or just plain ZigBee, is the original ZigBee stack profile. This stack profile supports not only mesh routing, but also tree routing as described in the previous section. Stack profile 0x01 tends to be used in applications where cost is one of the major concerns. Its main characteristics are:

- Predictable address assignment
- Tree routing
- Smaller code size
- Up to 10 hops in the network

Stack profile 0x02, also called ZigBee Pro, is a later addition to ZigBee, added in the 2007 specification. It does not include tree routing, but includes the feature of source routing for dealing with data concentrators (or gateways). Stack profile 0x02 tends be used where cost is less of an issue, and larger networks are more important. Its main characteristics are:

- Random (stochastic) address assignment

- Source routing

- Multicast

- Larger code size

- Up to 30 hops in the network

Both stack profiles support unicasting, broadcasting, groups, endpoints, clusters, the ZigBee public application profiles, and every ZDO and application-level feature described so far in this document. Really, the major differences between the two stack profiles are address assignment and optional routing methods, and the maximum number of hops.

The over-the-air beacon payload describes which stack profile on which the given network is running. You've seen plenty of examples of stack profile 0x01 beacons. Here is a stack profile 0x02 beacon:

```
IEEE 802.15.4
NWK Layer Information: 0x8422
    .... .... .... 0010 = Stack Profile (0x2)
    .... .... 0010 .... = nwkcProtocolVersion (0x2)
    .... ..00 .... .... = Reserved (0x0)
    .... .1.. .... .... = Router Capacity: True
    .000 0... .... .... = Device Depth (0x0)
    1... .... .... .... = End Device Capacity: True
    Extended PAN ID: 0x0050c237b0040001
```

An entire ZigBee network is either stack profile 0x01 or stack profile 0x02. Nodes from a stack profile 0x02 network may join a stack profile 0x01 network and vice versa, but the joining nodes will conform to the stack profile that they are joining. ZigBee Routers which are not capable of doing the proper routing methods for the given stack profile must join as ZigBee End Devices.

Some ZigBee stack vendors allow a run-time choice of stack profile, while others allow compile-time choice only. If run-time choice is important to your application (perhaps to reduce the number of manufactured parts), check with your stack and/or chip vendor.

> There are two ZigBee stack profiles: 0x01 (ZigBee) and 0x02 (ZigBee Pro).
>
> Stack Profiles define alternate routing methods and maximum number of hops.
>
> ZigBee nodes may join either stack profile.

CHAPTER 8

Commissioning ZigBee Networks

Commissioning is the process of connecting ZigBee applications to each other. This process, while simple in concept, can be fairly complicated to do well and can involve quite a few steps. Commissioning includes:

- Searching for appropriate networks to join

- Joining the correct network

- Determining which nodes on the network to talk to

- Determining how to talk to those other nodes (groups, bindings, or directly)

- Determining if communication has stopped and what to do about it, perhaps even searching for the network on a different channel

Before diving into commissioning, it's worthwhile to talk about a man who served on another commission, and is rumored to have been the inspiration for the name ZigBee.

Zbigniew Brzezinski, Born in Warsaw Poland in 1928, was the National Security Advisor to Jimmy Carter. He later served during Ronald Reagan's administration on the NSC-Defense Department Commission on Integrated Long-Term Strategy for the United States.

Always a human rights activist, in 1989 Brzezinski toured Russia and visited a memorial to the Katyn Massacre. He asked the Soviet government to acknowledge the truth about the event for which he received a standing ovation from the Soviet Academy of Sciences. The Berlin Wall fell not ten days later, significantly changing the political landscape in Eastern Europe.

Brzezinski participated in the formation of the Trilateral Commission during the 1970s and 1980s, in order to more closely cement U.S.-Japanese-European relations, just as the ZigBee Alliance works toward an international standard developed and supported by companies in all these regions.

Brzezinski is currently a professor of American foreign policy at Johns Hopkins University's School of Advanced International Studies.

The rumor goes that in his concern for national security he conceived of a network of sensor devices that could be easily deployed, and if a device was destroyed or removed, the network would continue to operate. It was his inspiration, along with a grant to MIT, which was the kernel of an idea that was later to be named for him. But since no one would be able to spell or pronounce Zbigniew Brzezinski, they shortened it to Zig B. or ZigBee.

8.1 Commissioning Overview

Commissioning is the process of configuring the nodes in the network so they can communicate data to each other. Imagine that you are a ZigBee node. You want to join a network, find an application to talk with, and get to work. So you send out a beacon request. Perhaps dozens of ZigBee networks respond, all within hearing range (see Figure 8.1). Which one should you join?

Figure 8.1: Which Network to Join?

Okay, so somehow, magically (to be explained in this chapter), you join the right network. You also need some security so you don't control the neighbor's network and they don't control yours. Somehow you obtain a security key for the network. Now, how do you find which other nodes in that network to talk to? Perhaps you are a light switch. Well, if you searched the network you would see there are 43 lights on it. Should you control them all? A group of them? Only one? Which one? How is a simple switch to decide?

ZigBee provides a robust set of primitives to accomplish all aspects of commissioning. To help understand these primitives in common usage, I'll describe three distinct scenarios which cover the majority of commissioning issues:

- Simple commissioning
- Butterfly commissioning
- Custom commissioning

The first two scenarios deal with devices that must interoperate between multiple manufacturers (or OEMs), without knowledge of which other devices may be in the network before they join. The last one assumes the manufacturer knows something about the network, and perhaps even defines all the nodes in it.

ZigBee commissioning follows a concept called the "Butterfly Model" (see Figure 8.2). The idea is that a device is "born" with whatever information can reasonably be configured in a static manner and at manufacturing time. At its most basic, and in the absence of any other configuration information, a fresh device will join the first network that offers itself. After joining, it "digests" information that prepares it for its next life, and so on. After some number of these phases it arrives at its stable operational state.

Figure 8.2: ZigBee Commissioning Uses Butterfly Mode

Every device has the ability to get back to the original "factory" state, in case something goes wrong, or the installer wants to start over (perhaps to move the device to another network). This is often called a factory reset. Remember, ZigBee devices store their current state in non-volatile memory to survive power outages, so just removing power and rebooting won't do it. The device has to forget what it has learned.

The commissioning primitives can be found in a variety of ZigBee components, including the NWK layer, APS, ZDO, ZDP, and the ZigBee Cluster Library. An application running in a node has full access to all of these primitives. Most commissioning primitives are available in both local and over-the-air forms:

- *ZigBee Device Object (ZDO)* contains methods for finding and joining the network in various ways, calling on the NWK layer to do some of this work.

- *ZigBee Device Profile (ZDP)* contains device (node) and service (application) discovery, as well as remote table-management functions.

- *ZigBee Cluster Library (ZCL)* provides over-the-air group and scene management.

- *The Commissioning Cluster* provides a standard over-the-air means for setting up security keys, PAN IDs, the channel mask, and manager addresses.

ZDP and ZDO were largely written by Don Sturek, currently of Texas Instruments. He was the technical editor for the APS, ZDP, and ZDO portions of the ZigBee specification, and was chair of The Architecture Committee (TAG) for ZigBee.

ZDO is really the portion of ZigBee that decides which network to join. It bases this decision on a set of fields from the various information bases found in the ZigBee layers. If a commissioning tool (such as a remote PC or a handheld) is used, the commissioning cluster can remotely instruct ZDO to start up in various ways, but it's always ZDO that does the work inside the node.

The fields used by ZDO to form or join the network are shown in Table 8.1.

For ZigBee public profiles, these are normally set at the factory to include nothing but a MAC address in a widget. Other fields are set to all 0x00s, or all 0xffs (to indicate that they are not set). But, in a private profile, they may be set.

ZDP, ZCL, and the Commissioning Cluster are used only after a node is on a network.

Table 8.1: ZDO Startup Fields

Field	Description
MAC Address	The MAC address is a unique 64-bit number assigned at manufacturing time. It is never changed.
apsChannelMask	The channel mask describes which channels should be scanned when forming or joining. The field is a 32-bit bit-mask of 802.15.4 channels. Only channels 11 through 26 are valid for ZigBee, which operates in the 2.4 GHz RF spectrum. Examples: 0x00000800 = channel 11 (bit 11) 0x04001000 = channels 12 and 26 (bits 12 and 26) 0x07fff800 = all channels (11–26)
apsUseExtendedPANID	This is a 64-bit address (similar to the MAC address). Set this to all 0x00s for ZigBee Application Profiles. It is sometime set to a specific number for private profiles.
nwkPANId	Always set to 0xffff (no PAN ID). This will choose a random PAN ID.
nwkNetworkAddress	Set to 0xffff to indicate no short address.
nwkStackProfile	Set this to the preferred stack profile (0x01 or 0x02). Most Application Profiles allow the node to join either stack profile.
TrustCenterAddress	Normally set to 0x0000. Can be set to a different short address if the TC is not on the ZigBee Coordinator.
NetworkKey	Set to all 0x00s if no network key. Set to a specific network key for preconfigured keys. Usually given to the node by the network in Home Automation.
NwkKeySeqNum	Only relevant to over-the-air commissioning tools.
TrustCenterMasterKey	Only relevant High Security networks. This is defined either by the Application Profile or a commissioning tool.

The major ZDP commissioning commands are shown in the following list. See Chapter 5, " ZigBee, ZDO, and ZDP," for a more detailed explanation:

- **ZDP-Bind, ZDP-Unbind**, and **ZDP-End-Device-Bind** add and remove entries from remote binding tables.

- **ZDP-Simple-Descriptor-Request** and **ZDP-Active-Endpoint-Request** and **ZDP-Match-Descriptor** determine which applications exist on remote nodes.

- **ZDP-IEEE-Address-Request** can find all the nodes on the network (just start at the ZC).

- **ZDP-Mgmt-Bind** can determine which applications are bound to which.

- **ZDP-Permit-Joining-Request** enables and disables permit-join in the network.

The major ZCL commissioning commands include the following. See Chapter 6 for a more detailed explanation of these commands.

- **ZCL-Add-Group-Request** and **ZCL-Remote-Group-Request** allow nodes to add and remove groups

Use a combination of **ZDP-Mgmt-Bind** and **ZCL-Get-Group-Membership** to determine which groups are used in the network.

ZCL-Add-Scene and **ZCL-Remove-Scene** and **ZCL-Remove-All-Scenes** to add and remove scenes for disparate devices (a switch, thermostat, and window shade can all react in some way to a scene).

Which brings us to the Commissioning Cluster, and the next section.

Commissioning is the process of setting up ZigBee nodes so that they can communicate data.

8.2   The Commissioning Cluster

The specification for ZDP did not include the ability to query or set the various startup parameters used by ZDO, including PAN ID, extended PAN ID, security key, and network address. This was done purposely, because in simple commissioning (see the next section) it is not needed. Since ZDP is required in every ZigBee device, it was decided to keep ZDP small and instead add this optional feature to the ZigBee Cluster Library as the Commissioning Cluster.

The Commissioning Cluster is very useful in commercial building environments. Network deployment and operation in these environments are often planned to a high degree and they often involve multiple, simultaneous installers during construction or retrofit. They often have more nodes in a given network and often contain many separate, but overlapping networks.

- Startup Attribute Set (SAS)

- Join Parameters Attribute Set

- End-Device Parameters Attribute Set

- Concentrator Parameters Attribute Set

The startup attribute set is used to find the right network and to set up any preconfigured security material. The join parameters control the frequency of joining. The end device parameters determine how often the end device polls its parent and when it searches for a new parent. The concentrator parameters are unique to data concentrators (or gateways as they are sometimes called).

The following table briefly describes the SAS parameters. Although the Commissioning Cluster is optional, every node contains these fields, as they are needed to form or join a network. They are set through the APSME-SET or NLME-SET functions, which affect the APS and NWK information bases.

The SAS parameters (see Table 8.2) only take effect after the node is reset by the Commissioning Cluster command Restart Device Request. Until then, they are merely settings in a table in RAM in the remote node.

The startup control is the most interesting field. It contains instructions on the way to reset the node. Mode 0x00 means *silent join*, which is the node contains everything it needs to know, just as if it had already joined that new network, including all security keys, etc. ... Mode 0x01 is for ZC capable nodes, and tells it to form a network. Mode 0x02 tells it to rejoin, as if the node had lost contact with its parent. The nice thing about this mode is permit-join can be disabled and the node can still "join." Mode 0x03 tells the node to join from scratch. Permit-join must be enabled for the node to join.

If the node will be joining, the Join Parameters attributes may also be set (see Table 8.3).

The rejoin interval determines how often the node will attempt the join or rejoin. Usually, the node will attempt this frequently at first, then gradually slow down so as not to flood the air with useless rejoin attempts, in essence, assuming that the network will come back later.

The End Device Parameters Attribute set (see Table 8.4) is used only on ZigBee end devices.

Sleepy (RxOnIdle = FALSE) ZigBee End devices (ZEDs) poll their parent at intervals for messages. This allows the Commissioning Cluster to determine the interval, which affects battery life. The ParentRetryThreshold applies to both sleepy and wakeful ZEDs. If they just can't communicate with their parent (their only link to the network), then they must eventually find a new parent.

Table 8.2: Startup Attribute Set

SAS Attribute	Description
ShortAddress	This is the 16-bit network short address. Set to 0x0000–0xfff7 for valid addresses, or 0xffff to indicate it is not yet established.
ExtendedPANId	Which extended PAN ID will the node form or join when reset? ExtendedPANId 0x00f0c27710000000 is the commissioning extended PAN. Set to all 0x00s to indicate any extended PAN.
PANId	Normally starting out as 0xffff, which means choose a random PAN ID. Most applications care only about the extended PAN ID. Both PANId and ExtendedPANId are found in beacons.
ChannelMask	A 32-bit mask for deciding which channels to search when forming or joining. Only bits 11–26 may be set. It takes time (about 1/3 to 1 second) for each channel scanned.
ProtocolVersion	Always set to 0x02.
StackProfile	Set to 0x01 or 0x02, the preferred stack profile.
StartupControl	0x00—silent join. 0x01—form a network. 0x02—rejoin a network. 0x03—associate join a network.
TrustCenterAddress	A short address to find the trust center. This is required in high security. Normally 0x0000 (the ZC).
TrustCenterMasterKey	The master key is used to establish a link key with the trust center through SKKE.
NetworkKey	The network key
UseInsecureJoin	Set to TRUE for standard security, FALSE for high security.
PreconfiguredLinkKey	Assumes SKKE has already been performed.
NetworkKeySeqNum	Key sequence number for the network key. A node may have more than 1 network key (old and new).
NetworkKeyType	Set to 0x01 for standard security, 0x05 for high security.
NetworkManagerAddress	Normally set to 0x0000 (the ZC). This node is in charge of frequency agility, if enabled.

ZigBee has the concept of a gateway. Many-to-one routing allows concentrators to easily operate as a gateway without consuming too many mesh-networking resources. This process is described in detail in Appendix A, "ZigBee 2007 and ZigBee Pro."

Table 8.3: Join Parameters Attribute Set

Join Parameters Attribute	Description
ScanAttempts	From 1 to 0xff. 0xff means forever.
TimeBetweenScans	From 1 to 0xffff, in milliseconds.
RejoinInterval	Lower bounds for rejoining, in seconds. Defaults to 60.
MaxRejoinInterval	Upper bounds for rejoining, in seconds. Defaults to 1 hour.

Table 8.4: End Device Parameters Attribute Set

End Device Parameters Attribute Set	Description
IndirectPollRate	The rate, in milliseconds, to poll the parent
ParentRetryThreshold	The number of failed attempts to contact a parent that will cause a "find new parent" procedure to be initiated

Any router node may serve as a gateway. Data concentrators, or gateways, have a few special parameters, as shown in Table 8.5.

The idea behind using the commissioning cluster is so that a commissioning tool (either a handheld or PC) can act as a commissioning network. This network then uses the Commissioning Cluster (via the ZCL-Set-Attributes command) to set up the new node. Then the commissioning tool tells the new node to reset using one of the Commissioning Cluster commands shown in Table 8.6.

Restart Device tells the remote node to restart using the parameters set up in the commissioning attributes. Some nodes can handle more than one attribute set for the commissioning cluster. If this is the case, then the save/restore save and restore to

Table 8.5: Concentrator Parameters Attribute Set

Concentrator Parameters Attribute Set	Description
ConcentratorFlag	After restarting, will this node be a concentrator or not? Assumes that the commissioning tool already knows.
ConcentratorRadius	To what radius will this discover the many-to-one route? The default (and maximum) is 5.
ConcentratorDiscoveryTime	Many-to-one route discovery can occur automatically (1 through 0xffff seconds) or manually (0x0000).

Table 8.6: Commissioning Cluster Commands

ID	Command	Mandatory/Optional
0x00	Restart Device Request	Mandatory
0x01	Save Startup Parameters Request	Optional
0x02	Restore Startup Parameters Request	Optional
0x03	Reset Startup Parameters Request	Mandatory

the primary commissioning cluster attribute sets. The Reset Startup Parameters restores the "factory" defaults.

> The Commissioning Cluster allows over-the-air setup of the startup procedure.
>
> The Commissioning Cluster is optional.

8.3   Example 1: Simple Commissioning

One of the simplest commissioning techniques is that employed by the ZigBee Home Automation Application Profile. The installer of this network is expected to be either a home owner or a professional installer. Either way, the installer is definitely *not* expected to be computer expert. In the simplest case, the installer turns on the devices and they just work. In more complicated situations, pressing a few buttons does the trick.

As an example, assume I just bought a ZigBee system from The Home Depot. The "Porch Light" starter pack comes with two battery-operated switches, four lights, and a remote control (see Figure 8.3). The purpose of this simple home automation system is to turn on the front and back porch lights when I drive home (for safety), and then to turn them off automatically after I've left (for energy savings). The porch lights can also be turned on or off from either of the switches or from the remote. One switch is to be placed just inside the front door, and the other is placed in my bedroom. The next time I go to bed and realize that I didn't turn out the porch lights (again), pressing a bedside button solves the problem.

Since this is a retro-fit system, the ZigBee lights are little gizmos that screw into an existing light socket (probably ceiling cans), and then the light bulb screws into the gizmos. The switches look like normal wall switches and can either be stuck on a cement or brick wall, or recessed into the drywall by cutting a small square hole. The remote unit clips onto the sun visor in my car.

Figure 8.3: ZigBee Simple Commissioning

The kit also comes with a small little box, called a ZigBee Network Controller. The ZNC plugs into my DSL modem, cable modem, or WiFi™ router, if I have one. If not, it just plugs into the wall.

The user instructions that might come with the kit could be quite simple:

1. Warning! Remove power to the porch lights (either at the circuit breaker or wall switch).

2. Plug the ZigBee Network Controller into a wall socket. A green LED comes on, to indicate that everything is okay.

3. Unscrew the existing light bulbs. Screw the ZigBee Light Base-Units into sockets. Screw the light bulbs back into the ZigBee Light Base-Units.

4. Remove the clear plastic tab from the battery holders on the switch units.

5. Press the "Add ZigBee Devices" button on the ZigBee Network Controller. The green LED will flicker. When it becomes solid again, all the ZigBee devices are connected to the network.

6. Test each light switch. Be sure each switch turns the lights on and off. If not, see the troubleshooting section.

7. Install one light switch by the door. Install one light switch in the bedroom. Place the remote control in the car.

Even if the purchaser lost the instructions, they'd probably try what's described above: plug everything in, turn it on, and start pressing buttons. It's that simple.

So how does this work from a ZigBee developer perspective? It starts with setting up the right commissioning parameters. The Home Automation Application Profile specifies this very clearly.

The node starts out knowing nothing but its own MAC address and Application Profile. It has no PAN ID (set to 0xffff), it has no extended PAN ID (set to 0x0000000000000000), it has no channel (channel mask is set to 0x07fff800), it has no network key, and it has no profile ID (it will join either stack profile 0x01 or 0x02).

When power is applied, the node begins scanning for networks. As soon as it finds one with permit-join enabled, it joins that network and receives its short address and network key (in the clear with a transport key) from its parent.

The only widget which forms a ZigBee network in the example above is the ZigBee Network Coordinator (ZC). The ZC will form a network using its MAC address as its extended PAN ID, and a random PAN ID. It will form the PAN randomly on one of the preferred channels for HA: 11, 14, 15, 19, 20, 24, or 25. The ZC will leave permit-join enabled for two minutes when it is first plugged in or after the user presses the "Add ZigBee Devices." Once it has formed the network, it saves this information to non-volatile memory, so that power can be interrupted with no harm to the network. It also includes a recessed "reset" switch that can put it back into factory reset.

The switches and remote control devices are sleepy ZigBee End devices (ZEDs). The lights are ZigBee Routers (ZRs), both of which are device types that attempt to join a ZigBee network. Remember, they will join any network, any PAN ID, any channel. The ZRs and ZEDs will scan for networks on all the channels and then attempt to join. They'll repeat this several times before backing off to trying once a minute. If the user presses a switch and it still hasn't joined a network, then it will attempt to do so right then (immediate user response).

The only potential problem with this easy-joining scheme occurs if two or more installers (perhaps neighbors in an apartment or condominium) attempt to add ZigBee devices to their network at the same time. This is unlikely, but possible. The troubleshooting solution is simply to reset the joining devices and try again.

Once on the network, the ZRs and ZEDs record the channel, PAN ID, extended PAN ID, network (short) address, and network security key into non-volatile memory. They can be powered off or the batteries can be changed, and they still retain the knowledge of which network they are on.

The ZED switches have been preconfigured in their binding tables to use a specific group ID for the destination of the HA:On/Off:Toggle command. A hard-coded group ID can be used because group IDs are unique to each network.

The smart automobile remote takes more explanation. How can a remote automatically turn on the lights when I drive home, without pressing any buttons? There is a fairly easy way to do this.

The device has two modes: "home" and "away."

Every 30 seconds, the device wakes up and attempts to communicate with its parent.

If it was "home" and can no longer communicate, it goes to "away" state. If it is "away" and it can now communicate, it goes to "home" state.

If "away," it will attempt a rejoin every 30 seconds, which may get the same parent or not. When rejoin is successful, the node goes to "home" state.

When the device goes from "away" to "home," it sends the HA:On/Off:On command. Voilà!

Later, this system could be expanded to include more lights and switches, heating and cooling control, and a complete home theater system. The devices are standardized, regardless of who the manufacturer might be. They all work the same way, and they can all be configured over-the-air with bindings, groups, and scenes.

Even the original devices from that starter pack can be reconfigured to include new bindings. Remember how the lights began as belonging to a group, and the switches sent to the group? Well, the switches can be set to control other devices with simple button presses. Press a button on a light, another on a switch, and they can be bound using ZDP-End-device-Bind. The same goes for a thermostat and a temperature sensor.

The ZigBee Network Coordinator, because it has an Ethernet connection to plug into a wireless router or DSL modem, can be monitored or controlled by a laptop, PC, or even over the Internet with a little bit of PC software. A nice drag-and-drop, menu-driven program on the PC (with a ZigBee USB dongle) or a television (ZigBee-enabled) makes reconfiguring the house easy, even for novices.

The primitives are all there. It just takes a little imagination to use them to great effect.

You've seen all these ZigBee primitives before. See Chapter 5, "ZigBee, ZDO, and ZDP," for ZDP (binding) commands; Chapter 6, "The ZigBee Cluster Library," for ZCL (groups and scene) commands; Chapter 5, "ZigBee, ZDO, and ZDP;" and Chapter 7, "The ZigBee Networking Layer," for ZDO and NWK (joining) commands.

To see the "Porch Lights" example in full source code, complete with an over-the-air capture, go to the http://www.zigbookexamples.com Web site. Keep in mind the source code is for the Freescale platform, but the concepts apply to all ZigBee platforms.

The example uses four ZigBee boards, all NVM-enabled so they will remember their settings even across battery changes or reboots:

- **ZcNcbZnc**: The ZigBee network controller. It starts up automatically. SW1 opens the network for one minute to add other devices.

- **ZrSrbLight**: A porch light. Uses a random MAC address, so multiple lights can be programmed. It has no switches.

- **ZedSrbSwitch**: The switch that is placed by the front door, or in the bedroom. SW1 toggles the remote light.

- **ZedSrbCarRemote**: The remote that operates in the same way as the switch, except that it automatically turns on the light when in range. To get it out of range, use a coffee can to cover it.

LSW4 puts all of these devices back to factory reset (forgetting what they learned in NVM).

> Simple commissioning nodes will join any network.
>
> Binding applications is often preconfigured or handled through button presses.

8.4   Example 2: Commercial Commissioning

Simple commission works well for the small, unplanned network. For larger networks, or large groups of multiple networks, such as those found in commercial buildings, hotels, and hospitals, simple commissioning is just, well, too simple.

Networks need to be more secure to operate in a commercial environment. Networks also must be planned, both for the installation process, and for operations and maintenance. Down time or delays in deployment can be very expensive.

For these applications, the network is planned out ahead of time. Blueprints are used to decide where the network components will go. Tests are run ahead of time to make sure there is sufficient ZigBee Router coverage. The rooms are usually built one-by-one as autonomous units, and the installer must not only verify that the network is functioning, but that all the equipment installed (lights, switches, thermostats, heating and cooling units, door locks, etc.) works as expected. A very detailed checklist is built and written with the training of the installer in mind who will usually be an electrician.

In larger installations, it is common to have multiple installers all working in the same area. Figure 8.4 shows a typical set of hotel rooms. Two installers are setting up adjacent rooms, taking devices from their boxes, installing them and then commissioning them to network together.

Each device is tested, repaired if there is a problem, and replaced if the problem can't be resolved. The installation tool may have a bar code scanner on it, so they can scan the device when getting it out of the box. In this way the commissioning tool knows about the device and can check it off a list, or do other intelligent behavior.

These devices (lights, switches, door locks, thermostats, etc.), which may originate from multiple OEMs (perhaps Philips lights, Trane thermostats, and Schneider Electric air conditioners) are set up out-of-the-box to automatically join the commissioning network on extended PAN ID 0x00f0c27710000000. ZigBee defines this special Extended PAN ID so that devices built by many OEMs can all be commissioned in the same way.

Think of it as two step process. First, get on the commissioning network and receive the commissioned data. Then, reset the node to join the operating network, and complete the commissioning process (see Figure 8.5).

If it is expected that there will only be multiple installers in any given vicinity, then the ZigBee standard commissioning extended PAN ID, 0x00f0c27710000000, is used only to

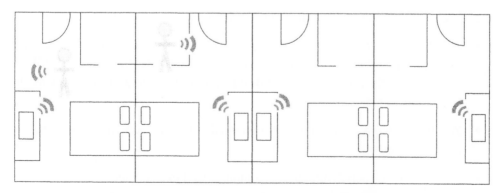

Figure 8.4: ZigBee Commercial Commissioning

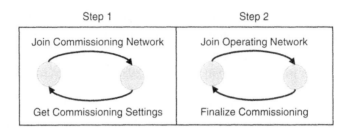

Figure 8.5: Commercial Commissioning Requires at Least Two Steps

get the extended PAN ID of the commissioning tool itself. Then the node is reset (via the Commissioning Cluster) to go onto the commissioning tool's PAN. ZigBee reserves the Extended PAN ID range of 0x00f0c27710000001 to 0x00f0c2771000ffff for the purpose of commissioning tools.

In ZigBee Stack Profile 0x01, the node usually receives its network short address when the node rejoins the new network via the Commissioning Cluster Reset Device command (with StartupControl set to mode 0x02).

But in stack profile 0x02, also called ZigBee Pro, it is possible to set the network short address explicitly. In this case, the commissioning tools would know the short address to assign to each node, perhaps based on some formula involving hotel room number. Then, when the node is reset, it can either use the rejoin as above, or even silent join (with StartupControl set to mode 0x00) so that it can be done a little faster (no waiting for negotiation with the trust center). It's assumed in this case that the tool has all the necessary keys and has communicated them to the nodes.

The example in this section, "CommercialCommissioning," uses the following nodes:

- *ZcNcbCommissioningTool*: Commissions (over-the-air) the light and switch. The LCD displays which node joined the commissioning network. A press of SW1 commissions the joined node and informs it to reset to the operating network. It has permit-joining on.

- *ZcSrbOperatingNetwork*: Pre-commissioned as the operating network. It has permit-joining off.

- *ZrSrbLight*: "Generic" light that is commissioned over-the-air.

- *ZrSrbSwitch*: "Generic" switch that is commissioned over-the-air.

To see the full source code and capture, go to the http://www.zigbookexamples.com website. As usual, the source code is for the Freescale platform. See Chapter 3, "The ZigBee Development Environment," for general instructions on using the source code. The basic steps for the demo are:

1. Download all the images. Turn all boards off.

2. Capture with Daintree (optional) on channel 25.

3. Boot the ZcSrbOperatingNetwork board. This lights all LEDs to indicate that it's the operating network.

4. Boot the ZcNcbCommissioningTool. This displays that information on the LCD.

5. Boot either the light or the switch. The LCD on the commissioning tool displays the fact that a light or switch has joined the commissioning network.

6. Note in the over-the-air capture that the node is associated with the commissioning cluster.

7. Press SW1. Watch as the new extended PAN ID is communicated over-the-air, and the node is instructed to reset. If the node is a switch, notice that the binding command is also given to it.

8. Boot the other board (either light or switch). Press SW1 on the commissioning tool again to provide the information to the other board, and to reset it.

9. Press SW1 on the ZrSrbSwitch. Notice that it toggles the light! This node is commissioned.

This is the basic process. Of course, there is much more to it, and commercial tools are beginning to come out from various vendors. At the time of this writing, both Daintree and Atalum sell commissioning tools.

> *Commercial Commissioning allows full planning of the network.*
>
> *A special commissioning tool and a temporary commissioning network complete the process.*

8.5   Example 3: Custom Commissioning

Devices that work with ZigBee public Application Profiles must work in a specific and generic way in order for them to interoperate with devices made by other OEMs. ZigBee specifies how commissioning works in each public profile very clearly. Private profiles do not have this restriction.

A private profile may commission the network in much more creative ways. For example, the entire network can work right out-of-the-box. Just order a kit: It forms a unique network on a unique PAN, and everything in the box is already connected and ready to communicate.

An application of this could be a manoverboard system for small pleasure craft and commercial fishing boats. Every router and battery-operated man-overboard device knows in advance exactly which network it is on, which extended PAN ID, which node address, which network key, everything. All the nodes essentially "silent join," and the network begins operating correctly the moment the devices are turned on.

The example for custom commissioning is a merchandise quality-tracking system. In this system, a variety of sensors monitor the physical environment inside tractor trailer trucks, perhaps including temperature, humidity, and shock. The goal is to ensure that a load of frozen salmon arrived frozen, and has stayed frozen the entire journey, or that a cargo of produce (such as lettuce) was never too warm, too cold, or too dry (see Figure 8.6).

Today, the majority of these systems depend on manual spot-check inspection. An inspector will open boxes, examine some of the fish or lettuce, and make an educated guess as to whether the goods are within specification or not. This process is both

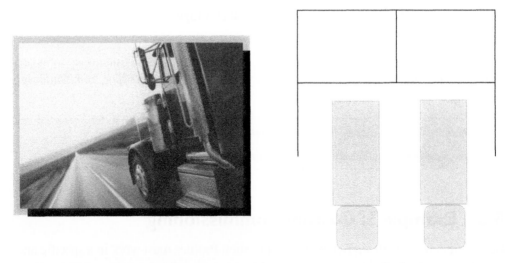

Figure 8.6: Loading Bays Could Use Custom Commissioning

time-consuming and is subject to error. Placing sensors in the vehicles to do this work makes sense. But how do we get that sensor data out of the truck? In this case, wireless monitoring makes perfect sense.

The sensors record data during the journey, and when the rear doors are opened, the sensor nodes automatically connect to a network in the docking bay, and download the data to a gateway, which is attached to a PC and a central tracking system.

All the nodes in this system are produced by the same manufacturer. So how do these nodes travel from one network to another? How do they get on the network in the loading bay and automatically download information? The procedure makes a few assumptions, an acceptable variation because it is a private profile:

- The gateway (to where the data will be sent) is assumed to be node 0x0000.

- Sensors, while in sensing mode, gather data once each minute.

- Once every three minutes, the sensors also send a beacon request (active scan) to scan for networks.

- Only networks in Extended PAN ID range 0x12345678xxxxABCD will be considered as potential networks to join.

- If the same network that was exited previously is found within 30 minutes, it is not joined.

- If a new network is found, or the same network is found 30 minutes or more after the sensor lost contact with all networks, the new network will be joined using the NWK-Rejoin command. This NWK-Rejoin is used so that permit-joining may always remain off, in the networks in the loading bays.

- Once joined, the sensors send all their data to the gateway (acknowledged). After the data is transmitted, it is forgotten, and the node leaves the network to go back into sensing mode.

Using these simple rules, the sensor nodes will always join a proper network for transmitting their data. They conserve power while not on the network, saving it for sensing operations and occasional data transmission.

Notice there is no binding. Nodes just send directly to the gateway (node 0x0000). Notice there is no over-the-air commissioning. The nodes know what they want to do before they are even started. These nodes don't even need the standard ZigBee non-volatile memory to be enabled, except perhaps as storage for the data. But probably another storage mechanism, such as serial flash, will be used for data storage. Everything was commissioned into the nodes at the time of manufacture.

The example in this section, "Commercial Commissioning," uses the following nodes:

- *ZcNcbGateway*: Gathers the data. It transmits the data over the serial port (simple ASCII).

- *ZedSrbSensor*: Gathers temperature data once each minute and transmits it when it gets in range of the gateway.

Simply set the ZedSrbSensor outside, down the hall, or anywhere away from the Gateway node. Attach the Gateway and use HyperTerminal on that USB/COM port (mine is COM4) at 38,400 baud, 8N1. When the sensor comes within range, it will download the gathered data.

I've accelerated the process so that it works better for a demo. The demo gathers data once every 20 seconds, and only needs to be away from the gateway for one minute before it will link to it again.

"Custom Commissioning" is very flexible. All the ZigBee Device Object (ZDO), ZigBee Device Profile (ZDP), and ZigBee Cluster Library (ZCL) commands are at your disposal. Use them creatively. If you like, tell me about them. I love hearing of new applications for ZigBee, and new creative ways to use it. My email address is drewg@sanjuansw.com.

> Custom Commissioning allows a wide range of commissioning options.
>
> Use Custom Commissioning in private application profiles.

CHAPTER 9

ZigBee Gateways

So far in this book, the chapters have centered on how ZigBee nodes communicate with other ZigBee nodes. However, many ZigBee networks are not stand-alone: They need to interact with other networks or at least with some data concentrator such as a PC. This chapter explains how gateways work in ZigBee, and describes some common pitfalls when deploying them in the field.

Before continuing with ZigBee gateways, consider this final story about how ZigBee got its name.

A narrow cobblestone alleyway winds its way through Amsterdam, leading past an ancient house built sometime in the 17th century. Through lighted windows, gleaming copper cookware, rustic beams, and romantic artwork can be seen adorning the interior, and fresh flowers cover the tables and entryway. A sign hangs on wrought-iron outside the door, which reads, "The Black Sheep." The old house, now a restaurant, boasts a menu encompassing both French-inspired dishes and updated Dutch fare. It is late at night, nearly closing time. This night is special, for it is the last night: the restaurant is closing for good.

This night is also special for another reason. Around one of the last occupied tables, a man named Bob Heile sits with four other men. The last of the plates have been cleared and the empty wine bottles have been pushed to one side. A shout of laughter erupts as the men shout out strange-sounding words. This group of men is trying to come up with a name for a new technology, a new wireless technology aimed at low power, low cost, and low speed networking. The late night and alcohol help. The original concept came from Philips, called Firefly. But the goal has moved far beyond Firefly, far beyond lighting, in fact, to general purpose wireless sensor and control networks.

What you are witnessing is the real story of the birth of ZigBee technology and the ZigBee Alliance. Eventually, over more wine and more laughter, a name emerges for this technology, a name that is unique, easy to say, and most importantly, can be trademarked. "To ZigBee," toasts the group. "To ZigBee!"

If you travel to a ZigBee Open House, you will see Bob Heile, the chairman of the ZigBee Alliance, wearing a business suit (one of the few times he does). Around his neck is a tie, adorned with many white sheep and one black sheep. Ask him about it, and you will hear an interesting tale.

For those of you who travel to Amsterdam, you may be interested to learn that the restaurant has reopened. The Black Sheep can be found at Korte Leidsedwarsstraat 24, Amsterdam, Netherlands 1017 RC · 020-622-3021.

A ZigBee gateway, quite simply, is a means of transferring data between a ZigBee network and devices on another network. In illustrations, the other networks are usually depicted as clouds, while the interested devices and the ZigBee network are attached to the cloud (see Figure 9.1).

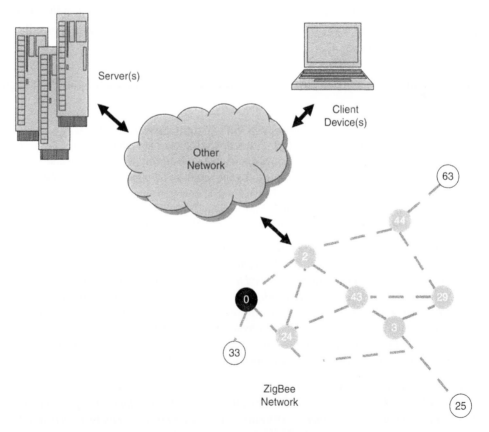

Figure 9.1: A ZigBee Gateway

One interesting example of an existing gateway is a device found in the Home Automation Profile, called a Combined Interface Device (device ID 0x0007 in the HA application profile). The hardware for this device is normally a USB ZigBee dongle, which connects to a PC, TV, or a similar display. Through a menu-driven program, the home owner can monitor or control devices in the ZigBee network: turn off the lights, check the status of the security system, or remotely allow someone into the home. Often, this PC program provides some kind of Internet experience, allowing the ZigBee network to be controlled or monitored over the Web, as well as from local PCs, wall displays, or the television.

Figure 9.2 is a screen shot that shows one view of a ZigBee network as seen inside the Control4 Composer Home Edition. Control 4 (http://www.control4.com) designs, manufactures, and sells ZigBee and other Home Automation products.

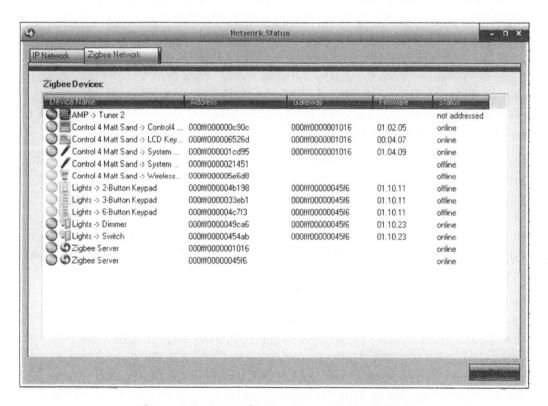

Figure 9.2: Control4 Composer Home Edition

Another example of a gateway would be a commercial building automation system. One that I've worked with is a building management system that provides energy savings and statistics through monitoring and control of energy-consuming devices within a hotel, such as the heating, ventilation, and air conditioning (HVAC) systems. These networks are deployed in hotels ranging from very small mom-and-pop businesses to the very large Mandalay Bay in Las Vegas, Nevada. Systems of this nature typically pay for themselves in a matter of months.

One simple application of a gateway that interacts with ZigBee is the automatic printing of a list each morning of all the battery-operated ZigBee devices with low battery charge. For a large network, with many thousands of nodes, this can be a huge time and cost-saver for the maintenance personnel. The ZigBee gateway in this case is connected to an Ethernet backbone, which in turn communicates to one or more PCs sitting in a maintenance room in the basement of the hotel.

A third example (which actually requires two gateways) is automatic meter reading (AMR). The utility meters may be networked with ZigBee, but ZigBee won't make it all the way back to the utility company. For this other networks are used, sometimes power-line networks, in other cases cellular networks, WiFi™, or proprietary networks. A gateway is used between the ZigBee network and this back-haul. Usually the gateway takes the form of serial communication from the ZigBee processor to some other host processor that is attached to the other, longer-range network. The data is then sent back to the central office for processing and even billing.

The other gateway required in the AMR system is the one from ZigBee to ZigBee. The AMR system, in addition to reading the meter, may need to monitor or control a Home Automation (HA) or Commercial Building Automation (CBA) network. That home network isn't likely to be part of the AMR network, but a stand-alone network that the home owner controls. To connect to that network in a way that allows for the disparate security protocols (light security in HA, heavy security for AMR), two separate networks are used (see Figure 9.3).

The California Energy Commission (doesn't it always start in California?) is putting together standards that will ensure significant energy savings for consumers, and also allow the electric companies the ability to control electrical usage in homes that have been signed up for an optional program to reduce peak usage. Reducing peak usage means not having to build extra power plants. Not having to build extra power plants means a huge savings for the utility companies, and reduced power bill for consumers. If you are interested in this effort, you can read more by googling "Title 24."

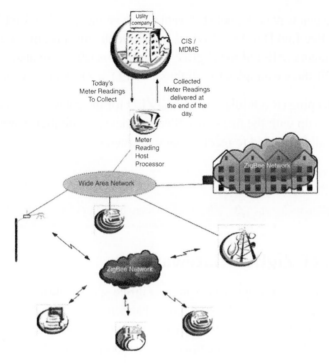

Figure 9.3: AMR Gateway

Within the topic of Gateways, there are three specific areas which are important to understand:

- Lack of current ZigBee gateway standards

- Data concentration

- Bandwidth problems at the data concentrator

All three of these problems are discussed in detail in the following sections.

In this chapter, forgive me if I use the terms *data concentrator* and *gateway* somewhat interchangeably. The difference is subtle: A data concentrator collects data from a ZigBee network, while a gateway is a data concentrator that also communicates this data to another network. Either can inject data into the ZigBee network.

One common misconception is that the gateway must be the ZigBee Coordinator (ZC). The truth is, any node in the network can be a gateway, and in fact multiple nodes in the network can be gateways at the same time. If you need room for a more complex application in the gateway node, but still want to use an inexpensive 8-bit MCU, choose an RxOnIdle ZigBee End Device (ZED), as they have the most room for application space. ZigBee Routers (ZRs) or the ZigBee Coordinators (ZCs) are also a good choice. In this chapter, I'll show examples for all three: a ZED, a ZR, and a ZC gateway.

Do NOT use a sleeping (RxOnIdle = FALSE) ZED for the gateway, as its parent would not be able to keep up with the messages. Sleeping devices poll their parent for messages and are not normally expected to receive many messages.

> Gateways communicate between a ZigBee network and a PC or other network.
>
> Any ZigBee node type (ZC, ZR, or ZED) may be a gateway.

9.1   A UART ZigBee Gateway

So, with all this talk of gateways, where in the ZigBee specification is the gateway section? Well, there isn't one. At least, it hasn't happened yet. The ZigBee Alliance is continuing to improve the technology, but gateways have not yet been standardized. (There is a ZigBee Bridge Device specification that has passed initial ballot as of the time of this writing, but a bridge is not a gateway.)

Every ZigBee implementation, from both stack and silicon vendors, includes some way to interface between the ZigBee board and a development PC via a serial, USB or Ethernet port, to allow the node to be programmed and debugged. In this section, I'll propose a small but effective ZigBee Serial Gateway Interface (ZSGI) that takes advantage of this serial port.

The main goal behind this gateway interface is, first of all, to keep it simple and easy to implement on an 8-bit MCU, which is typical of ZigBee nodes. For example, the Freescale, Texas Instruments, Renesas, Atmel, and Integration platforms offer a serial port (or a USB port that acts like a serial port). There is nothing to prevent this protocol from being used on Ethernet-based systems such as Ember's, with a little bit of IP wrapper around it.

Typically, in a serial gateway the small 8-bit ZigBee device communicates to a host processor or PC. It may be that the host is a regular desktop or laptop, and the communication stops there. It may be that the host processor is an ARM or other embedded processor, residing on a motherboard within the embedded widget. (I've seen this in WiFi routers and AMR meters.) The communication to the host processor may

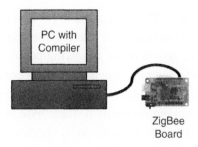

Figure 9.4: A Common ZigBee Serial Gateway

be a serial or USB cable, as shown in Figure 9.4, or it may simply use the Tx/Rx pins between two MCUs on the same motherboard.

This proposal assumes only Tx/Rx lines, and does not specify any hardware flow control. Note that on the Freescale platform, the MCU serial port goes through an FTDI USB chip so that it can be connected to a PC as a USB-serial device, because real serial ports are getting to be scarce these days on laptop or desktop computers.

On the host side, the ZSGI protocol could be extended to work over any network, including TCP/IP, WiFi, a cellular connection, or any other network which may want to communicate with nodes in a ZigBee network. Those extensions are not discussed here, however.

9.1.1 The ZigBee Serial Gateway Interface

The ZigBee Serial Gateway Interface is a half-duplex protocol, in which a serial message is sent, and a response is received. The over-the-serial-port frame is shown in Table 9.1.

The same frame format is used either for *commands* (packets sent from the host processor to the ZigBee processor) or *events* (packets sent from the ZigBee processor to the host processor).

Every packet begins with an STX (byte value 0x02). Next follows a GroupID, which indicates which category of message the packet belongs to (for example, which SAP

Table 9.1: ZigBee Serial Gateway Interface (Over-the-Serial-Port Frame)

Bytes:1	1	1	1	Variable	1
STX (0x02)	GroupID	MsgID	Length	Payload	Checksum

handler must respond to this call). The GroupID can be seen below in the Gateway Commands section. The MsgID indicates the specific message within that message group. The length indicates the length of the following payload field, in bytes. Next follows the actual payload, which varies depending on the combination of GroupID and MsgID. For example, RegisterEndpoint.req requires a full simple descriptor for data, whereas DeregisterEndpoint.req requires only a single byte of data, the EndPoint number. Finally a checksum follows, which is a simple XOR checksum (all bytes, from STX through checksum byte must checksum to 0x00, when all bytes are added).

The protocol can be expanded to support up to approximately 65,000 different commands (the combination of the 8-bit Group ID and 8-bit Message ID). Don't confuse the Group ID in the gateway interface with ZigBee groups. The Group ID in the gateway is simply a way of associating similar commands in the serial protocol.

9.1.2   Gateway Commands

Table 9.2 lists the commands and events for the ZigBee Serial Gateway Interface (ZSGI). The parameters of the interface match what's in the ZigBee specification. This keeps it vendor-independent and makes it easy to implement. Remember, the implementation in the 8-bit ZigBee MCU will be different for each stack vendor, because ZigBee does not specify APIs, only over-the-air behavior.

Table 9.2: ZigBee Serial Gateway Interface

Name	GroupID	MsgID	Parameters
RegisterEndpoint.req	A3	0B	See SimpleDescriptor (Table 2.35)
RegisterEndpoint.rsp	A4	0B	Status
DeregisterEndpoint.req	A3	0A	Endpoint #
DeregisterEndpoint.rsp	A4	0A	Status
StartNetwork.req	A3	E0	See NLME-JOIN.request (Table 3.16)
StartNetwork.rsp	A4	E0	Status
StopNetwork.req	A3	DC	See NLME-LEAVE.request (Table 3.22)
StartNetwork.rsp	A4	DC	Status
APS_Data.req	9C	00	APSDE-DATA.request (Table 2.2)
APS_Data.cnf	9D	00	APSDE-DATA.confirm (Table 2.3)
APS_Data.ind	9D	01	APSDE-DATA.indication (Table 2.4)

You'll note that this simple gateway interface does not include the ZigBee Cluster Library (ZCL) commands. These commands are higher-level than this gateway, and are expected to be implemented on the host processor side of this gateway connection, if needed by the Gateway application, by using the APS_Data.req command.

You'll also notice that there are no ZDP or configuration commands. This interface assumes that endpoint 0 can be used for sending and receiving ZDP commands, and that any configuration (such as channel list, adding groups, etc.) can be accomplished by using over-the-air commands. The interface also assumes that a node can send an over-the-air style command to itself, for example, by using ZDP to retrieve a node's own IEEE or NWK address.

9.1.3   Using the ZigBee Serial Gateway Interface

So, how does the host processor "application" actually use this set of commands to create a gateway, and to monitor and control the network? The typical order of events is described below.

1.   The host processor registers an application endpoint (0x01 – 0xF0), using the appropriate simple descriptor for the application. The example in this chapter uses the Home Automation OnOffLight.

2.   The host processor starts the network by using the StartNetwork.req command. Note that "starting the network" may actually mean joining a network (as opposed to forming a network), depending on whether the gateway node is a ZC, ZR, or ZED. (Yes, a gateway can be an RxOnIdle ZED!)

3.   ZigBee data can then be communicated on the ZDP and application endpoint. Note that there can be multiple application endpoints, supporting a variety of Application Profiles. Optionally, the gateway node may leave that network, and possibly even join a new one. The leave command resets all the layers.

This following example uses a PC to toggle an HA OnOffLight, through a ZigBee Serial Gateway Interface. The PC is connected to the gateway (a Freescale SRB board) via a USB port. From the PC side the USB connection appears to be a COM port, as seen from Windows Device Manager.

The ZigBee Serial Gateway Example comes with two projects, a Gateway and an HA OnOffLight. The Gateway image was created using BeeKit, by selecting an SRB board, ZigBee Coordinator with the ZigBee Test Client (ZTC) enabled. ZTC is Freescale's

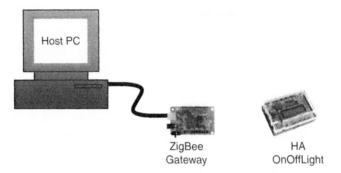

ZigBee
Gateway

HA
OnOffLight

Figure 9.5: ZigBee Serial Gateway Example

version of the ZigBee Serial Gateway Interface. The HA OnOffLight image was also created with BeeKit, selecting an NCB board with display enabled. Both boards are on channel 25, PAN ID 0x0f00.

If you wish to follow along with actual hardware, then program the SrbZcGateway.mcp project into the SRB board, and program the NcbZedHaOnOffLight.mcp project into the NCB board. For a description of the boards and development environment, see Chapter 3, "The ZigBee Development Environment."

To run the example, you must first enable a USB port as a serial port. This entails a few steps, but is only required once, just like enabling any new hardware in Windows:

1. Connect the SRB board which will be the gateway to the PC using a USB cable.

2. Turn on the SRB board.

3. Windows will indicate that it has found new hardware. Install the drivers from "C:\Program Files\Freescale\Drivers."

4. Windows will ask twice.

5. After the new hardware is connected, open Windows Device Manager to determine which COM port was assigned. On my system, it is COM7.

Next, download and install the Freescale TestTool. This tool is a GUI that understands the ZigBee Serial Gateway Interface, and will interact on the PC side of the serial gateway connection.

To ensure TestTool understands the ZSGI, follow these steps. These steps only need to occur once during setup.

1. Be sure that ZigBee2006.xml is in the "C:\Program Files\Freescale\Test Tool\ Xml" directory.

2. Run Test Tool.

3. From the menus, select Tools/Communication Settings… You must type in the COM number and choose 38,400 baud.

4. From the menus, select View/Command Console. This brings up the main screen for interacting with ZSGI.

5. On the Command Console screen is a drop-down box that is titled "Loaded Command Set." Choose ZigBee2006.

Test Tool is now set up.

To run the gateway example, reset both boards. LED1 should blink on both of them, indicating the normal Freescale "waiting to start" state. Press SW1 on both boards. This will start the network.

In Test Tool, there is a button labeled All Commands. Press this button, and select "APS Layer commands/APSDE-DATA.request." This is shown in Figure 9.6.

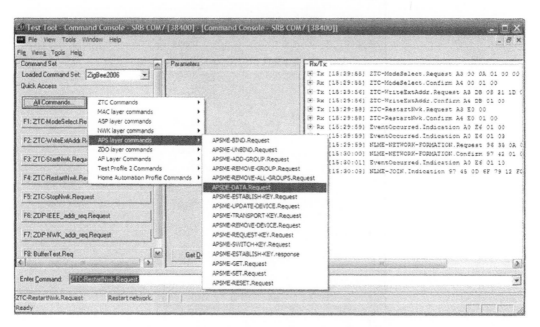

Figure 9.6: Test Tool Command Console

So what happened to commissioning? What happened to finding active endpoints, to looking for the particular HA OnOffLight to control? For the sake of clarity, I made some assumptions. I know that the gateway is the ZigBee Coordinator, so it will be node 0. I know that the default Freescale endpoint is endpoint 8. And I know that the HA OnOffLight is the first ZED in the network (0x796f). So, I simply tell the gateway node, through the serial port, to send an APSDE-DATA.request directly to node 0x796f. The payload is three octets: 0x01 0x42 0x02, which is the ZCL Toggle command. Send this command and the light toggles!

The actual screen in Test Tool is shown in Figure 9.7.

Any ZigBee command can be sent from the gateway. For example, change the DstEndpoint to 0x00, the ProfileId to 0x0000, the ClusterId to 0x0001, the asduLength to 0x05, and the Asdu contents to 0x00 0x6f 0x79 0x00 0x00, and the gateway will issue an IEEE address request to the ZDP of the HA OnOffLight. The results will come back to the gateway through endpoint 0x00.

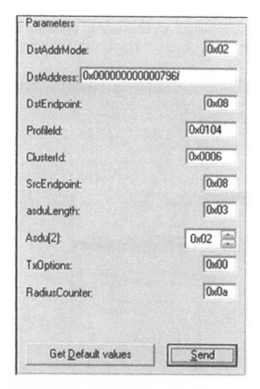

Figure 9.7: APSDE-DATA.req in Test Tool

The source code to the Freescale serial gateway (called ZTC), which implements the ZigBee Simple Gateway Interface plus many more commands, is too complicated to go into in detail here. The concept, however, is fairly simple.

The program in the gateway enables a serial port in the 8-bit MCU (in this case, at 38,400 baud, 8N1). The serial port is interrupt-driven, both on transmit and receive. When data is received over the serial port, it is decoded. If an entire ZSGI packet complete with accurate checksum is received, it is formatted and passed to the appropriate Service Access Point (SAP) handler, in this case, the APSDE SAP handler.

When data flows up through a SAP handler (data that comes from the radio), then it is formatted into a ZGSI packet and sent out the serial port. There are quite a few BeeKit options available to fine tune ZTC, enabling and disabling various features, but the simple way is to just check the ZTC box on the Advanced Features pane of the BeeKit Project Wizard.

ZigBee stack vendors which run on MCUs with serial ports tend to have serial support built into the stack. They usually have an initialization function, a serial transmit function, and a serial receive function.

> ZigBee currently lacks gateway standards.
>
> A serial gateway is easy to implement and use.

9.2 The Data Concentrator Problem

You may remember that ZigBee uses routing tables to route packets. If you don't remember this, go back and read Chapter 7, "The ZigBee Networking Layer."

Each routing table entry contains two significant fields: the final destination, and the next hop to use to get there. For example, if node 43 wants to communicate to node 19, its final destination would be node 19 and the next hop toward node 19 would be 2 (as shown in Figure 9.8).

What if all the nodes in the network want to communicate with node 19 (if node 19 was the gateway, for example)? Node 2 in the figure above would have a next hop of zero to get to node 19. Node 43 and node 44 would both have a next hop of two, but node 2 still only needs the one routing table entry to get to node 19, the next hop of zero. In fact, if a hundred nodes were all communicating through node 2 to get to node 19, it still needs

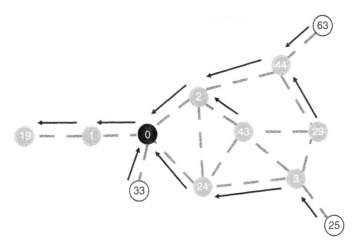

Figure 9.8: Sending Data to the Concentrator

only one routing table entry. Regardless of how many nodes there are in the ZigBee network, each node needs only one entry to communicate to the data concentrator.

So what's the problem? Ah, but what if the data concentrator needs to communicate to all those nodes in the ZigBee network, and not just receive data from them?

In Figure 9.9, node 19 would need to have enough routing table entries for every node in the network! And so would nodes 1 and 0. If the ZigBee network only has a dozen, or even a few dozen nodes, that's no problem, but what if the ZigBee network contains

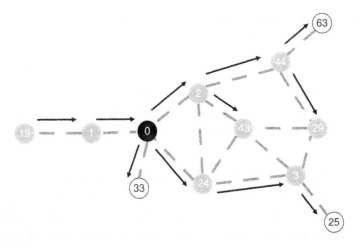

Figure 9.9: The Data Concentrator Sending Data to the Network

6,000 nodes? Then node 19 would need 6,000 × 5 bytes per routing table entry, or 30 K of RAM just for the routing tables. Considering ZigBee is targeted at 8-bit MCUs which have only 4 K RAM, you can see that this approach is not very practical.

One solution is of course to run ZigBee on a processor with more RAM, say some 32-bit processor such as an Intel x86, or ARM 9 with eight megabytes of RAM. But ZigBee is designed to be inexpensive, and that would mean every router would need enough RAM for any routes which pass through it. This could significantly increase the price of the ZigBee network.

ZigBee Pro, an upcoming version of the ZigBee specification, also has a solution to this problem, called *many-to-one route discovery* which employs the strategy of source routing in addition to table-driven mesh routing. The ZigBee Pro method assumes that the data concentrator (or the PC it talks to) has enough RAM to store a large set of source routes to all the nodes. I'll describe this ZigBee Pro concept and others in more detail in Appendix A, "ZigBee 2007 and ZigBee Pro."

Another practical solution that uses the current ZigBee specification that has been adopted in ZigBee networks employs "Island Controllers." This solution assumes that the network designer has control over at least some of the routers in the system. In this approach, communication is from the data concentrator to islands of nodes, rather than to each node directly. Again the host PC must have enough RAM to store the information about which islands lead to which nodes, Standard ZigBee mesh routing is used between islands, and between the last island and the final destination nodes.

To understand how this works, consider a hotel. In this example, each hotel room may contain up to 16 ZigBee nodes, all of which communicate to a room controller. The room controller is used so that each room can operate independently of every other room, regardless of what happens to the ZigBee network. Assume that there are 3,000 nodes in the ZigBee network. Perhaps there are many ZigBee networks in a large hotel, one per floor, but each one has a gateway. The networks are considered independent networks for the purposes of this discussion.

To communicate to all nodes in this network, the gateway would require 3,000 routing table entries, if using the brute force of communicating directly from the gateway to each node in the network. That clearly won't work with only 4 K of RAM in the gateway. So instead, a set of islands are defined on the PC side of the gateway. Assuming just 14 routing table entries per router (using 70 bytes of RAM), each "Island Controller" can handle up to 14 "island children." Each Island Controller has an application that will examine packets delivered on a special cluster, and deliver it to the next island in the

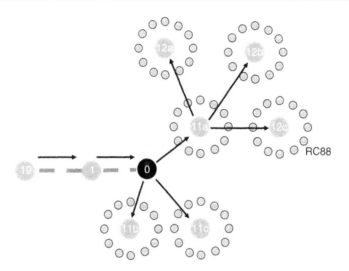

Figure 9.10: Islands Within the Network

chain, and eventually to the proper cluster on the proper node. With 3,000 nodes, two tiers of islands would be needed:

$$(16 \times 14 \times 14 = 3,136)$$

Gateway → IslandController1 → IslandController2 → Room Controller → End-Node

As an example, if the back-end building management system needs to deliver a packet to Room Controller 88 (RC88), or to one of its children, it would first deliver the packet to Island1a (I1a), which would in turn deliver it to Island2c (I2c), which would then deliver it to RC88, which would then deliver to one of its children.

It's interesting to note this looks something like the tree networking you learned about in Chapter 7, "The ZigBee Networking Layer." But the system actually uses ZigBee mesh networking to get to each Island Controller in the most efficient manner. This is not tree routing, but it does assume that the Island Controllers are fairly stable units, and can be accessed, somehow, through a path in the mesh network. The islands do not need to be a single hop away from each other, as the ZigBee mesh networking will find the most efficient path between islands.

This method also requires a special application in the Island Controllers, but none of the other ZigBee nodes need to know anything special about the delivery of packets from the gateway. In the next section, I'll provide simple source code that delivers the packets as described through an Island Controller system.

There is one more thing to be aware of. The ZigBee specification has no specific policy about retiring routes. To quote the spec, "The aging and retirement of routing table entries in order to reclaim table space from entries that are no longer in use is a recommended practice; it is, however, out of scope of this specification." This means your application should do some sort of aging if the stack implementation does not. Ask your stack vendor how this is accomplished in their implementation. If you don't age routes, the network could eventually stop routing packets if links have broken enough times, which, as a consequence of Murphy's Law, wouldn't be discovered until your network has been in the field for months, or even years.

9.2.1   Island Controller Example Program

This section describes a simple example of an "Island Controller" application. It uses a unique cluster, 0x15AD (a number that looks kind of like the word "island" if you stare at it long enough), to send data through a series of islands. It's up to the sender (gateway, or PC on the other side of the gateway) to determine which Island Controllers to use and in what order.

In this example, I'll use four nodes, as shown in Figure 9.11. The Island Gateway node (an SRB) is connected to a PC via the USB port. A PC program from Freescale, called Test Tool, sends a special packet into the network through the Island Gateway to inform the Home Automation OnOffLight to toggle.

This same configuration will scale to over 3,000 nodes with just 14 routing table entries, and the devices being monitored or controlled (the OnOffLight) do not need to know anything about the "Island Controller" concept.

I'll show you the relevant lines from the over-the-air capture, but the plan works like this. Whatever is to be sent by the Island Gateway to any node in the network is preceded by a header, and sent to cluster 0x15AD. The header format is shown in Table 9.3.

The first field, NumIslands, indicates how many "NextIsland" short address fields follow. If this is 0, then there are no more islands and the packet is delivered directly to the final node, cluster, and endpoint indicated by the NwkAddr, Cluster, and EP fields. The Island Controllers all use endpoint 8 for simplicity's sake (no need to discover active endpoints and read simple descriptors).

Each island controller that receives a packet decrements the number of Island fields and strips off the first "NextIsland" field in the island chain, then delivers the packet to that next Island Controller.

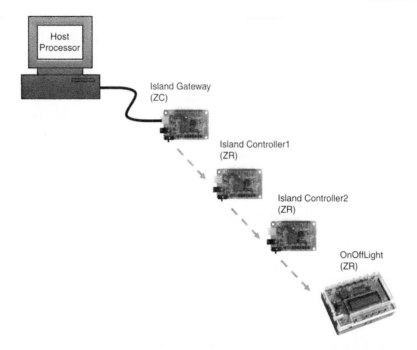

Figure 9.11: Island Controller

Note that when the actual end node receives the command, it will look like it came from the last island in the chain (also called the room controller). When the end node wants to send something back to the gateway, it would send directly to the gateway. Remember that sending information upstream toward the gateway uses very few routes; it is only downstream communication that needs to be addressed by the application.

The over-the-air message sent to the end node is an HA OnOff Toggle command to toggle the light. In the real world, this would probably be some configuration information such as informing the node not to go into power-saving mode even if the room is unoccupied. The information sent is irrelevant for the example.

So, what does this example look like over-the-air? The capture is shown in Table 9.4. The .dcf version of the capture can be found on-line at http://www.zigbookexamples.com.

Table 9.3: Island Controller Header

Octets:1	2 x Variable	2	2	1	Variable
NumIslands	NextIsland	NwkAddr	ClusterId	EP	actual packet

Table 9.4: Island Controller Capture

Seq No.	Time Delta	MAC Src	MAC Dst	NWK Src	NWK Dst	Protocol	Packet Type
1	–	–	0xffff	–	–	IEEE 802.15.4	Command: Beacon Request
2	+02.776	–	0xffff	–	–	IEEE 802.15.4	Command: Beacon Request
3	+00.004	0x0000	–	–	–	IEEE 802.15.4	Beacon: BO: 15, SO: 15, PC: 1, AP: 1, Nwk RC: 1, Nwk EDC: 1
4	+02.020	0x0050 c203dc 0f4411	0x0000	–	–	IEEE 802.15.4	Command: Association Request
5	+00.001	–	–	–	–	IEEE 802.15.4	Acknowledgment
6	+00.493	0x0050 c203dc 0f4411	0x0000	–	–	IEEE 802.15.4	Command: Data Request
7	+00.001	–	–	–	–	IEEE 802.15.4	Acknowledgment
8	+00.002	0x0050c2050005c800	0x0050c203dc0f4411	–	–	IEEE 802.15.4	Command: Association Response
9	+00.001	–	–	–	–	IEEE 802.15.4	Acknowledgment
10	+12.383	–	0xffff	–	–	IEEE 802.15.4	Command: Beacon Request
11	+00.002	0x0000	–	–	–	IEEE 802.15.4	Beacon: BO: 15, SO: 15, PC: 1, AP: 0, Nwk RC: 1, Nwk EDC: 1
12	+00.005	0x0001	–	–	–	IEEE 802.15.4	Beacon: BO: 15, SO: 15, PC: 0, AP: 1, Nwk RC: 1, Nwk EDC: 1
13	+02.018	0x0050c20d10046400	0x0001	–	–	IEEE 802.15.4	Command: Association Request
14	+00.001	–	–	–	–	IEEE 802.15.4	Acknowledgment
15	+00.493	0x0050c20d10046400	0x0001	–	–	IEEE 802.15.4	Command: Data Request
16	+00.001	–	–	–	–	IEEE 802.15.4	Acknowledgment
17	+00.002	0x0050c203dc0f4411	0x0050c20d10046400	–	–	IEEE 802.15.4	Command: Association Response
18	+00.001	–	–	–	–	IEEE 802.15.4	Acknowledgment

(*continued*)

Table 9.4: *continued*

Seq No.	Time Delta	MAC Src	MAC Dst	NWK Src	NWK Dst	Protocol	Packet Type
19	+50.906	-	0xffff	-	-	IEEE 802.15.4	Command: Beacon Request
20	+00.003	0x0002	-	-	-	IEEE 802.15.4	Beacon: BO: 15, SO: 15, PC: 0, AP: 1, Nwk RC: 1, Nwk EDC: 1
21	+00.004	0x0001	-	-	-	IEEE 802.15.4	Beacon: BO: 15, SO: 15, PC: 0, AP: 0, Nwk RC: 1, Nwk EDC: 1
22	+03.055	-	0xffff	-	-	IEEE 802.15.4	Command: Beacon Request
23	+00.003	0x0000	-	-	-	IEEE 802.15.4	Beacon: BO: 15, SO: 15, PC: 1, AP: 0, Nwk RC: 1, Nwk EDC: 1
24	+00.004	0x0002	-	-	-	IEEE 802.15.4	Beacon: BO: 15, SO: 15, PC: 0, AP: 1, Nwk RC: 1, Nwk EDC: 1
25	+00.001	0x0001	-	-	-	IEEE 802.15.4	Beacon: BO: 15, SO: 15, PC: 0, AP: 0, Nwk RC: 1, Nwk EDC: 1
26	+03.063	0xaaaaaaaaaaaaaaaa	0x0002	-	-	IEEE 802.15.4	Command: Association Request
27	+00.001	-	-	-	-	IEEE 802.15.4	Acknowledgment
28	+00.496	0xaaaaaaaaaaaaaaaa	0x0002	-	-	IEEE 802.15.4	Command: Data Request
29	+00.001	-	-	-	-	IEEE 802.15.4	Acknowledgment
30	+00.004	0x0050c20d10046400	0xaaaaaaaaaaaaaaaa	-	-	IEEE 802.15.4	Command: Association Response
31	+00.001	-	-	-	-	IEEE 802.15.4	Acknowledgment
32	+47.731	0x0351	0x0002	0x0351	0x0002	Zigbee APS Data	HA:Reserved
33	+00.002	-	-	-	-	IEEE 802.15.4	Acknowledgment
34	+00.003	0x0002	0x0001	0x0002	0x0001	Zigbee APS Data	HA:Reserved
35	+00.002	-	-	-	-	IEEE 802.15.4	Acknowledgment
36	+00.003	0x0001	0x0000	0x0001	0x0000	Zigbee APS Data	HA:On/off
37	+00.001	-	-	-	-	IEEE 802.15.4	Acknowledgment

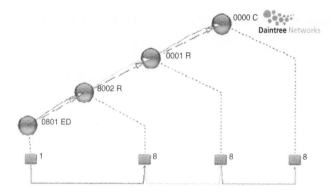

Figure 9.12: Island Gateway as Seen in Daintree

Packets 1 through 31 are the usual association requests and responses required to join all the nodes to the network.

Where things get interesting is at packet 32. Before examining the packets in detail, it's useful to understand the sequence of events. Look at Figure 9.12. The gateway is node 0x0351 (ZED) as seen on the left. A command issued from the gateway island hops through island 0x0002 (ZR), and then island 0x0001 (ZR) to reach the final destination, an HA OnOffLight at node 0x0000 (ZC) on the right.

It is irrelevant which routers in the network are Island Controllers, and those routers can even have other functions (in this case, they also act as an HA OnOffSwitch). ZigBee will use its mesh capability to find a route from island to island, and the overall scheme keeps the number of routes to a minimum, even when there are thousands of nodes which the gateway must communicate with.

Examining the details, notice that the packet 32 is initiated from the gateway. The destination address is not the final node it wants to speak to. Instead, the destination is the next island in the chain (0x0002):

```
Packet 32 Decode

IEEE 802.15.4
ZigBee NWK
  Frame Control: 0x0048
  Destination Address: 0x0002
  Source Address: 0x0351
  Radius = 5
  Sequence Number = 113
```

Table 9.5: Island Controller Header from Island Gateway

Octets:1	2 x Variable	2	2	1	Variable
NumIslands	NextIsland	NwkAddr	ClusterId	EP	actual packet
0x01	0x0001	0x0000	0x0006	0x08	0x01 0x42 0x02

```
ZigBee APS
   Frame Control: 0x00
   Destination Endpoint: 0x08
   Cluster Identifier: Reserved (0x15ad)
   Profile Identifier: HA (0x0104)
   Source Endpoint: 0x01
   Counter: 0xbf
ZigBee APS Data
   0000: 61 88 12 00 0f 02 00 51 03 48 00 02 00 51   a......Q.H...Q
   000e: 03 05 71 00 08 ad 15 04 01 01 bf 01 01 00   ..q..-...?...
   001c: 00 00 06 00 08 01 42 02                     ......B...
```

Pay attention also to the contents of the ZigBee APS Data portion of the frame in packet 32 highlighted above: It is an Island Controller frame header. Look at the Island Controller frame header below with the numbers plugged into the Island Controller Header table (see Table 9.5). As with all ZigBee communications, all 2-byte fields are little Endian.

At the next hop, packet 34, notice the NumIslands field has been reduced to 0. This means that the receiving node will be the last island in the chain.

```
Packet 34 Decode

IEEE 802.15.4
ZigBee NWK
   Frame Control: 0x0048
   Destination Address: 0x0001
   Source Address: 0x0002
   Radius = 10
   Sequence Number = 245
ZigBee APS
   Frame Control: 0x00
   Destination Endpoint: 0x08
   Cluster Identifier: Reserved (0x15ad)
   Profile Identifier: HA (0x0104)
   Source Endpoint: 0x08
   Counter: 0x6b
```

```
ZigBee APS Data
0000: 61 88 26 00 0f 01 00 02 00 48 00 01 00 02a.&......H....
000e: 00 0a f5 00 08 ad 15 04 01 08 6b 00 00 00..u..-...k...
001c: 06 00 08 01 42 02                        ....B...
```

And finally, with the last hop, notice that this looks like any other HA OnOffToggle command:

```
Packet 36 Decode

IEEE 802.15.4
ZigBee NWK
   Frame Control: 0x0048
   Destination Address: 0x0000
   Source Address: 0x0001
   Radius = 10
   Sequence Number = 93
ZigBee APS
   Frame Control: 0x00
   Destination Endpoint: 0x08
   Cluster Identifier: On/off (0x0006)
   Profile Identifier: HA (0x0104)
   Source Endpoint: 0x08
   Counter: 0xe7
ZigBee ZCL
   Frame Control: 0x01
   Transaction Sequence Number: 0x42
   Command Identifier: Toggle (0x02)
```

And voilà! The light toggles.

There are three images created in BeeKit for this project:

- IslandGateway (node 0x0351, project SrbZedIslandGateway.mcp)

- IslandController (nodes 0x0002, 0x0001, project SrbZrIslandController.mcp)

- Application (node 0x0000, project NcbZcHaOnOffLight.mcp)

The IslandGateway and Application use standard BeeKit templates. Only the IslandController contains any special source code. As discussed previously, the Island Controller implements a special cluster (0x15AD) and a special header for that cluster. The C code is pretty straightforward:

```
/* Island cluster ID, 0x15AD in little endian order */
#define gIslandCluster_c 0xAD15
```

```
/* header for island hopping */
typedef struct islandHeader_tag
{
  uint8_t numIslands;
  zbNwkAddr_t nextIsland[1];
} islandHeader_t;

/* header for final destination node */
typedef struct islandFinalHeader_tag
{
  uint8_t numIslands;
  zbNwkAddr_t nwkAddr;
  zbClusterId_t clusterId;
  zbEndPoint_t endPoint;
  uint8_t aData[1];
} islandFinalHeader_t;
```

In C it is common for variable length fields to be listed as if they were an array of [1], such as nextIsland[1] above, which really represents the list of next islands, the number of which is defined by numIslands. The length of the aData[1] field is really indicated by the APSDE-DATA.indication asduLength field. The data indication is shown below:

```
void BeeAppDataIndication
  (
  void
  )
{
  apsdeToAfMessage_t *pMsg;
  zbApsdeDataIndication_t *pIndication;
  zbStatus_t status;
  afToApsdeMessage_t *pMsgOut;
  islandHeader_t *pIslandHeader;
  islandFinalHeader_t *pFinalHeader;
  uint8_t *pAsdu;
  static afAddrInfo_t addrInfo;
  uint8_t iPayloadLen;

  while(MSG_Pending(&gAppDataIndicationQueue))
  {
    /* Get a message from a queue */
    pMsg = MSG_DeQueue(&gAppDataIndicationQueue);

    /* ask ZCL to handle the frame */
    pIndication = &(pMsg->msgData.dataIndication);

    /* sent to island cluster. Either send it on to next island,
    or directly to node. */
    if(IsEqual2BytesInt(pIndication->aClusterId,
```

```
gIslandCluster_c))
  {
  /* allocate a message for forwarding */
  pMsgOut = AF_MsgAlloc();
  if (!pMsgOut)
  {
    MSG_Free(pMsg); /* can't forward, no memory */
    continue;
  }

  /* prepare message */
  pAsdu = ((uint8_t *)pMsgOut + ApsmeGetAsduOffset());

  /* set up default destination */
  Set2Bytes(addrInfo.aClusterId, gIslandCluster_c);
  addrInfo.dstAddrMode = gZbAddrMode16Bit_c;
  addrInfo.radiusCounter = afDefaultRadius_c;
  addrInfo.srcEndPoint = addrInfo.dstEndPoint = appEndPoint;
  addrInfo.txOptions = gApsTxOptionNormal_c;
  iPayloadLen = pIndication->asduLength;

  /* more island hops to go? */
  pIslandHeader = (void *)pIndication->pAsdu;
  if(pIslandHeader->numIslands)
  {
    /* decrement the count and copy in entire packet minus
      one island hop */
      pAsdu[0] = pIslandHeader->numIslands-1;
      FLib_MemCpy(pAsdu + sizeof(pIslandHeader->numIslands),
        &(pIslandHeader->nextIsland[1]),
        pIndication->asduLength-sizeof(islandHeader_t));

    /* send to next hop */
    Copy2Bytes(addrInfo.dstAddr.aNwkAddr,
      pIslandHeader->nextIsland[0]);

    /* payload has shrunk by 2 bytes */
    iPayloadLen -= szeof(zbNwkAddr_t);
}
/* send directly to application */
else
{
  /* send payload to final destination node */
  pFinalHeader = (void *)pIslandHeader;
  addrInfo.dstEndPoint = pFinalHeader->endPoint;
  Copy2Bytes(addrInfo.aClusterId, pFinalHeader->clusterId);
  Copy2Bytes(addrInfo.dstAddr.aNwkAddr,
    pFinalHeader->nwkAddr);
```

```
   /* copy in the data */
    iPayloadLen -= (sizeof(islandFinalHeader_t)-
sizeof(pFinalHeader->aData));
    FLib_MemCpy(pAsdu, pFinalHeader->aData, iPayloadLen);
}

/* send data on to next node in the island chain
  (which may be final node) */
  (void)AF_DataRequestNoCopy(&addrInfo, iPayloadLen,
    pMsgOut, NULL);
}

/* this is still a switch. Handle switch code */
else
{
  status = ZCL_InterpretFrame(pIndication);

  /* not handled by ZCL interface, handle cluster here... */
  if(status == gZclMfgSpecific_c)
    {
      /* insert manufacturer specific code here... */
    }
  }

  /* Free memory allocated by data indication */
  MSG_Free(pMsg);
  }
}
```

This application simply checks for the cluster ID 0x15AD. If it's any other cluster the data indication is passed to the ZigBee Cluster Library (ZCL) to handle. If the cluster is 0x15AD (the IslandController cluster) then the first field is checked to see whether the packet should be forwarded to the next island in the chain, or sent to the final destination.

The determination of whether it should be forwarded depends on whether there are any remaining islands in the chain, determined by this statement:

```
if(pIslandHeader->numIslands)
```

Notice that when forwarded to the next island, the number of islands remaining is decremented, and that island hop is removed:

```
pAsdu[0] = pIslandHeader->numIslands - 1;
```

As usual, the complete source code can be found on http://www.zigbookexamples.com.

The advantage of this "Island Controller" technique is that very few routes are needed to support very large networks (thousands of nodes) from a single gateway. The disadvantage is that a human must plan the network and determine which nodes will be Island Controllers. Planning of this nature typically takes place in commercial buildings, in any case. Another disadvantage is that the receiving node can't tell who sent the packet, as the packet appears to come from the last Island Controller. Commissioning can make sure the data is sent to the proper gateway.

> Inbound traffic to a gateway requires only one route per hop.
>
> Outbound traffic from gateway requires many routes, Island Hopping, or ZigBee Pro.

9.3 Bandwidth and the Gateway

Okay, so now there are enough routes to communicate from the gateway to any node in the network and back again. But there is something else to consider besides just routes. Is there enough bandwidth?

2.4 GHz 802.15.4 radios communicate at 250 kbps (kilobits per second). This is considered the maximum bandwidth of any given radio. But all radios in a given vicinity share that same bandwidth, so don't expect applications to communicate at 250 kbps.

For determining average bandwidth available for applications, consider the following:

- Interferers
- Density of the network
- ZigBee protocol overhead
- Communication patterns

9.3.1 Interferers

All radios experience some form of interference. In fact, wireless often has a reputation as an unreliable medium, one which ZigBee spends much effort to correct. A basic rule of thumb is to assume that 50% of the bandwidth is taken up by interferers at any given moment in time. Of course, this formula is much too general to apply in all situations, but it provides a starting point for discussion.

One of the most common interferers is WiFi. WiFi is prevalent in many areas where ZigBee is installed, and will only get more so with time. By using CSMA-CA and retries, ZigBee is able to continue to communicate, even if the WiFi traffic is

WiFi™ speaks at less than 100% duty cycle

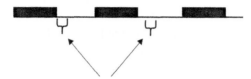

ZigBee uses CSMA-CA to speak during the quiet periods

Figure 9.13: ZigBee and WiFi interference

particularly heavy. Take a look at Figure 9.13. There are periods of silence, even when WiFi is communicating continually. ZigBee takes advantage of these silent periods to communicate or to retry packets. Tests conducted by the ZigBee Alliance had a 0% packet error rate, even with the ZigBee radio within one foot of the WiFi router! That's not to say that ZigBee didn't need to use its retry mechanism, but no packets were lost from an application perspective.

It is worth noting, however, that ZigBee did not test 802.11.n (only a/b and g). A copy of this white paper will be available on the ZigBee Web site (http://www.zigbee.org).

One of the other interesting things about WiFi and ZigBee is that usually WiFi channels 1, 6, or 11 are used, which means that many ZigBee channels will be free, given any single implementation of WiFi. As seen in Figure 9.14, ZigBee (802.15.4) channels 15, 20, 25, and 26 are always free from WiFi interference, regardless of which WiFi channel is used.

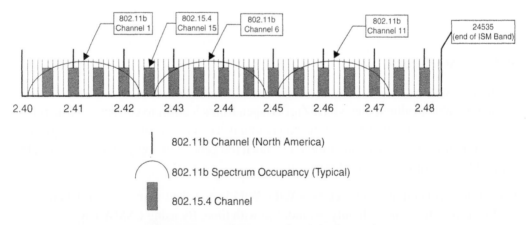

Figure 9.14: ZigBee and WiFi Channels

Many radio technologies, such as Bluetooth™, also share the 2.4 GHz space. Empirically, ZigBee has shown itself to be very robust in noisy RF environments. At trade shows I routinely see many wireless technology demos break or have trouble in the chaotic environment. Not so with ZigBee. The ZigBee demos just keep working.

Other common interferers are cordless telephones, microwave ovens (yes, microwaves operate in the 2.4 GHz band), and general RF noise. I'm told even sunspots can cause some interference. Modern microwave ovens are built to screen most of their RF interference. In fact, at San Juan Software, we use microwave ovens as a way of isolating ZigBee nodes. (Yes, we are careful not to turn the oven on while a ZigBee board is inside!)

> Assume 50% of the available bandwidth is used by interferers.

9.3.2 ZigBee Protocol Overhead

ZigBee consumes bandwidth to enhance reliability, extend network range, and to commission the network.

For example, the 802.15.4 MAC will retry up to three times (for a total of four transmissions) to send a message to the next hop. If ZigBee APS retries are used (one of the TxOptions on an APSDE-DATA.request), then ZigBee will retry up to three times as well, for a total of 16 possible transmissions of a single packet.

ZigBee uses unicasts to send data along a route, but every node in the vicinity of each hop can hear that unicast. It is rejected by all but the intended node, but the transmission still consumes bandwidth.

ZigBee broadcasts may repeat up to two times (a total of three times) and each node in the network or within the radius of that broadcast repeats it. Broadcasts can consume a lot of bandwidth. Most reasonably sized networks (100-plus nodes) can only handle about four to five broadcasts at any given time. Where practical, try to use unicasts, either mesh or along the tree, instead of broadcasts. Unicasts use far less resources in terms of bandwidth and RAM. But remember that commands like ZDP-NWK-ADDR.request *require* a broadcast. So does discovering a route. Likewise, groupcasts are a form of broadcast.

One of the most difficult times for ZigBee functioning is during commissioning. Many nodes are competing for the same resources, and applications tend to be a lot chattier during this period. Ideally, it is best to commission the network one device at a time. This includes joining, as well as determining which of the various nodes any given node needs to speak with.

One thing that consumes bandwidth that is often forgotten is ZigBee End-Device polling. Sleeping (RxOnIdle = FALSE) ZEDs poll their parent to see if their parent is keeping any messages for them while they were asleep. This poll rate by default is often set up to something like two seconds, because that's a useful rate during commissioning. Set this poll rate to as long as possible. The Home Controls Stack Profile (stack profile 0x01) allows this poll rate to be as slow as once per hour. It's really up to what the application can bear. Remember, this does not affect the transmission rate of the device (the light switch can still send the on/off command immediately when switched), only the receiving rate. And most implementations poll for a message shortly after transmitting. Polling once every 6.5 seconds is a good poll rate because it matches well with the MAC purge rate.

> Use unicasts, not broadcasts (or groupcasts) where possible.
>
> Commission nodes one at a time.
>
> Reduce poll-rate for ZigBee End-Devices. Once every 6.5 seconds is a good poll rate.

9.3.3   Density of the Network

Too many ZigBee nodes in the same vicinity can interfere with each other. In a typical home, the ZigBee network may contain 50 to 100 nodes, all within hearing range of each other. In a commercial network, such as building automation, this number can be significantly higher. The bandwidth in any given vicinity is limited (only one node may be speaking at any given time) and is shared by all the nodes in that vicinity. Each node uses Carrier-Sense-Multiple-Access Channel-Assessment (CSMA-CA) before speaking, so there must be some silence between packets.

Network density is determined by the number of nodes that a given node can hear, based on the transmit power of the other nodes, and the receiving node's ability to hear those transmissions.

802.15.4, including ZigBee, requires from 1 to 4 milliseconds, depending on the size of the packet. Assuming a 50% duty cycle (which allows enough quiet time for CSMA-CA to work) this allows an average of about 250 packets per second. Because ZigBee is an acknowledged protocol, this is really a maximum of 125 application packets per second, shared by all nodes. So if there are 50 nodes in the vicinity, this equates to two packets per second per node.

Remember, that some of this bandwidth will be taken up by other portions of the ZigBee protocol. For example, if a node in this or another part of the network is performing a route discovery, the packets are repeated three times by each router in the network. A group or multicast likewise eats up bandwidth. ZigBee end devices poll their parents for information, which also consumes bandwidth.

My rule of thumb is to allow 50% of the bandwidth average for ZigBee protocol overhead. So I don't write applications that send data any more frequently than one packet per second, on average. I also work to minimize over-the-air traffic wherever possible, keeping in mind the density of the network.

> Reduce network density, or reduce over-the-air traffic for a healthy network.

9.3.4   Communication Patterns

One of the aspects of bandwidth consumption that is often forgotten is communication pattern. Even if there is a very noisy RF channel (lots of bandwidth consumed by interferers), and a very dense network (hundreds of nodes in the vicinity), ZigBee applications can still continue to communicate reliably. How? They do this by reducing the traffic, and randomizing *when* the traffic occurs.

Consider this scenario. Perhaps 100 nodes must report their status to a data concentrator or gateway. If all of them attempt to communicate once every minute, once every hour, or even once every day, they can overload the ZigBee network should they try to communicate at the same time.

For the purposes of most applications, it is not necessary to specify the precise moment at which to transmit data; a measurement from anywhere during a specified interval is sufficient. For example, sending in a temperature reading can occur at any time within a given minute, for home or commercial building automation. The temperature won't change that rapidly even if the heating or cooling units come on right away.

Adding a little bit of randomization (called jitter) can go a long way toward helping an application be a team player in larger networks. ZigBee uses randomization in repeating broadcasts or sending unicast retries. The Smart Energy application profile, which monitors and regulates electrical power usage, contains a whole section (E.3.5) on randomization.

The following code shows what a random jitter in a temperature sensor application might look like:

```
void TimerCallback (tmrTimerID_t timerID)
{
  if(timerID == gTemperatureTimerId)
  {
  GetCurrentTemperature();        //read from temp sensor
  TransmitTemperatureReading(); //send reading to gateway
  StartRandomTemperatureReadingTimer();
  }
}

void StartRandomTemperatureReadingTimer(void)
{
  uint16_t randomJitter;
  randomJitter = 1000 * (50 + GetRandomRange(0, 20));
  TMR_StartSingleShotTimer(mTimer, randomJitter, TimerCallback);
}
```

This procedure tends to spread the messages out over time very well. Just a jitter of a few milliseconds can be enough to make even a large network function better.

Another way to help messages flow more smoothly in a ZigBee network is to combine data when possible. Take that same example of a temperature sensor. The sensor might take 10 or even 100 readings, average them, and then send out a single number at the end of a minute, rather than sending out many readings at shorter intervals. Or perhaps the application might only send a reading when the temperature has changed by five degrees, or after two minutes have passed, whichever comes first.

In another application I was involved with, battery-operated ZigBee nodes were worn by crew members on commercial ships as "man overboard" detectors. These nodes, called "tags," sent a message to the gateway once every two seconds to indicate that the node (that is, the person) was still on the network. If any node didn't check in, it could mean that the crew member was in the water. Even for very small networks (under 50 nodes) the system would sometimes overload and give false man overboard readings (no check-in) over a 24-hour period. The solution was to collate the incoming information on the receiving routers, and only send one message to the gateway for every 50 nodes. This greatly reduced the traffic, allowing the network to scale much higher (into the hundreds of nodes), while still allowing a very quick transmit rate (in ZigBee terms) from each of the battery-operated nodes.

> Reduce over-the-air traffic when possible.
>
> Use jitter to communicate in a more distributed pattern.

9.4 Custom Gateways

The ZigBee Serial Gateway Interface introduced in Section 9.1 works well if simply communicating ZigBee messages, but what if the gateway itself must have special functionality?

In this final example, I'll revisit controlling the iPod over ZigBee, this time with a PC interface to the iPod, as well as a remote control. In this example, an NCB is used for the gateway, which will show commands as they are sent to the iPod, and indicate volume on the LEDs (see Figure 9.15).

A PC program written in Visual Basic for .Net brings up an iPod image on the screen where the following can be controlled:

- Play/Pause

- Next Song

- Previous Song

- Volume Down

- Volume Up

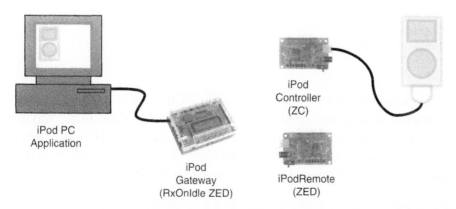

Figure 9.15: iPod Gateway

As with most examples in this book, the channel selected is channel 25, PAN ID 0x0f00, if you wish to capture this with Daintree SNA.

The iPod can be controlled either by the remote or through the gateway. The whole thing is shown in the figure. It's interesting to note, this three-node example will work even in a full Home Automation network with any number of nodes, with just the following changes:

- The iPod Controller would need to be changed from the ZigBee Coordinator to an RxOnIdle ZED.

- The channel mask must be set to all channels (currently it's channel 25 only).

- The PAN ID must be set to 0xffff to allow joining any PAN.

If desired, enable NVM in BeeKit and export the properties again (not the project, or you'll overwrite the BeeApp.c), to remember the PAN settings after joining a network.

I won't show an over-the-air capture of this again, because it looks exactly the same as the capture in Chapter 4, "ZigBee Applications." The interesting part is how the PC application interfaces with the gateway.

To enable the gateway, I enabled the ZigBee Test Client (ZTC) in the project, just as I did when making the ZigBee Serial Gateway Interface. The way to extend ZTC is, as usual, through the use of a register call and a callback. Take a look at the code fragment below, found in BeeAppInit():

```
...

/* indicate the app on the LCD */
LCD_WriteString(2, "iPodGateway");

/* register with ZTC */
ZTC_RegisterAppInterfaceToTestClient(ReceiveZtcMessage);

...
```

A string is written to the LCD to indicate that this NCB board is an iPodGateway. Then the callback function, ReceiveZtcMessage() is registered with ZTC. In this way, any command that ZTC doesn't already know about will go here. The PC application takes advantage of this, and uses a special group ID (0x60) to indicate that this is an iPod command. Group IDs 0x80 and above are reserved for BeeStack.

The header information, in addition to the iPod over-the-air messages, includes a prototype for ReceiveZtcMessage(). Note the enum iPodRemoteCommand_t. This enum has values for Play/Pause, Skip Forward, etc. We use the values from this enum to be message IDs from the gateway. Recall when the ZigBee Serial Gateway Interface was described earlier. The interface included both a Group ID and a Message ID. For Play/Pause, that would be 0x60 0x00:

```
/* also used for Msg IDs */
typedef enum {
  gPlayPause_c = 0,
  gSkipForward_c,
  gSkipReverse_c,
  gVolumeDown_c,
  gVolumeUp_c,
  ButtonRelease_c
} iPodRemoteCommand_t;

/*
  iPod Request/Response Message
  field size value
  header 2 0xff 0x55
  length 1 size of mode + command + parameter
  mode 1 the mode the command is referring to
  command 2 the two byte command
  parameter 0..n optional parameter, depending on the command
  checksum 1 0x100-( (sum of all length/mode/command/parameter bytes)
  & 0xFF)
*/

uint8_t PlayPause[] = {0xFF, 0x55, 0x03, 0x02, 0x00, 0x01, 0xFA};
uint8_t SkipForward[] = {0xFF, 0x55, 0x03, 0x02, 0x00, 0x08, 0xF3};
uint8_t SkipReverse[] = {0xFF, 0x55, 0x03, 0x02, 0x00, 0x10, 0xEB};
uint8_t VolumeDown[] = {0xFF, 0x55, 0x03, 0x02, 0x00, 0x04, 0xF7};
uint8_t VolumeUp[] = {0xFF, 0x55, 0x03, 0x02, 0x00, 0x02, 0xF9};
uint8_t ButtonRelease[] = {0xFF, 0x55, 0x03, 0x02, 0x00, 0x00, 0xFB};

/* poor man's delay, assumes 16MHz */
#define DELAY_100US() { {uint8_t i = 200; do {} while(--i);} }

/* Group ID for ZTC interface */
#define gZtcGroupID_c  0x60

void SendiPodRemoteCommandOTA(iPodRemoteCommand_t cmd);
void ReceiveZtcMessage(ZTCMessage_t * pMsg);
```

The actual code to send to the iPod Controller is very simple. See SendiPodRemote CommandOTA() below. (By the way, OTA stands for over-the-air.)

Notice that the destination is always the ZigBee Coordinator (node 0x0000). This code is okay for an example in a book or a test in the lab, but would not be sufficient in a commercial product. A commercial product would use correct binding to find the right services. The exception to this rule would be if all the nodes in the ZigBee network will be controlled by you or the OEM. In that case, a shortcut such as assuming that the iPod Controller is the ZC is perfectly acceptable. This saves over-the-air traffic, and keeps the application simple.

Also notice that the cluster used is not even specified here. The code simply picks the first cluster out of the application SimpleDescriptor's output cluster list. This allows the developer to change his or her mind about the cluster ID without having to change code. Simply adjust the cluster in BeeKit:

```
/****************************************************************
SendIpodRemoteCommand
*
*This function sends a simple remote control command to the ipode
****************************************************************/
void SendiPodRemoteCommandOTA(iPodRemoteCommand_t cmd) {
  afAddrInfo_t addrInfo;
  uint8_t TransmitBuffer[1];

  /*copy iPod command to TransmitBuffer */
  TransmitBuffer[0] = cmd;

  /* set up address information */
  addrInfo.dstAddrMode = gZbAddrMode16Bit_c;
  /*Always send packet to coordinator*/
  /* should be updated to use binding */
  addrInfo.dstAddr.aNwkAddr[1] = 0x00;
  addrInfo.dstAddr.aNwkAddr[0] = 0x00;
  addrInfo.dstEndPoint = appEndPoint;
  addrInfo.srcEndPoint = appEndPoint;
  addrInfo.txOptions = gApsTxOptionNone_c;
  addrInfo.radiusCounter = afDefaultRadius_c;

  /* set up cluster */
  addrInfo.aClusterId[0] =
    endPointList[0].pEndpointDesc->pSimpleDesc
    ->pAppOutClusterList[0];
  addrInfo.aClusterId[1] =
    endPointList[0].pEndpointDesc->pSimpleDesc
  ->pAppOutClusterList[1];

  /* send the data request */
  (void)AF_DataRequest(&addrInfo, sizeof(TransmitBuffer),
    TransmitBuffer, NULL);
}
```

```
/***************************************************************
ReceiveZtcMessage
 *
 *Receive the ZTC message and send it over the air.
 ***************************************************************/
 void ReceiveZtcMessage
 (
     ZTCMessage_t* pMsg /* IN: message from ZTC/UART (Test Tool)*/
 )
 {
   uint8_t messageID;
   static const char *sziPodCmd[] =
   {
     "Play/Pause",
     "Skip Forward",
     "Skip Backward",
     "Volume Down",
     "Volume Up"
   };

   /* ignore invalid opcodes */
   if(pMsg->opCode != gZtcGroupID_c)
     return;

   /* display command on NCB screen */
   messageID = pMsg->opCodeId;
   LCD_WriteString(2, sziPodCmd[messageID]);

   /* determine which message to send to iPod Controller */
   SendIpodRemoteCommandOTA(messageID);
 }
```

The final function, ReceiveZtcMessage() actually takes the incoming serial message, which has already been validated by ZTC, and uses the Message ID as the command to be sent to the iPod Controller. The command is displayed on the LCD screen, so it's obvious which commands came through the gateway. ZTC will pass along any message it doesn't already understand.

> Custom gateways are easy to create and extend through ZTC.
>
> Binding is not necessary if the OEM controls all the nodes in the ZigBee network.

ZigBee 2007 and ZigBee Pro

This appendix covers the features added to the new ZigBee 2007 specification.

A.1 ZigBee 2006, 2007, and ZigBee Pro

ZigBee is really three, three, yes three stacks in one: ZigBee 2006, ZigBee 2007 (often called just ZigBee) and ZigBee Pro. Originally, there was also ZigBee 2004, but that stack is considered deprecated, and is no longer in use (except in legacy systems).

The three ZigBee stacks have more similarities than they do differences, and indeed they can join networks that were started with the other stacks. Some stack vendors offer only one, while others offer all three stacks with a choice at compile and/or run-time. When evaluating a stack or a stack vendor, be sure you know which stack configuration you are looking for. The three ZigBee stacks are compared in Table A.1.

Why choose one stack over the other? The simple answer is size, cost, Application Profile, and vendor.

When using 128 K flash parts or less, not much room is left over for an application using ZigBee Pro. This varies, of course, between vendors. On the other hand, if you are using a hardware platform with plenty of room, ZigBee Pro does have the most features and the highest security model. I suggest during evaluation that you build one of the sample applications to see how much ROM (flash) and RAM is taken up by the ZigBee stack on a particular platform. You can also ask the vendor for a map file (the output of the compile/link process), which will show the size of the stack vs. the application.

Table A.1: ZigBee Stacks Compared

Feature	ZigBee 2006	ZigBee 2007	ZigBee Pro
Size in ROM/RAM	Smallest	Small	Bigger
Stack Profile	0x01	0x01	0x02
Maximum hops	10	10	30
Maximum nodes in network	31,101	31,101	65,540
Mesh networking	✓	✓	✓
Broadcasting	✓	✓	✓
Tree routing	✓	✓	–
Frequency Agility	–	✓	✓
Bandwidth Used By Protocol	Least	More	Most
Fragmentation	–	✓	✓
Multicasting	–	–	✓
Source routing	–	–	✓
Symmetric Links	–	–	✓
Standard Security (AES 128 bit)	✓	✓	✓
High Security (SKKE)	–	–	✓

If cost is a significant factor in your product, you may be better off using ZigBee 2007 or ZigBee 2006. These are the smallest of the stacks, and so a less expensive part with less RAM and less Flash memory may be used.

Some Application Profiles require link-keys, or fragmentation. If so, the OEM must choose a stack profile that supports the feature. For example, Smart Energy (formerly Automatic Metering) requires fragmentation. Only ZigBee 2007 and ZigBee Pro support this feature, so ZigBee 2006 is not an option if the OEM is making an automatic meter or energy service portal.

The final factor is your choice of vendor. We all have our favorite vendors. Perhaps we have worked with their parts for years, and so we trust them. Perhaps we already have the necessary tools, or they have a really great support rep. Perhaps their part meets our hardware requirements in voltage range, supply availability, etc., better than the one from the other vendors. Whatever the reason, we all have vendors we prefer. Sometimes that's good enough. Any vendor that offers a certified stack is likely to have a good product. If the particular vendor does not offer a ZigBee certified stack, well, I would simply consider that product a proprietary solution.

A.2   Stochastic Addressing

Address assignment in stack profile 0x02 is very different than in stack profile 0x01. The artificial number-of-hop-limitation imposed by the tree/Cskip address assignment scheme was deemed too restricting for ZigBee Pro networks, so a new address scheme was chosen: Stochastic Addressing.

The algorithm is simple. The parent of a joining node picks a random address for the joining child. When the child joins, a device announce is broadcast across the network, and if a conflict is detected, the conflict detection notification is broadcast back. Then the process is attempted again.

The statistics for collisions are the same as for hashing tables, and for the Birthday Problem. The address space is 64 K. The probability of a pair of nodes choosing the same address reaches 50% at around 300 nodes, and 99% at 777 nodes, so you're pretty much guaranteed a conflict within about 800 nodes or so. That's alright. ZigBee will take care of the conflict using the address conflict resolution procedure. A good explanation of this effect can be found on Wikipedia. Look up the "Birthday Paradox."

For sleepy children, the parent will take care of the address conflict resolution, and will send a rejoin response to the child (just as if the child had asked to rejoin to a new parent), but with the new (and hopefully non-conflicting) address. All devices that are awake (RxOnIdle = TRUE) will take care of the address conflict themselves.

Generally, address conflicts are rare. They occur only when a node joins (or rejoins) a network, because nodes keep their addresses for their entire life on the network. When conflicts do occur, the address conflict resolution takes care of the problem fairly quickly.

The biggest issue is the number of potential broadcasts in the system. This means that every joining node must broadcast at least once. There is the potential for another broadcast or two detecting an address conflict, with another broadcast resolving it. So plan for the worst case: four broadcasts every time a node joins (or rejoins) the network.

Another common way to assign addresses in ZigBee Pro that circumvents the address conflict altogether is to choose unique addresses ahead of time, with a commissioning tool. That way, all addresses are unique by design. See Chapter 8, "Commissioning ZigBee Networks," for more on commissioning.

A.3 Source Routing

Mesh routing is a great method for distributed, ad-hoc routing. One of its weaknesses is that routes require tables in each node. This is not a problem if the device has sufficient RAM, but in ZigBee devices, RAM is often very precious.

A routing table entry is pretty small: six bytes per route in most implementations. If a given router needs to keep track of 50, or even 100 routes, that's not too bad in a device with 4 K RAM. But what if a node needs to keep track of 1,000 routes? As 1,000 routes times 6 bytes equals 6,000 bytes, it suddenly becomes problematic.

In ZigBee, this problem usually only comes into play with gateways (sometimes called data concentrators). Most normal communication occurs between devices that are fairly near, so the distributed route scheme works really well.

But consider the gateway. For example, say that ZigBee devices in hotel rooms need to send status information continuously back to some back-end service, so that the hotel management can better supervise the cleaning staff, monitor energy, and determine room occupancy. (By the way, there are a number of ZigBee installations in hotels around the world.)

As shown in Figure A.1, many nodes might be sending data into the gateway. This is not a problem for ZigBee mesh routes, as there is only one route table entry in each node to the single destination, the gateway. Remember, routes in ZigBee only store the next hop to the final destination in each node, so even if 2,000 different nodes are all routing to the gateway, only one entry is needed to find that single destination. If the gateway (the large node at the top) was NwkAddr 0x1111, then every router in the network need only use one routing table entry inbound to find the next hop toward node 0x1111.

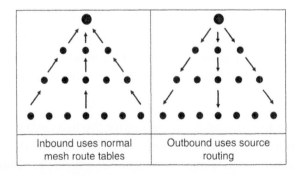

| Inbound uses normal mesh route tables | Outbound uses source routing |

Figure A.1: Source Routing Solves the Gateway Problem

The problem comes in if the gateway needs to communicate back to every node in the network, outbound. Now there are many routes required, because there are many destinations. If the gateway needs to communicate with 2,000 nodes, there needs to be 2,000 routes in the gateway, and something close to that in the routers near the gateway. This can simply require too much memory for a typical ZigBee device.

ZigBee source routing solves this problem. What ZigBee source routing does is make the assumption that the gateway (or data concentrator) has plenty of memory and can store the entire route (each hop) to all those 2,000 destination nodes.

The basic procedure for setting up source routing is this:

1. The gateway (concentrator) broadcasts a many-to-one route request out to a certain radius (limited to five in ZigBee Pro).

2. All the routers that hear the many-to-one route request record the shortest path to the gateway with a route table entry. Effectively, this means all routers within a certain radius now have a route back to the gateway.

3. When a device initiates a unicast to the gateway, the routing table entry is checked, and if this is the first time, the device will initiate a route record command to record all the hops to the gateway.

4. The gateway receives the route record command prior to receiving the data indication.

5. Now anytime the gateway needs to communicate to the device, it can send the unicast with source routing enabled.

For example, assume node "A" is a gateway. It sends out a many-to-one route request, radius 5. This many-to-one route request reaches all nodes within this radius, in particular, node "E" which now has a route back to node "A":

$$E \rightarrow D \rightarrow C \rightarrow B \rightarrow A$$

Now "E" communicates something to "A," unicast. Because this is the first time "E" has spoken to "A," some bits in the route entry indicate "E" should send a NWK route record command prior to sending the data. The application doesn't need to do anything special, as the route record occurs automatically within the NWK layer of the ZigBee stack.

Format of Individual Frame Types

NWK data frame format with source routing

Octets:2	2	2	1	1	1	1	2-n	Variable
Frame Control	DstAddr	SrcAddr	Radius	Seq. number	Relay Cnt	Relay Idx	Relay List	Data payload
	Routing fields							
	NWK header							NWK payload

Figure A.2: Source Routing Frame

The NWK route record command records the route along the way, which results in a route record indication at the gateway. Now, the gateway knows the route back to node "E."

$$A \rightarrow B \rightarrow C \rightarrow D \rightarrow E$$

When the gateway transmits data to node "E," it will indicate a source route in the data request. As usual, ZigBee specifies the exact over-the-air form of a source routed data request. A bit in the NWK frame control field indicates the packet is source-routed. The term "relay" merely refers to the intermediate nodes between the gateway and the destination node (see Figure A.2). They are in the same order as recorded with the route record command, and the index just indicates the "next-hop."

The API is less clear. It is assumed that if a gateway is sending to a device and it already has a source route from a NWK route record command, then it will use the source route. Some vendors may offer the option at the gateway to use source routing or not, or to detect whether it has a source route to the desired destination node.

Note that there can be multiple gateways. It's perfectly reasonable to have four or five gateways in a network, and for nodes to use the nearest one for communicating to the outside world. This decision really lies with the application. Many-to-one and source routing are only available in stack profile 0x02 (ZigBee Pro).

A.4 Multicast

Multicasting is a form of broadcasting. Like broadcasting, multicasting allows a single node to transmit packets to many nodes using a single APSDE-DATA.request. Unlike

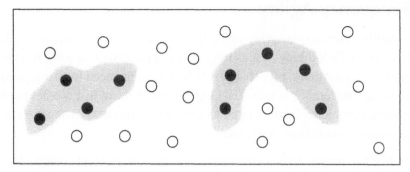

Figure A.3: Multicast Groups Can Be Disconnected Shapes

broadcasting, multicasting is not limited to a circular radius. Multicasting can be any shape.

Multicast accomplishes this by not decrementing the radius as long as a repeating node is part of the multicast group.

Another field, called the non-member radius, helps with disconnected multicast groups. This field indicates how far the broadcast will travel beyond nodes that are in the group. The group can then be disjointed (not connected), as shown in Figure A.3, while still working with the multicast.

The apsNonmemberRadius field was originally called multicast fuzz, but clearer heads with less of a sense of humor renamed the field.

The nodes in black with a gray background are in the multicast group, while the other nodes are not. Only nodes in the group repeat the broadcast, so if the apsNonmemberRadius is set to 0 or 1, a multicast initiated in the set of nodes in the shape at the left would not reach the nodes in the shape at the right. But, if the apsNonmemberRadius was set to 2 or more, the multicast could span the disconnected group and would be received and repeated by nodes in both disconnected shapes.

Essentially, consider a multicast to be a more efficient broadcast. Also, the multicast, like a groupcast, only goes to those application endpoints that are members of the group.

These groups are the same groups shared by the APS layer, so commands such as the APSME-ADD-GROUP.request work with these groups (see Chapter 4, "ZigBee Applications"). The over-the-air ZigBee Cluster Library group commands (see Chapter 6, "The ZigBee Cluster Library") also work with multicast groups. Multicasting is only available in stack profile 0x02 (ZigBee Pro).

A.5   Frequency Agility

Frequency Agility refers to the ability of ZigBee to change physical channels after the network has started. Don't think of it as frequency hopping, such as what Bluetooth™ does, where the channel constantly changes. In 802.15.4, the need to change frequencies (channels) due to interference is rare. The O-QPSK mechanism just doesn't run into the same interference problems as other radio technologies, and the retry mechanisms built into ZigBee resolve most other interference issues.

Instead, ZigBee Frequency Agility concentrates on the rare channel change. The channel change might be needed due to regulations. For example, some hospitals do not permit interferers on the same frequency as their WiFi™ networks, so if a new WiFi network were installed, the existing ZigBee network might need to be set to a different, non-overlapping channel. It's possible that one channel is particularly noisy. Perhaps a channel is suddenly experiencing significant RF noise from new machinery, forcing significant numbers of retries, and slowing the response time of the ZigBee nodes.

It's interesting to note that the ZigBee Alliance could not cause enough interference to actually force a channel change with WiFi and microwave interferers. Multiple WiFi networks were set going, downloading movies, and using up as much bandwidth as possible over the same RF frequencies as ZigBee. Microwave ovens were turned on near the ZigBee nodes, and ZigBee just didn't fail. Sure, there were a few more retries, but understand that it can be difficult to generate enough RF noise to stop ZigBee. There is a ZigBee whitepaper on this topic.

The basics of Frequency Agility are this:

- The routers keep track of failure counts.

- If the failure count exceeds a threshold (25% of messages fail, when the message count is 20 or greater), then the router scans its set of channels and reports the interference on the channels to the channel master (sometimes called network manager).

- The routers will only do this a maximum of four times per hour, so as not to flood the network with reports.

- If the channel master decides to change channel (due to enough reports or another reason, such as the user desiring a channel change), it issues a Mgmt_NWK_Update_req to indicate the channel change.

- Sleepy end devices, if they lose track of their parents, will automatically scan the channels if they can't find a parent on the current channel, and will look for them on the other channels in their apsChannelMask. This is done because it's possible the sleepy ZEDs will miss a channel-change command.

- Wakeful nodes (routers and RxOnIdle = TRUE end-devices) that don't receive the Mgmt_NWK_Update_req will look for the network on one of the other channels, if they reach their error threshold and can't communicate with the channel master after a certain period of time. In doing so they make the assumption that interference is too high to communicate on the current channel.

The Mgmt_NWK_Update_req can also set the *apsChannelMask* over-the-air, the set of channels to be used when changing channels automatically. The command below is instructing the network to change to channel 14.

```
Seq    Channel       NWK Dest       Packet Type
---------------------------------------------------------
45     11            0xfffd         ZDP:MgmtNwkUpdateReq

Frame 45 (Length = 34 bytes)
IEEE 802.15.4
ZigBee NWK
  Frame Control: 0x0048
  Destination Address: 0xfffd
  Source Address: 0x0000
  Radius = 10
  Sequence Number = 8
ZigBee APS
  Frame Control: 0x08
  Destination Endpoint: 0x00
  Cluster Identifier: MgmtNwkUpdateReq (0x0038)
  Profile Identifier: ZDP (0x0000)
  Source Endpoint: 0x00
  Counter: 0x4b
ZigBee ZDO
  Transaction Seq Number: 0x06
  Mgmt Network Update Request
    Channel Mask: 0x00004000 (channel 14)
    Reserved: 0x00
    Scan Duration: Channel Change Request (0xfe)
    Network Update Id: 0x01
```

The channel change can happen fairly quickly if all goes well. The nodes receive the command, and after they are done rebroadcasting it (about 500 to 800 ms), they may switch channels. This can propagate across a large network (perhaps 2,000 nodes) on the order of about 10 seconds or so. But if things don't go well (if a router misses the three broadcast retries due to interference, perhaps), then it can take some time for those nodes to switch channels.

It is completely up to the policy of the channel master (aka network manager) to decide when to switch channels. The usual case is when a network administrator decides to change channels based on reports. Only those networks that will be completely unattended by a human should be considered for automatic channel change.

Frequency Agility is available both in ZigBee 2007 (stack profile 0x01) and ZigBee Pro (stack profile 0x02).

A.6   Fragmentation

The 802.15.4 PHY is 127 octets (bytes) in length. After protocol and security overhead is taken into account, the application payload ranges between 72 to 100 bytes.

Sometimes applications want to send more data than this as a single unit. Perhaps the application has collected a set of sensor or logged data, and wants to communicate it all at once. Perhaps even a new code image is being sent over-the-air.

The application could split up the data itself, taking into account that the data may arrive out of order, and require reassembling on the other end. Or the application can let ZigBee do it, using an optional feature called Fragmentation (see Figure A.4).

ZigBee fragmentation takes care of all the details, making sure the packets are reassembled in correct order and checking that the fragmented packet was received in its entirety, complete with end-to-end retries and acknowledgment. The application has control over the following fragmentation parameters:

- blockSize
- apsInterframeDelay
- apscMaxWindowSize

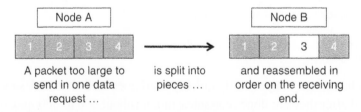

Figure A.4: Fragmented Packets Are Split into Pieces and Reassembled

The packet is split up into a set of fragments (called blocks). The size of each fragmented block may be up to the maximum size available to APSDE-DATA.request(), which depends on the current security parameters and transmission options. If no security is used, the block size can be up to 98 bytes, which results in a full 127-byte PHY packet. In the Freescale solution, a call to ApsmeGetMaxAsduLength() will return this maximum length.

The apsInterframeDelay determines how much time passes between blocks within a window. The apscMaxWindowSize defines how many blocks are sent before an end-to-end acknowledgment is expected.

But here is a word of warning! I can't tell you how many times this has caught me: thinking that the ZigBee stack was malfunctioning, only to find it was user error. Make sure the window size is the same on both the sending and receiving nodes! Yes, it's listed as a constant, but if you compile the constant as window size 1 on one set of nodes, and window size 3 on another set of nodes, the resulting ACK chaos results in a failed, fragmented transmission every time. I wish ZigBee had sent the window size over-the-air along with the first packet, as it does for block size … (Sigh). There is room in the fragmentation header for window size … Perhaps in some future version.

Note that the window size is irrelevant for the nodes between (routers). Only the nodes at either end need to agree on this parameter. In the fragmented data request shown in Figure A.5, the window size has been set to 3 (that is, 3 blocks in a window

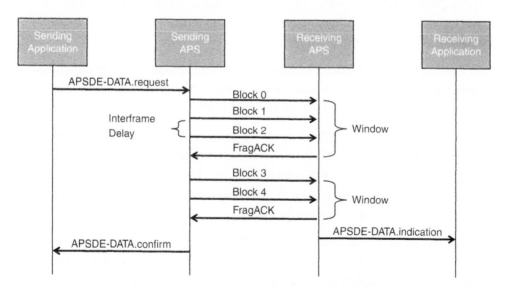

Figure A.5: A Fragmented Transmission

before an ACK is expected). The FragACK indicates to the sender which blocks in the window were received. The sender will resend any missed blocks. The interframe delay determines the time between blocks. Unless this application "owns" the network set this value to 100 ms or so, so that the network has time to process packets while a large, fragmented packet is being sent.

The APSDE-DATA.confirm is not returned until the sending application has received acknowledgment that all blocks were delivered to the receiving application, unless the transmission failed because of too many retries. The APSDE-DATA.indication on the receiving end will only be sent up to the application after the entire set of blocks has been received.

Note that fragmentation is only available in unicast. Due to the automatic retry and acknowledgment mechanism, fragmented packets cannot be broadcast, groupcast, or multicast.

Fragmentation is available both in ZigBee 2007 (stack profile 0x01) and ZigBee Pro (stack profile 0x02).

A.7   Symmetric Routes and Link Status

In ZigBee 2006 and ZigBee 2007, all routes are unidirectional. That is, if node "A" discovers a route to node "B," and node "B" wants to reply, it requires two route discoveries: "A" to "B," and "B" to "A."

Not so with ZigBee Pro. ZigBee Pro automatically sets up routes in both directions, assuming the nodes will communicate bidirectionally. A single route request causes a route entry to be entered both forward and backward.

As part of this process, ZigBee Pro keeps track of route symmetry by using link status commands. Every 15 seconds (plus some jitter time, defined by the NIB field nwkLinkStatusPeriod) a router will send out a radius 1 broadcast to indicate link status. All the neighboring routers keep track of these messages to determine which routers can hear them and which routers they can hear.

What does this mean to your application? ZigBee Pro networks consume more bandwidth for the network protocol, leaving less for your application. However, they do prevent that "asymmetric link" problem where one node can shout really loudly, but can't hear its neighbors.

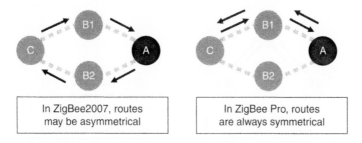

Figure A.6: Routes are Symmetrical in ZigBee Pro

Link Status commands and symmetrical links are available in ZigBee Pro (stack profile 0x02) only (see Figure A.6).

A.8 Link Keys, SKKE, and High Security

One very interesting feature added to ZigBee Pro is High Security.

The ZigBee Standard Security model is very secure, using AES 128-bit encryption and authentication, but it has a very important rule: All nodes on the network trust each other.

With ZigBee High Security, the model still uses AES 128-bit encryption and authentication, but the rule is very different: Only the applications that are specifically designed to talk together trust each other.

As an analogy, think of your house versus a hotel. In your house, you trust everyone. You don't expect family members or friends to steal the best china. Now think of a hotel. Although many people are in the same building, only some of them trust each other. You may trust the front desk clerk to provide you with a new room key, and to take your credit card. You may trust others that are staying in the same hotel to come into your room, perhaps other members of your family, but the trust certainly doesn't extend to everyone in the hotel.

The way in which ZigBee accomplishes the higher-trust model involves a number of mechanisms. One of them involves significantly more interaction with the Trust Center. In the Standard Security model, the trust center is used to allow or refuse a node on the network, with only the MAC address for information. In the High Security model, the node must establish an authentication sequence.

In Standard Security, the network can (optionally) be set up so the network key is given to the application in the clear from the node's parent. Not so with High Security. Transactions are always secured.

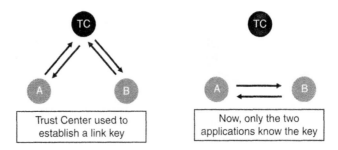

Figure A.7: Link Keys Establish Private Information Between Applications

When a High Security application wishes to speak with another application, it establishes a Link Key with that other application. The link key is used so that no other nodes on the network can listen in to that conversation. In the Smart Energy profile, multiple customers (with sensitive customer data) are on the same ZigBee network. The public utility companies want to be *very sure* customer A won't be able to read customer B's data.

This link key is negotiated using ephemeral data. The Trust Center is used as part of this process, using a special (essential link) key that only a given node and the Trust Center know. At the completion of the process, the two nodes now share a private key (the link key) that only they know, as shown in Figure A.7.

The only trusted node in the entire network is the Trust Center. All other nodes are treated with suspicion. It's no longer enough merely to have the network key.

So what does the application need to do to be able to use this high level of security? Very little, in fact. Two bits in the TxOptions field of the APSDE-DATA.request determine what type of security is applied:

```
TxOptions
0x01 = Security enabled transmission
0x02 = Use NWK key
```

MAC HDR	NWK HDR	NWK AUX HDR	APS HDR	APS AUX HDR	Payload	APS AUX MIC	NWK AUX MIC	FCS

Figure A.8: With Link Keys, the Frame Is Secured Twice

MAC HDR	NWK HDR	NWK AUX HDR	APS HDR	Payload	NWK AUX MIC	FCS

Figure A.9: Without Link Keys, the Frame Is Secured Once

If both options are false, only the NWK key is used, just as in Standard Security. If bit 0 (value 0x01) is set, it means that a link key needs to be established for the two nodes (if it hasn't been established already). If Bit 1 (value 0x02) is set, then the network key will be used instead of a link key for the APS key.

If either bit is set, then the APS frame will be secured in addition to the network frame. Yes, that does mean the packet is secured twice: the APS data is secured with the link key, then the NWK payload is secured with the NWK key. The over-the-air frame will have an APS AUX header and MIC, in addition to the NWK AUX header and MIC, as seen in Figure A.8.

The only use I've found for security with the network frame is if the application is out of memory and can't establish a link key. If the data is just not sensitive, and can be known by all nodes on the network, don't bother to secure APS at all.

If only the network key is used (as it is in Standard Security, or with both security options off), then only the NWK payload is secured (see Figure A.9).

Remember, that the entire frame is *always* authenticated, that is, no modified, false, or repeated data will be accepted by the receiving ZigBee nodes.

ZigBee 2007 includes fragmentation and frequency agility.

ZigBee Pro includes multicast, source routing, symmetric routes, and High Security.

Use ZigBee 2007 for a smaller image.

Use ZigBee Pro for more features.

ZigBee Quick Reference

This chapter is a quick reference for the ZigBee architecture and commands. For a more detailed explanation, see the main body of this book. This quick reference focuses on primitives available to the application, not the entire ZigBee specification.

B.1   The ZigBee Architecture

You've seen the architecture diagram shown in Figure B.1 many times. It's repeated here for convenience.

The next sections provide a quick reference to the portions of ZigBee an application interacts with directly: the APS and ZDO/ZDP layers.

B.2   APS Commands

The Application Support Sub-layer (APS) provides the application interface to the network data services and various application services. See Chapter 4, "ZigBee Applications," for more details.

Table B.1 lists the commands and primitives supported by ZigBee 2007 and ZigBee Pro. The primitives, parameters, or options in the description in *italics* are for ZigBee Pro only.

B.3   ZDP Commands

The ZigBee Device Profile (ZDP) is an application on endpoint 0 supporting over-the-air commands to get and set a variety of information regarding ZigBee nodes and applications. See Chapter 5, "ZigBee, ZDO, and ZDP," for details.

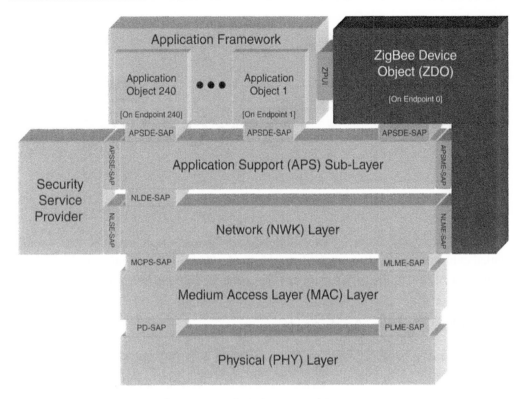

Figure B.1: The ZigBee Architecture

I have purposefully left out the discovery cache commands. There is discussion within the ZigBee Alliance of deprecating this set of commands altogether, as they are not used in the industry.

The Mode field describes whether or not the request should be unicast, and whether it is optional or mandatory on the server or client side. U = unicast only, B = broadcast only, E = either unicast or broadcast, M = mandatory, O = optional. For example, a ZDP command that should be unicast, is only mandatory on the server side, but optional on the client side would be the UMO. The mode field does not apply to responses.

If a response is warranted, it is always returned unicast to the node requesting the information.

All commands shown in Table B.2 operate in the same way on ZigBee 2007 or Pro.

Table B.1: APS Primitives

Primitive	Parameters	Description
APSDE-DATA.request	DstAddrMode, DstAddress, DstEndpoint, ProfileID, ClusterID, SrcEndpoint, asduLength, asdu, TxOptions, RadiusCounter	Send data from one node to one or more others. DstAddrMode = 0 = use binding table 1 = groupcast, *multicast(pro)* 2 = direct 16, broadcast (0xfffc = ZR, 0xfffd = RxOnIdle, 0xffff = all) 3 = direct 64 (MAC address) TxOptions = 1 = Security enabled at APS 2 = Use NWK key for APS key 4 = APS end-to-end ACK 8 = Fragmented (ACK implied)
APSDE-DATA.confirm	DstAddrMode, DstAddress, DstEndpoint, SrcEndpoint, Status, TxTime	Results of a data request. The confirm is not sent until a route is discovered and the packet is out of the radio to the next hop. The confirm indicates end-to-end acknowledgment if APS ACK is enabled. Status = SUCCESS, NO_SHORT_ADDRESS, NO_BOUND_DEVICE, SECURITY_FAIL, NO_ACK, ASDU_TOO_LONG
APSDE-DATA.indication	DstAddrMode, DstAddress, DstEndpoint, SrcAddrMode, SrcAddress, SrcEndpoint, ProfileID, ClusterID, asduLength, asdu, Status, SecurityStatus, LinkQuality, RxTime	Data has been received from another node. DstAddrMode = 1 = groupcast or *multicast (pro)* 2 = unicast or broadcast Status indicates if security or fragmentation failed. SecurityStatus indicates which key was used to secure the asdu.

(continued)

Table B.1: *continued*

Primitive	Parameters	Description
APSME-BIND.request	SrcAddr, SrcEndpoint, ClusterID, DstAddrMode, DstAddr, DstEndpoint	Binds a local endpoint to a remote node/endpoint or group. DstAddrMode = 1 = groupcast, *multicast(pro)* 3 = direct 64 (MAC address) Confirm Status = SUCCESS, ILLEGAL_REQUEST, TABLE_FULL or NOT_SUPPORTED
APSME-UNBIND.request	SrcAddr, SrcEndpoint, ClusterID, DstAddrMode, DstAddr, DstEndpoint	Clears a local binding table entry. DstAddrMode = 1 = groupcast, *multicast(pro)* 3 = direct 64 (MAC address) Confirm Status = SUCCESS, ILLEGAL_REQUEST, INVALID_BINDING or NOT_SUPPORTED
APSME-GET.request	AIB_Attribute	Get a local APS Information Base attribute. Confirm Status = SUCCESS, INVALID_PARAMETER, UNSUPPORTED_ATTRIBUTE
APSME-SET.request	AIB_Attribute, AIB_AttrLength, AIB_AttrValue	Set a local APS Information Base attribute. Confirm Status = SUCCESS, INVALID_PARAMETER, UNSUPPORTED_ATTRIBUTE
APSME-ADD-GROUP.request	GroupAddress, Endpoint	Adds a local endpoint to a group. Confirm Status = SUCCESS, INVALID_PARAMETER, TABLE_FULL
APSME-REMOVE-GROUP.request	GroupAddress, Endpoint	Removes a local endpoint from a group. Confirm Status = SUCCESS, INVALID_PARAMETER, INVALID_GROUP
APSME-REMOVE-ALL-GROUPS.request	Endpoint	Removes a local endpoint from all groups. Confirm Status = SUCCESS, INVALID_PARAMETER

Table B.2: ZDP Commands and Responses

Primitive	Mode	Parameters	Description
NWK_addr_req	BMO	IEEEAddress, RequestType, StartIndex	From an IEEEAddress, find the 16-bit NwkAddr. RequestType = 0 = the device only 1 = the device and its children
NWK_addr_rsp		Status, IEEEAddress, NwkAddress, [NumDevices], [StartIndex]	Results of a NWK_Addr_req. Status = SUCCESS, INV_REQUESTTYPE NumDevices and StartIndex only present if RequestType 1.
IEEE_addr_req	UMO	NwkAddress, RequestType, StartIndex	From a NwkAddress, find the IEEEAddress. RequestType = 0 = the device only 1 = the device and its children
IEEE_addr_rsp		Status, IEEEAddress, NwkAddress, [NumDevices], [StartIndex]	Results of an IEEE_Addr_req. Status = SUCCESS, INV_REQUESTTYPE NumDevices and StartIndex only present if RequestType 1.
Node_Desc_req	UMO	NwkAddrOfInterest	Request a node descriptor.
Node_Desc_rsp		Status, NwkAddrOfInterest, NodeDescriptor	Status = SUCCESS (cannot fail)
Power_Desc_req	UMO	NwkAddrOfInterest	Request a power descriptor.
Power_Desc_rsp		Status, NwkAddrOfInterest, PowerDescriptor	Status = SUCCESS (cannot fail)
Simple_Desc_req	UMO	Status, NwkAddrOfInterest, Endpoint	Request a simple descriptor from a remote node and endpoint.
Simple_Desc_rsp		Status, NwkAddrOfInterest, Length, SimpleDescriptor	Status = SUCCESS, INVALID_EP, NOT_ACTIVE
Active_EP_req	UMO	NwkAddrOfInterest	Request a list of active endpoints from a remote node.

(continued)

Table B.2: *continued*

Primitive	Mode	Parameters	Description
Active_EP_rsp		Status, NwkAddrOfInterest, ActiveEPCount, ActiveEPList	Lists all active endpoints. Status = SUCCESS (cannot fail)
Match-Desc_req	EMO	NwkAddrOfInterest, ProfileID, NumInClusters, InClusterList, NumOutClusters, OutClusterList	Look for remote endpoints that match the profile and cluster list.
Match-Desc_rsp		Status, NwkAddrOfInterest, NumMatchingEPs, MatchEPList	Lists each matching endpoint. Status = SUCCESS (cannot fail).
Complex_Desc_req	UOO	NwkAddrOfInterest	Request a complex descriptor.
Complex_Desc_rsp		Status, NwkAddrOfInterest, Length, ComplexDescriptor	Status = SUCCESS, NOT_SUPPORTED
User_Desc_req	UOO	NwkAddrOfInterest	Request a user descriptor.
User_Desc_rsp		Status, NwkAddrOfInterest, Length, UserDescriptor	Status = SUCCESS, NOT_SUPPORTED
Device_annce	BMM	NwkAddress, IEEEAddress, Capability	Indicate to the network that a node has changed NwkAddr or capability information.
			No response to Device_annce
User_Desc_set	UOO	NwkAddrOfInterest, Length, UserDescriptor	Set the contents of a remote user descriptor.
User_Desc_rsp		Status, NwkAddrOfInterest	Status = SUCCESS, NOT_SUPPORTED
System_Server_ Discovery_req	BMO	ServerMask	Find one of the system server nodes. ServerMask = 0x01 = primary trust center 0x40 = network manager

Table B.2: *continued*

Primitive	Mode	Parameters	Description
System_Server_ Discovery_rsp		Status, ServerMask	Mask indicates which server was found. Multiple responses may be returned, depending on request.
Extended_Simple_ Desc_req	UOO	NwkAddrOfInterest, Endpoint, StartIndex	Request a large simple descriptor from a remote node and endpoint.
Extended_Simple_ Desc_rsp		Status, NwkAddrOfInterest, Endpoint, InClusterCount, OutClusterCount, StartIndex ClusterList	Status = SUCCESS, INVALID_EP, OT_SUPPORTED ClusterList starts at StartIndex.
Extended_Active_EP_ req	UOO	NwkAddrOfInterest, StartIndex	Request a large list of active endpoints from a remote node.
Extended_Active_ EP_rsp		Status, NwkAddrOfInterest, ActiveEPCount, StartIndex, ActiveEPList	Status = SUCCESS, INVALID_EP, NOT_SUPPORTED ActiveEPList starts at StartIndex.
End_Device_Bind_req	UMO	SrcNwkAddr, SrcIEEEAddr, SrcEndpoint, ProfileID, NumInClusters, InClusterList, NumOutClusters, OutClusterList	Binds the node(s) with matching output clusters to the other node. Two nodes must issue this request to be successful. NwkAddress and IEEEAddress are the addresses of the requesting node.
End_Device_Bind_rsp		Status	SUCCESS, INVALID_EP, TIMEOUT, NO_MATCH
Bind_req	UOO	SrcIEEEAddr, SrcEndpoint, ClusterID, DstAddrMode, DstAddress, DstEndpoint	Add an entry in a remote binding table. See APS-BIND.request.
Bind_rsp		Status	SUCCESS, NOT_SUPPORTED, INVALID_EP, TABLE_FULL, NOT_AUTHORIZED

(continued)

Table B.2: *continued*

Primitive	Mode	Parameters	Description
Unbind_req	UOO	SrcIEEEAddr, SrcEndpoint, ClusterID, DstAddrMode, DstAddress, DstEndpoint	Remove an entry in a remote binding table. See APS-UNBIND. request.
Unbind_rsp		Status	SUCCESS, NOT_SUPPORTED, INVALID_EP, NO_ENTRY, NOT_AUTHORIZED
Mgmt_NWK_Disc_req	UOO	ScanChannels, ScanDuration, StartIndex	Request a remote device scan for networks on the list of channels.
Mgmt_NWK_Disc_rsp		Status, NetworkCount, StartIndex, ListCount, NetworkList	SUCCESS, NOT_SUPPORTED NetworkList = ExtendedPanID, LogicalChannel (11–26), StackProfile(4 bits: 0x1 or 0x2) ZigBeeVersion(4 bits: 02), BeaconOrder(4 bits: 0xf), SuperframeOrder(4 bits: 0xf), PermitJoining(0 or 1)
Mgmt_Lqi_req	UOO	StartIndex	Request the neighbor table from a remote node.
Mgmt_Lqi_rsp		Status, TotalEntries, StartIndex, ListCount, NeighborTableList	SUCCESS, NOT_SUPPORTED
Mgmt_Rtg_req	UOO	StartIndex	Request the routing table from a remote node.
Mgmt_Rtg_rsp		Status, TotalEntries, StartIndex, ListCount, RoutingTableList	SUCCESS, NOT_SUPPORTED
Mgmt_Bind_req	UOO	StartIndex	Request the binding table from a remote node.

Table B.2: *continued*

Primitive	Mode	Parameters	Description
Mgmt_Bind_rsp		Status, TotalEntries, StartIndex, ListCount Binding TableList	SUCCESS, NOT_SUPPORTED
Mgmt_Leave_req	UOO	IEEEAddress, ControlField	Request that a remote device and/or its children leave the network. ControlField = 0x80 = rejoin after leaving 0x40 = remove children
Mgmt_Leave_rsp		Status	SUCCESS, NOT_SUPPORTED, NOT_AUTHORIZED
Mgmt_Direct_Join_req	UOO	IEEEAddress, Capability	Adds an entry to a remote node's neighbor table to allow direct join.
Mgmt_Direct_Join_rsp		Status	SUCCESS, NOT_SUPPORTED, NOT_AUTHORIZED
Mgmt_Permit_ Joining_req	EMO	PermitDuration, TC_Significance	Enable or disable joining in a remote node or the entire network.
Mgmt_Permit_ Joining_rsp		Status	SUCCESS, INVALID_REQUEST, NOT_AUTHORIZED
Mgmt_NWK_Update_ req	EOO	ScanChannels, ScanDuration, [ScanCount], [nwkUpdateID], [nwkManagerAddr]	Used for frequency agility. ScanDuration = 0x00 – 0x05 = energy scan 0xfe = change channel 0xff = update AIB with new apsChannelMask and nwkManagerAddr
Mgmt_NWK_Update_ notify		Status, ScannedChannels, TotalTransmissions, TransmissionFailures, ListCount, EnergyValues	Results of a Mgmt_NWK_Update_req

ZigBee Cluster Library Quick Reference

This chapter is a Quick Reference for the ZigBee Cluster Library. For a more detailed explanation, see Chapter 6, "The ZigBee Cluster Library." This Quick Reference does not list profile-specific clusters.

C.1   ZCL Foundation

This section describes ZigBee Cluster Library commands that are cross-cluster. All commands operate in the same way on ZigBee 2007 or ZigBee Pro. [Brackets] denote optional fields. M/O denotes mandatory or optional support, if ZCL is supported (Table C.1).

C.2   ZCL General Clusters

This section describes the ZigBee Cluster Library General (all-purpose) clusters in quick reference format. I've listed only those clusters used in multiple application profiles.

All commands operate in the same way on ZigBee 2007 or ZigBee Pro. See Table C.2.

Table C.1: ZigBee Cluster Library Foundation Comma

Command Name	M / O	Parameters	Description
Read attributes	O	AttrId, [AttrId...]	Read one or more attributes.
Read attributes response	M	AttrRecord, [AttrRecord...]	Return value of one or more attributes. AttrRecord = AttrId, Status, DataType, Data
Write attributes	O	AttrRecord, [AttrRecord...]	Write one or more attributes. AttrRecord = AttrId, DataType, Data
Write attributes undivided	O	AttrRecord, [AttrRecord...]	Write one or more attributes as a set. AttrRecord = AttrId, DataType, Data
Write attributes response	M	StatusRecord, [StatusRecord...]	Return success status of write attributes. StatusRecord = Status, AttrId
Write attributes no response	M	AttrRecord, [AttrRecord...]	Write one or more attributes, no response. AttrRecord = AttrId, DataType, Data
Configure reporting	O	CfgRecord, [CfgRecord...]	Configure attributes for reporting.CfgRecord = Direction, [DirSpecifcFields] Direction(0) = DataType, MinReportInterval, MaxReportInterval, ChangeVal Direction(1) = TimeOut
Configure reporting response	M	CfgRspRecord, [CfgRspRecord...]	Status of configure attributes. CfgRspRecord = Status, Direction, AttrId
Read reporting configuration	O	RdReport, [RdReport...]	Read current reporting configuration. RdReport = Direction, AttrId
Read reporting configuration response	M	RdReportRsp, [RdReportRsp...]	Return current reporting configuration. RdReportRsp = Status, Direction, AttrId, [DirSpecifcFields] Direction(0) = DataType, MinReportInterval, MaxReportInterval, ChangeVal Direction(1) = TimeOut
Report attributes	O	AttrId, DataType, Data	Attribute report, depends on configuration.
Default response	M	CmdID, Status	Unsupported command response.

Table C.1: *continued*

Command Name	M / O	Parameters	Description
Discover attributes	O	StartAttrID, MaxAttrIDs	Determine supported attributes on remote node.
Discover attributes response	M	DiscoveryDone, AttrInfo, [AttrInfo...]	Results of discover attributes command. AttrInfo = AttrId, DataType

Table C.2: ZigBee Cluster Library General Clusters

Cluster Name	Attributes	Commands	Description
Basic	ZCL Version, App Version, Stack Version, HW Version, MfgName, ModelID, DateCode, PowerSource	FactoryReset	Attributes for determining basic information about a device, setting user device information such as location, and enabling a device.
Power Configuration	MainsVoltage, MainsFrequency, MainsAlarmMask, MainsMinThreshold, MainsMaxThreshold, MainsTripPoint, BatteryVoltage, BatteryMfg, BatterySize, BatteryAHrRating, BatteryQuantity, BatteryRatedVoltage, BatteryAlarmMask, BatteryMinThreshold	—	Attributes for determining more detailed information about a device's power source(s), and for configuring under/over voltage alarms.
Identify	IdentifyTime	Identify, IdentifyQuery	Attributes and commands for putting a device into Identification Mode (e.g., flashing a light).

(continued)

Table C.2: *continued*

Cluster Name	Attributes	Commands	Description
Groups	NameSupport	AddGroup, ViewGroup, GetGroupMembership, RemoveGroup, RemoveAllGroups, AddGroupIfIdentifying	Attributes and commands for group configuration and manipulation.
Scenes	SceneCount, CurrentScene, CurrentGroup, SceneValid, NameSupport LastConfiguredBy	AddScene, ViewScene, RemoveScene, RemoveAllScenes, StoreScene, RecallScene, GetSceneMembership	Attributes and commands for scene configuration and manipulation.
On/Off	OnOff	Off, On, Toggle	Attributes and commands for switching devices between "On" and "Off" states.
Level Control	CurrentLevel, RemainingTime, TransitionTime, OnLevel	MoveToLevel, Move, Step, Stop, MoveToLevelOnOff, MoveOnOff, StepOnOff, StopOnOff	Attributes and commands for controlling devices that can be set to a level between fully "On" and fully "Off."

Index

Printed and bound by CPI Group (UK) Ltd, Croydon, CR0 4YY

03/10/2024

01040338-0001